Lecture Notes in Artificial Intelligence 10101

Subseries of Lecture Notes in Computer Science

LNAI Series Editors

Randy Goebel
University of Alberta, Edmonton, Canada
Yuzuru Tanaka
Hokkaido University, Sapporo, Japan
Wolfgang Wahlster
DFKI and Saarland University, Saarbrücken, Germany

LNAI Founding Series Editor

Joerg Siekmann
DFKI and Saarland University, Saarbrücken, Germany

Christian Bessiere · Luc De Raedt
Lars Kotthoff · Siegfried Nijssen
Barry O'Sullivan · Dino Pedreschi (Eds.)

Data Mining and Constraint Programming

Foundations of a Cross-Disciplinary Approach

 Springer

Editors

Christian Bessiere
Université Montpellier 2
Montpellier
France

Luc De Raedt
KU Leuven
Heverlee
Belgium

Lars Kotthoff
University of British Columbia
Vancouver, BC
Canada

Siegfried Nijssen
Université Catholique de Louvain
Louvain-la-Neuve
Belgium

Barry O'Sullivan
University College Cork
Cork
Ireland

Dino Pedreschi
University of Pisa
Pisa
Italy

ISSN 0302-9743   ISSN 1611-3349   (electronic)
Lecture Notes in Artificial Intelligence
ISBN 978-3-319-50136-9   ISBN 978-3-319-50137-6   (eBook)
DOI 10.1007/978-3-319-50137-6

Library of Congress Control Number: 2016959176

LNCS Sublibrary: SL7 – Artificial Intelligence

© Springer International Publishing AG 2016
This work is subject to copyright. All rights are reserved by the Publisher, whether the whole or part of the material is concerned, specifically the rights of translation, reprinting, reuse of illustrations, recitation, broadcasting, reproduction on microfilms or in any other physical way, and transmission or information storage and retrieval, electronic adaptation, computer software, or by similar or dissimilar methodology now known or hereafter developed.
The use of general descriptive names, registered names, trademarks, service marks, etc. in this publication does not imply, even in the absence of a specific statement, that such names are exempt from the relevant protective laws and regulations and therefore free for general use.
The publisher, the authors and the editors are safe to assume that the advice and information in this book are believed to be true and accurate at the date of publication. Neither the publisher nor the authors or the editors give a warranty, express or implied, with respect to the material contained herein or for any errors or omissions that may have been made.

Printed on acid-free paper

This Springer imprint is published by Springer Nature
The registered company is Springer International Publishing AG
The registered company address is: Gewerbestrasse 11, 6330 Cham, Switzerland

Preface

In industry, society, and science, advanced software is used for planning, scheduling, and allocating resources in order to improve the quality of service, reduce costs, or optimize resource consumption. Examples include power companies generating and distributing electricity, hospitals planning their surgeries, and public transportation companies scheduling their time-tables. This type of problem is often referred to as *constraint satisfaction and combinatorial optimization problems*.

Despite the availability of effective and scalable *solvers* that are applicable to a wide range of applications, current approaches to this problem are still unsatisfactory. The reason is that in all these applications it is very hard to acquire the constraints and criteria (that is, the *model*) needed to specify the problem, and, even if one has succeeded in capturing the model at one point, it is likely that it needs to be to changed over time to reflect changes in the environment. Therefore, there is an urgent need for optimizing and revising a model over time based on data that should be continuously gathered about the performance of the solutions and the environment they are used in.

Exploiting gathered data to modify the model is difficult and labour intensive with state-of-the-art solvers, as these solvers do not support data mining (DM) and machine learning (ML). However, existing frameworks for constraint satisfaction and combinatorial optimization problems do not support ML/DM techniques. In current ICT technology, DM and ML have almost always been studied independently from solving technology such as constraint programming (CP). On the other hand, a growing number of studies indicate that significant benefits can be obtained by connecting these two fields.

This led us to believe – almost five years ago – that it was the right time to develop the foundations of an integrated and cross-disciplinary approach to these two fields. A successful integration of CP and DM has the potential to lead to a new ICT paradigm with far-reaching implications that would change the face of DM/ML as well as CP technology. It would not only allow one to use DM techniques in CP to identify and update constraints and optimization criteria, but also to employ such constraints and criteria in DM and ML in order to discover models compatible with such prior knowledge. This book reports on the key results obtained on this research topic within the European FP7 FET Open project no. 284715 on "Inductive Constraint Programming" and a number of associated workshops and Dagstuhl seminars.

The book is structured in five parts. Part I contains an introduction to CP by Barry Hurley and Barry O'Sullivan and an introduction to DM by Valerio Grossi, Dino Pedreschi, and Franco Turini.

The next two parts address different challenges related to using ML and DM in a CP context. The first of these is the *model acquisition problem*, which aims at learning the different components of the CP model. This includes the identification of the domains to use, the constraints and possibly the preference or optimization function to be used. This is the topic of Part II. The first contribution, by Christian Bessiere, Abderrazak

Daoudi, Emmanuel Hebrard, George Katsirelos, Nadjib Lazaar, Younes Mechqrane, Nina Narodytska, Claude-Guy Quimper, and Toby Walsh, discusses an algorithm that acquires constraints by querying the user. The contribution by Nicolas Beldiceanu and Helmut Simonis describes a system for generating finite domain constraint models based on a global constraint catalog. The contribution by Luc De Raedt, Anton Dries, Tias Guns, and Christian Bessiere investigates the problem of learning constraint satisfaction problems from an inductive logic programming perspective. The contribution by Andrea Passerini discusses Learning Modulo Theories, a novel learning framework capable of dealing with hybrid domains.

The second challenge is that once the model is known, it needs to be solved. Reformulating models, optimizing the parameters of the solver, or considering alternative solvers is needed to solve the problem efficiently. Hints on how to improve a model and the best technique for solving it can be obtained by analyzing data collected during the run of solvers, or data collected from user studies. Part III reports on a number of techniques for model reformulation and solver optimization, that is: techniques for learning how to find solutions faster and more easily. In this part, a contribution by Lars Kotthoff provides a survey of algorithm selection techniques. Subsequently, Barry Hurley, Lars Kotthoff, Yuri Malitsky, Deepak Mehta, and Barry O'Sullivan present the Proteus portfolio solver and several improvements to portfolio techniques. Finally, Amine Balafrej, Christian Bessiere, Anastasia Paparrizou, and Gilles Trombettoni present techniques that adapt the level of consistency ensured by a solver during the search.

Part IV reports on the use of constraints and CP within a DM and ML context. This is motivated by the observation that many DM and ML tasks are essentially constraint satisfaction and optimization problems and that, therefore, they may benefit from CP principles and techniques. By specifying the constraints and optimisation criteria explicitly, DM and ML problem specifications become declarative and can potentially be solved by CP systems. Furthermore, several high-level modeling languages have been developed within CP that can potentially be applied or extended to ML and DM. The contribution by Anton Dries, Tias Guns, Siegfried Nijssen, Behrouz Babaki, Thanh Le Van, Benjamin Negrevergne, Sergey Paramonov, and Luc De Raedt introduces MiningZinc, a unifying framework and modeling language with associated solvers for DM and CP. Subsequently, Valerio Grossi, Tias Guns, Anna Monreale, Mirco Nanni, and Siegfried Nijssen show how many clustering problems can be formalized as constraint optimization problems.

Finally, Part V takes a more practical perspective. The first chapter by Christian Bessiere, Luc De Raedt, Tias Guns, Lars Kotthoff, Mirco Nanni, Siegfried Nijssen, Barry O'Sullivan, Anastasia Paparrizou, Dino Pedreschi, and Helmut Simonis reports on the iterative approach to inductive CP. The key idea is that the CP and ML components interact with each other and with the world in order to adapt the solutions to changes in the world. This is an essential need in problems that change under the effect of time, or problems that are influenced by the application of a previous solution. It is also very effective for problems that are only partially specified and where the ML component learns from observation of applying a partial solution, e.g., in the case of constraint acquisition. In addition, it reports on a number of applications of inductive CP in the areas of carpooling (with a contribution by Mirco Nanni, Lars Kotthoff,

Riccardo Guidotti, Barry O'Sullivan, and Dino Pedreschi), health care (with a contribution by Barry Hurley, Lars Kotthoff, Barry O'Sullivan, and Helmut Simonis), and energy (with a contribution by Barry Hurley, Barry O'Sullivan, and Helmut Simonis).

The editors would like to thank the European Union for supporting the EU FET FP7 ICON project, the reviewers of the project, Alan Frisch, Bart Goethals, and Francesca Rossi, and the project officer Aymard de Touzalin for their constructive feedback and support, all participants of the ICON project for their contributions (Behrouz Babaki, Amine Balafrej, Remi Coletta, Abderrazak Daoudi, Anton Dries, Valerio Grossi, Riccardo Guidotti, Tias Guns, Barry Hurley, Nadjib Lazaar, Yuri Malitsky, Younes Mechqrane, Wannes Meert, Anna Monreale, Benjamin Negrevergne, Anastasia Paparrizou, Sergey Paramanov, Andrea Romei, Salvatore Ruggiero, Helmut Simonis, and Franco Turini), as well as the participants of the Dagstuhl seminars 11201 and 14411 and the Cocomile workshop series. Furthermore, they are grateful to all reviewers of chapters in this book.

September 2016

Christian Bessiere
Luc De Raedt
Lars Kotthoff
Siegfried Nijssen
Barry O'Sullivan
Dino Pedreschi

Reviewers

Blockeel, Hendrik   KU Leuven, Belgium
Dao, Thi-Bich-Hanh   Université d'Orléans, France
Davidson, Ian   UC Davis, USA
De Causmaecker, Patrick   KU Leuven, Belgium
Dries, Anton   KU Leuven, Belgium
Frisch, Alan   University of York, UK
Gent, Ian   University of St. Andrews, UK
Guns, Tias   KU Leuven, Belgium
Hoos, Holger   University of British Columbia, Canada
Lazaar, Nadjib   Université de Montpellier, France
Mauro, Jacopo   University of Oslo, Norway
Nanni, Mirco   ISTI - CNR, Italy
Nightingale, Peter   University of St. Andrews, UK
Passerini, Andrea   Università degli Studi di Trento, Italy
Pearson, Justin   Uppsala University, Sweden
Simonis, Helmut   University College Cork, Ireland
Tack, Guido   Monash University, Australia
Vanschoren, Joaquin   Technische Universiteit Eindhoven, The Netherlands
Vrain, Christel   Université d'Orléans, France
Zelezny, Filip   Czech Technical University in Prague, Czech Republic

Contents

Showcases

Background

Introduction to Combinatorial Optimisation in Numberjack

Barry Hurley[✉] and Barry O'Sullivan

Insight Centre for Data Analytics, Department of Computer Science,
University College Cork, Cork, Ireland
{barry.hurley,barry.osullivan}@insight-centre.org

Abstract. This chapter presents an introduction to combinatorial optimisation in the context of the high-level modelling platform, Numberjack. The process of developing an effective model for a combinatorial problem is presented, along with details on how such problems can be solved using three of the most prominent solution paradigms.

1 Introduction

Combinatorial optimisation problems arise in many important real-world applications such as scheduling, planning, configuration, rostering, timetabling, vehicle routing, network design, bioinformatics, and many more. Intelligent, automated approaches to these problems can provide high quality solutions from a number of perspectives such as sustainability, energy efficiency, cost, time, etc., and can scale to tackle large problems far beyond the reach of manual methods. Optimisation technologies have been used to design a fibre optical network for entire countries, minimising the amount of cable to be laid, while also maintaining certain levels of redundancy [40]; to design electricity, water, and data networks [50]; to schedule scientific experiments on the Rosetta-Philae mission [49]; assign gates to airplanes [51]; as well as numerous timetabling, scheduling, and configuration applications [47,54].

There exist a number of alternative approaches to solve combinatorial problems, three of the most prominent methods being *Constraint Programming* (CP) [47], *Boolean Satisfiability* (SAT) [10], and *Mixed Integer Programming* (MIP) [56]. These techniques provide a generic platform to tackle a broad range of problems, from simple puzzles to large scale industrial applications. They provide a framework upon which real-world problems can be specified declaratively, largely relieving the user of the task of specifying how a solution should be found.

It is generally possible to solve the same problem with any of these methods, however they differ in terms of problem representation and solution methodology. In a nutshell, in the constraint programming paradigm variables take their values from finite sets of possibilities, with solutions typically found using a combination of systematic backtracking search and polynomial-time inference algorithms that reduce the size of the search space. A satisfiability problem is defined in terms of Boolean variables and a single form of constraint, namely a disjunction of

© Springer International Publishing AG 2016
C. Bessiere et al. (Eds.): Data Mining and Constraint Programming, LNAI 10101, pp. 3–24, 2016.
DOI: 10.1007/978-3-319-50137-6_1

Boolean variables or their negations. Instances are also solved using backtracking search, using unit-propagation for inference, as well as learning new clauses when failures are encountered. The mixed integer programming problem is defined by a set of linear expressions over integer and real-valued variables. Solutions are typically found by branch and bound search, using linear relaxations to make decisions, and the generation of cutting planes to prune the search space.

It is often not clear which approach is best for a particular problem, thus we may employ a higher-level modelling language to aide in the process. Modelling platforms such as Numberjack[1] [29], MiniZinc [42], and Essense [21] offer the user the ability to declare their model in a high-level language and the framework will handle the interface and encodings to the various paradigms. This enables rapid prototyping of different models with numerous solvers. This chapter presents an introduction to combinatorial optimisation, with examples given in Numberjack's modelling language.

The following sections introduce various facets of combinatorial optimisation. First, Sect. 2 introduces the basic building blocks of variables, constraints, and inference, as well as giving examples of how effective models can be declared in a modelling framework such as Numberjack. Section 3 introduces three of the most prominent approaches for solving combinatorial problems. Some underlying details of systematic search methods and their associated heuristics are detailed in Sect. 4.

2 Modelling Using Numberjack

Numberjack is a library, written in Python, which allows the user to model and solve combinatorial optimisation problems. It provides a common interface to a number of underlying C/C++ solvers seamlessly and efficiently. The remainder of this section details the process of building an effective solution to combinatorial problems.

A combinatorial optimisation model of a problem consists of a declarative specification, separating in so far as possible the formulation and the search strategy. However, modelling a problem effectively can be seen as an art in itself. The difficulty lies in producing a *solvable* model, i.e. one that quickly finds optimal solutions or determines that none exist. Naturally, there are many alternative models for a single problem, often it is not clear which one is best.

The basic process of developing a model consists of first defining what constitutes the variables and their corresponding domains, i.e. what decisions need to be made and what are the possible outcomes that can be taken for each one. Next, the constraints on the relationships between the variables must be defined. If some criterion is to be optimised, an objective function needs to be specified. Finally, if the model is well defined it can be passed off directly to a solver which will search for a (optimal) solution. Often, we may need to specify some heuristics for how the solver should perform the search, such as the variable or value ordering before the solver can effectively solve the problem.

---

[1] http://numberjack.ucc.ie/.

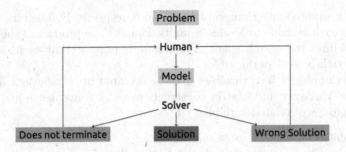

Fig. 1. Abstract process of modelling a problem.

For a user, the process of developing a suitable model often requires a number of iterations, depicted in Fig. 1. Two common issues arise in the development of a solution: the model either does not accurately represent the problem, or a solution is not found by the solver in reasonable time. The former is more of a real-world problem requiring the assistance of a domain expert, where eliciting the true constraints of the problem, which may not even be well understood, is a challenge. The latter can pose a larger challenge from a number of perspectives and may require the input of an expert in combinatorial optimisation.

Many different viewpoints can be taken in modelling a problem, so it can be easy to come up with a single model, but it may not necessarily be an efficient model. Important choices are to be made such as what are the variables, what are their domains, and what restrictions should be stated between them. Such decisions naturally affect the form of constraints which can be applied and unquestionably the effectiveness of the solver in finding a solution. Empirical performance may not be clear until it is actually evaluated.

Furthermore, solvers vary in terms of their capabilities, e.g. despite the hundreds of *global constraints* that have been developed [6], each solver typically implements a relatively small subset. Thus, the choice of which solver to use may be dictated by the global constraints required for a problem. Modelling languages lift the limitation of developing a model using a single solver. Instead, the model is implemented in a high-level solver-independent language that can be translated or encoded for a number of solvers. Nevertheless, these systems still rely on the user to produce a good model of their problem.

The next sections describe in more detail the primary components of a combinatorial optimisation model.

2.1 Variables

The variables constitute a fundamental component of a combinatorial problem. They are each represented by the finite set of values from which they can be assigned, often defined by a lower and upper bound. Typically these are restricted to integer values but some extensions do consider real-valued, set [24,58], and graph [16,17] variables. Boolean variables can take the values *true* or *false*, but are often interpreted interchangeably as 1 and 0, respectively. The ultimate task is to assign each variable to a value from its domain. The product of the variable domains defines the search space. Thus, it is important that each domain is defined as tightly as is permissible.

Some examples of how variables may be declared in Numberjack are given below. The VarArray and Matrix constructs serve as convenience methods for declaring groups of related variables.

Variable()	# *Boolean variable*
Variable('x')	# *Boolean variable called 'x'*
Variable(u)	# *Variable with domain of [0..u−1]*
Variable(l, u)	# *Variable with domain of [l..u]*
Variable(alist)	# *Variable with domain specified as a list*
VarArray(N)	# *Array of N Boolean variables*
VarArray(N, u)	# *Array of N variables with domains [0..u−1]*
VarArray(N, l, u)	# *Array of N variables with domains [l..u]*
Matrix(N, M)	# *N x M matrix of Boolean variables*
Matrix(N, M, l, u)	# *N x M matrix of variables with domains [l..u]*

2.2 Constraints

Constraints define relationships between the variables, forbidding invalid solutions to the problem. A *unary constraint* is the simplest form of constraint involving a single variable and is satisfied by preprocessing the domain of the variable. *Binary constraints* relate two variables, such as saying they cannot be equal, and *global constraints* [6] involve a larger set of variables, modelling more complex relations. The remainder of this section presents some common binary constraints, whereas Sect. 2.4 is devoted to the presentation of global constraints.

One of the most basic binary constraints is the disequality constraint which simply states that two variables must not be assigned the same value, for example $X \neq Y$. Inequalities such as $\langle <, \leq, \geq, > \rangle$ state a relationship which must hold between the respective assignments. In terms of their respective abilities to narrow the search space, these inequality constraints are stronger than the disequality constraint.

The *tightness* of a constraint is a measure of how many assignment tuples are forbidden, and subsequently how much of the search space is pruned. In particular, for a disequality constraint, with a tightness of $\frac{1}{d}$, we may only infer that a value may be removed from the domain of the opposite variable when one of the variables has been assigned, whereas for the inequalities, changes in the bounds or absence of certain values may reduce the domain of the other variable in the constraint. Such constraints are trivially specified in Numberjack using operator overloading, examples of which are presented in Table 1.

The expressivity of these binary constraints may be augmented by using expressions of the form $X + c < Y$, where c is a constant. Here, the expression

Table 1. Example binary constraint definitions in Numberjack.

Constraint	Numberjack code
Disequality	x != z
Greater than	x > y
Less-or-equal	y <= z
Logical-or	x | y
Logical-and	y \& z

$X + c$ becomes a *view* on the variable X, mirroring the offset domain without increasing the search space. Such constraints are useful in many scenarios, for example in scheduling if we would like to express the constraint that task2 starts after task1 has finished, we might specify a constraint of the form:

$$\text{task1}_{start} + \text{task1}_{duration} < \text{task2}_{start}$$

In many applications the model requires knowledge of the satisfiability of a particular constraint. In this case, we may *reify* the truth value of a constraint to a Boolean variable by writing something of the form:

$$z == (x < y) \qquad\qquad z <= (x < y) \qquad\qquad (x == y)\ != (a == b)$$

The first statement reifies the less-than relationships between x and y, enforcing that z is 1 iff x is less than y and 0 otherwise. The second example ensures that z is 0 if x is not less than y, if z is 1 then the less than relationship must hold, and other relationships are undefined. Finally, the third statement constrains the two pairs (x, y) and (a, b) such that exactly one pair must be assigned the same value and one pair must be assigned different values.

2.3   Inference

A central component in solving a CSP involves inferring variable information based on the constraints and the current state of the search, removing values from the domains that cannot possibly participate in any solution [38]. Based on the current partial assignment to variables during search, a value in a domain of an unassigned variable may, if assigned, violate a constraint then it is said to be *inconsistent*. Therefore it can be removed from the domain. No possible extension of the current assignment allows such a value to participate in a solution. These values are said to be *pruned* from the domain and consequently parts of the search tree will not be explored.

Figure 2 depicts the outcome of performing inference on a Sudoku problem which has been modelled as a CSP. Figure 2(a) shows the initial state of the CSP, where each cell corresponds to a single variable and its domain is the values from 1 to 9. Some cells have been pre-assigned with clues from the input.

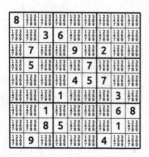

(a) Initial domains with preset clues.

(b) After propagating the preset clues.

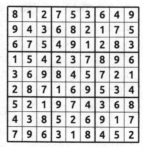

(c) Complete solution.

Fig. 2. Example of CP propagation on a Sudoku instance.

Constraints of the problem enforce that cells within each row, column, and 3 by 3 block take unique values, i.e. a series of all-different constraints. Evidently, where an initial clue is given as input, no cell in the corresponding row, column, or block may take this value, and so these values can be removed from their domains. Before we start any search, inference can be performed based on the all-different constraints, and the information given by the present clues, to remove inconsistent values in corresponding variables. Figure 2(b) depicts the result of propagating this knowledge, and values that cannot participate in any solution are removed from the domains of variables, resulting in a smaller search space. During search, this process is repeated in circumstances such as when a new variable has been assigned or backtracking has occurred.

Note that iteratively propagating the constraints to the domains is typically enough to solve quintessential Sudoku problems. However, the example Sudoku presented in Fig. 2 requires a combination of search, albeit a very small amount, and inference to find the complete solution depicted in Fig. 2(c). We must remark that the Sudoku example depicts a rather simple aspect of consistency, nevertheless it serves to illustrate the concept. Constraint programming and other combinatorial optimisation systems offer the ability to perform much more sophisticated reasoning, some of which is discussed in the following section.

Enforcing consistency during search reduces the search space but comes at an increased computational cost at each node. A trade-off must be made between pruning the search space and searching at a faster rate. Thus, constraint programming offers different levels of consistency that can be enforced, from constraint level *local consistency* to *global consistency* [9]. Local consistency concerns individual constraints in isolation, whereas global consistency equates to a complete solution satisfying all constraints. Generally speaking, each additional level of consistency has the capability to prune larger parts of the search space but entails a higher computational complexity.

The following definition, from [9], formally defines the concept of generalised arc consistency for a constraint network.

Definition 1 ((Generalised) Arc-Consistency ((G)AC)). *Given a CSP network $\mathcal{N} = \langle \mathcal{X}, \mathcal{D}, \mathcal{C} \rangle$, a constraint $c \in \mathcal{C}$, and a variable $X_i \in \mathcal{X}(c)$,*

- *A value $v_i \in D(X_i)$ is consistent with c in $\mathcal{D}$ iff there exists a valid tuple τ satisfying c such that $v_i = \tau[X_i]$. Such a tuple is called a* support *for (X_i, v_i) on c.*
- *The domain $\mathcal{D}$ is (generalised) arc consistent on c for X_i iff all the values in $D(X_i)$ are consistent with c in $\mathcal{D}$.*
- *The network $\mathcal{N}$ is (generalised) arc consistent iff $\mathcal{D}$ is (generalised) arc consistent for all variables in $\mathcal{X}$ on all constraints in $\mathcal{C}$.*
- *The network $\mathcal{N}$ is arc inconsistent if $\emptyset$ is the only domain tighter than $\mathcal{D}$ which is (generalised) arc consistent for all variables on all constraints.*

2.4   Global Constraints

Global constraints define constraints over an arbitrarily sized set of variables, presenting many benefits for constraint programming [53]. Notably, they can succinctly convey complex relationships between variables, allowing for a concise specification of a problem. More importantly, from a pragmatic perspective, this enables higher levels of reasoning to be performed by dedicated inference algorithms, reducing the search space significantly. For example, propagation for global constraints such as *all-different* and *cardinality* constraints can be achieved in low polynomial time using flow-based algorithms [45,46], much more efficiently than general purpose consistency algorithms.

To illustrate an example of such reasoning, consider an all-different constraint over the variables $\mathcal{X} = \{X_1, \ldots, X_5\}$, with initial domains $D(\mathcal{X}) = \{1, \ldots, 5\}$, declaring that each variable in the set must be assigned a unique value. Suppose that the domains have been reduced during search to those listed is Fig. 3(a). Note that the domain of variables $\{X_1, X_2, X_3\}$ constitute the Hall set $\{1, 2, 5\}$, whereby these three variables must each be assigned a unique value from the Hall set. Thus, any assignment of these values to other variables in the constraint can never result in a satisfying assignment, so they can be removed from the domains of the remaining variables, $\{X_4, X_5\}$. Had the all-different constraint been decomposed into a clique of dis-equalities, then such reasoning could not have been performed.

The Global Constraint Catalogue [6] collects definitions for all global constraints defined in the CP literature, at the time of writing this listing contains

$$
\begin{array}{llcll}
X_1 & \in & \{1, 2, \quad 5\} & X_1 & \in & \{1, 2, \quad 5\} \\
X_2 & \in & \{1, 2, \quad 5\} & X_2 & \in & \{1, 2, \quad 5\} \\
X_3 & \in & \{1, 2, \quad 5\} & X_3 & \in & \{1, 2, \quad 5\} \\
X_4 & \in & \{ \quad 2, 3, 4 \quad \} & X_4 & \in & \{ \quad 3, 4 \quad \} \\
X_5 & \in & \{1, 2, 3, 4, 5\} & X_5 & \in & \{ \quad 3, 4 \quad \}
\end{array}
$$

(a) Initial domains.   (b) After propagating the Hall set.

Fig. 3. Example of propagation on a Hall set $\{1, 2, 5\}$.

over 400 constraints and is continually increasing. Such a vast catalogue provides many opportunities for the application of constraint programming, however one practical issue faced by users is in identifying which one is appropriate for their problem.

2.4.1 Example Global Constraints

In practice, most constraint solving libraries only provide implementations for a small number of those listed in the global constraint catalogue. This section describes some of the most prominent and widely used global constraints.

Linear Sum. This general expression constrains the dot-product linear combination of a vector of variables and a vector of coefficients. Mathematically, these constraints take the form:

$$\sum_i w_i \cdot x_i \quad \triangle \quad c$$

where w is a vector of integer or real valued weights, x is a vector of variables, $\triangle$ is a relational operator from the set $\langle <, \leq, =, \geq, > \rangle$, and c is a constant.

This is the only constraint type expressible in integer linear programming but it provides a flexible representation since a number of high-level constraints can be decomposed or encoded in this form. For example, the constraint $x > y$ can be written in linear form as $x - y > 0$. Additionally, since they only deal with problems in a standard form it enables integer programming solvers to perform high-levels of reasoning, proving extremely powerful [56].

A linear sum of variables can be expressed in a number of ways in Numberjack, for example each of the following are equivalent:

```
2*a + b + 0.5*c + 3*d        == e
Sum([2*a, b, 0.5*c, 3*d])    == e
Sum([a,b,c,d], [2,1,0.5,3])  == e
```

In general, it is expensive and difficult for a constraint programming solver to perform a large amount of reasoning on linear sum constraints, particularly if there is a large number of variables or their domains are large. For example, in a linear sum with a large number of variables, there is a huge number of possible assignment permutations in which to check for supports, at least until a number of variables are fixed. Thus, in practice, their use with constraint programming solvers is often limited to cases with a small number of variables and small domains.

All-Different. One of the most widely known, intuitive, and well studied global constraints is the *all-different* constraint [36, 45] which simply specifies that a set of variables must be assigned distinct values. Such a relation arises in many practical applications such as resource allocation, e.g. to state that a resource may not be used more than once at a single time point. An all-different constraint may be specified in Numberjack simply by passing a list of variables (or a VarArray) as follows:

```
AllDiff([x1, x2, x3, x4])
AllDiff(vararray)
```

An intuitive application of the all-different constraint is the Sudoku problem, as illustrated in Fig. 2, whereby each row, column, and 3×3 cell is constrained to take distinct values. Such a condition can be modelled using an all-different for each row, column, and cell, giving a model with a total of 27 global constraints.

The all-different constraint may also be decomposed into a clique of disequalities between every pair of variables ($\forall i < j : X_i \neq X_j$). This decomposition requires $\binom{n}{2}$ binary constraints for each all-different, equating to a total of 972 (810 unique) binary disequality constraints for the Sudoku problem. However, this formulation looses the strong propagation that all-different enables, resulting in a larger search space to be explored.

Global Cardinality. The global cardinality constraint [1] places lower and upper bounds on the number of occurrences of certain values amongst a set of variables. The global cardinality constraint models restrictions in applications such as timetabling when there may be a limit on the number of consecutive activity types. For example in Numberjack, we can write the following:

> myvariablearray = **VarArray**(10, 1, 5)
> **Gcc**(myvariablearray, {3: [2, 2], 2: [0, 3], 4: [1, 10]})

to state that amongst the variables in 'myvariablearray', the value 3 must occur exactly twice, the value 2 at most three times, and the value 4 at least once.

Element. The element constraint [30] allows indexing into a variable or value array, at solving time, by the value of another variable. This can provide a very powerful modelling construct. A simple example of its use in Numberjack is:

> myvariablearray = **VarArray**(10, 1, 20)
> indexvar = **Variable**(10)
> y == Element(myvariablearray, indexvar)

This uses the value assigned to 'indexvar' as an index into the variable array 'myvariablearray', binding the resulting variable to be equal to the variable 'y'.

Cumulative. The cumulative constraint [2] proves extremely useful in many scheduling and packing problems. Two significant and important application areas for constraint programming. For example, in a scheduling scenario with a given set of tasks, each requiring a specific quantity of resource, the cumulative constraint restricts the total consumption of the resource to not exceed a predefined limit at each time point. Tasks are allowed to overlap but their cumulative resource consumption must not exceed a predefined fixed limit. Figure 4 illustrates an example schedule of five overlapping tasks on a resource with a capacity of 5. Given the scheduling of task 1 at time point 0, the earliest task 2 can start is 3 since its resource consumption is 2. Task 4 on the other hand can also start at 0, since its resource consumption of 1 fits within the remaining capacity. The cumulative constraint may also be viewed as modelling the packing of two-dimensional rectangles.

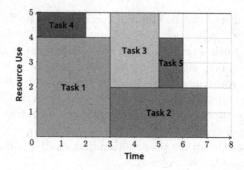

Fig. 4. Example task assignment on a cumulative resource.

2.5   Optimisation

Numerous industrial applications of combinatorial optimisation require going beyond a single satisfiable solution. Frequently the interest is in finding good, or the absolute best, quality solution. For example, we might wish to define the *objective function* to minimise cost, wastage, loss, or to maximise profit, yield, customer satisfaction, and so on. These expressions can intuitively be specified in Numberjack as follows:

Minimise(openingcosts + supplycosts)
Maximise(**Sum**(items, weights))

Different approaches are taken to solve such optimisation problems. Constraint programming can treat the objective function as another variable, performing branch and bound search on its range. It solves a series of satisfaction sub-problems, searching for a solution with an objective value below a certain threshold. On each subsequent call, the threshold is reduced until the problem is proven unsatisfiable or a resource limit has been exceeded. A satisfiability solver can similarly be used to solve some optimisation approaches, although its practicality is limited to problems where the domain of the objective function is small. Graphical model solvers perform sophisticated reasoning on the feasibility of bounds and values of local cost functions to tighten bounds on the objective. The application of the technology tends to be targeted at small, highly non-linear objective functions. Mixed integer programming solvers are most naturally suited to solving (linear) optimisation problems. The linear relaxations at their core yields effective lower-bounds. Critically, a MIP solver also examines the dual of the problem, yielding an upper-bound. Combining the two gives a precise indication of the range within which the optimal solution lies; when the two bounds are equal, optimality has been proven.

3   Solving Technologies

This section presents a more formal description of the aforementioned approaches to solving combinatorial problems.

3.1   Constraint Programming

Constraint programming problems are defined by a tuple $\langle \mathcal{X}, \mathcal{D}, \mathcal{C} \rangle$, defining the variables, domains, and constraints respectively. A variable $X_i \in \mathcal{X}$ has a domain of possible assignments from $\mathcal{D}$, denoted by $D(X_i) \in \mathcal{D}$. Constraints in $\mathcal{C}$ restrict the set of values which can be assigned to interconnected variables of the problem. For a given constraint $c \in \mathcal{C}$, we will refer to the relevant variables by the set $X(c)$, i.e. the *scope* of the constraint. In a binary constraint satisfaction problem, whereby $\forall c \in \mathcal{C} : |X(c)| \leq 2$, we may refer to a constraint between variables X_i and X_j by c_{ij}. The graph composed of nodes representing the variables and (hyper-)edges between the nodes representing the scopes of each constraint is often referred to as the *constraint network*.

In so far as is possible, constraint programming attempts to separate the definition of a problem from the solving process, to the extent that it is said to represent the holy grail of programming: "the user states the problem, the computer solves it" [19]. A solution to a CSP consists of a mapping from each variable to one of the values in its domain such that all constraints are satisfied. Solutions are typically found using a combination of backtracking-style search and inference; which are covered in Sects. 4 and 2.3 respectively.

3.2   Satisfiability

The satisfiability problem (SAT) [10] is one of the most prominent and long-standing areas of study in computer science, most notably by being the first problem to be proven $\mathcal{NP}$-complete and lying at the heart of the $\mathcal{P} \overset{?}{=} \mathcal{NP}$ question [14]. The problem consists of a set of Boolean variables and a propositional formula over these variables. The task is to decide whether or not there exists a truth assignment to the variables such that the propositional formula evaluates to true, and, if this is the case, to find this assignment.

SAT instances consist of a propositional logic formula, usually expressed in conjunctive normal form (CNF). The representation consists of a conjunction of clauses, where each clause is a disjunction of literals. A literal is either a Boolean variable or its negation. Each clause is a disjunction of its literals and the formula is a conjunction of each clause. The following SAT formula is in CNF:

$$(x_1 \lor x_2 \lor \neg x_4) \land (\neg x_2 \lor \neg x_3) \land (x_3 \lor x_4)$$

This instance consists of four SAT variables. One assignment to the variables which would satisfy the above formula would be to set $x_1 = true$, $x_2 = false$, $x_3 = true$, and $x_4 = true$.

3.3   Mixed Integer Programming

The mixed integer programming (MIP) [56] problem consists of a set of linear constraints over integer and real-valued variables, where the goal is to find an assignment to the variables minimising a linear objective function. More formally, a MIP problem takes the form:

$$min \quad cx + dy \tag{1}$$
$$s.t. \ Ax + By \geq 0 \tag{2}$$
$$x, y \geq 0 \tag{3}$$
$$y \text{ integer} \tag{4}$$

where x and y are two vectors of real-valued and integer variables, respectively. c and d are vectors of coefficients defining the objective function to be minimised. The matrices A and B represent coefficients of a set of linear constraints.

Analogous to the constraint and satisfiability solving techniques seen in previous sections, modern techniques for solving a mixed integer programming problems consist of a combination of search and various forms of inference. Firstly, a number of pre-solving techniques are applied which rewrite and reduce some parts of the constraints. This maintains the same form of problem, while generally resulting in a reduced, tighter problem.

Subsequently, the space of solutions is explored using branch and bound search. At each node in the search tree, the integrality constraints on variables in y are relaxed, the resulting formulation, namely the *LP relaxation*, is solved to optimality using linear programming techniques such as the simplex algorithm [41]. If it happens that the solution also satisfies the integrality constraints, then a feasible solution has been found. The best integer solution found during search is called the *incumbent* and its objective value provides an upper-bound on the optimal solution value.

In practice however, an integer solution to the LP relaxation rarely occurs and so the fractional solution is used to guide the search. Furthermore, the objective value of the non-integral solution also provides a lower-bound on the solution of the integral problem. The distance between the best lower and upper bound is deemed the *optimality gap*, when its value reaches zero, optimality has been proved. The search procedure then branches on one of the y variables for which a non-integral value was assigned. For example, if integer variable y_i was assigned the value 2.8 in the LP relaxation solution, then two sub-problems are created with constraints $y_i \leq 2$ and $y_i \geq 3$ respectively. If the solution to the LP relaxation in any of the resulting sub-problems is infeasible or is greater than the incumbent, then that node can be dropped and another node explored. This process is repeated recursively until optimality is proven or the problem is proved infeasible.

3.4   Choice Is Good

As the previous sections have outlined, the solution technologies for constraint programming, satisfiability, and mixed integer programming problems are all operationally different. Specifically: CP uses constraint propagation with back-tracking search; SAT utilises unit-propagation, clause learning, and search; and

MIP exploits linear relaxations, cutting planes, with branch and bound search. Often, it is not clear which solution technology is best suited for a particular problem so it can be worthwhile to experiment with different approaches. Fortunately, the user does not need to manually produce a different model for each approach since many problems can be encoded between CP, SAT, and MIP; a process which can be significantly simplified by using modelling frameworks. The following sections illustrate the performance differences between approaches on some example problems.

3.4.1   Example: Warehouse Location Problem

The Warehouse Location Problem [31] considers a set of existing shops and a candidate set of warehouses to be opened, the problem is to choose which warehouses are to be opened and consequently the respective shops which each one will supply. There is a cost associated with opening each warehouse, as well as a supply cost for each warehouse-shop supply pair, the objective being to minimise the total cost of warehouse operations and supply costs. A complete Numberjack model for the warehouse location problem is given in Fig. 5.

Table 2 compares the performance of a mixed integer programming solver and a constraint programming solver, namely SCIP and Mistral respectively, on some instances of the Warehouse Location Problem. SCIP is able to solve each of instance to optimality very quickly, whereas the CP solver takes over one hour of CPU-time to find solutions of worse quality. In this case, the CP solver is not able to perform much reasoning on the objective function for this problem, a weighted linear sum, whereas the MIP solver is able to produce tight bounds very quickly and narrow the search.

3.4.2   Example: Highly Combinatorial Puzzles

We compare a constraint programming, a satisfiability, and a mixed integer programming solver on some benchmarks of two arithmetic puzzles. Specifically, constructing a Costas Array and constructing a Golomb ruler of minimal size. Both of these problems are parameterised by a single value specifying the size of the instance. The Costas Array problem [15] is to place n points on an $n \times n$ board such that each row and column contains only one point, and the pairwise distances between points is also distinct. This can be modelled using a vector of n variables to decide the column of each point, and enforcing all-different constraints on the vector of variables and on the triangular distance matrix. A Golomb ruler [52] is defined by placing a set of m marks at integer positions on a ruler such that the pairwise differences between marks are distinct. The objective is to find rulers of minimal length. Numberjack models for the Costas Array and Golomb Ruler problems are presented in Figs. 6 and 7 respectively. Problems such as these are not limited to academic interest but do map to many real world applications.

Table 3 illustrates the empirical performance differences between CP, SAT, and MIP approaches on these problems. Here, the constraint programming solver (Mistral) is very effective. The satisfiability solver performs comparably well on

```
1   model = Model()

3   # 0/1 for each warehouse to decide which ones to open
4   WareHouseOpen = VarArray(data.NumberOfWarehouses)

6   # 0/1 matrixfor each shop (row) decide which warehouse (col) will supply it
7   ShopSupplied = Matrix(data.NumberOfShops, data.NumberOfWarehouses)

9   # Cost of running warehouses
10  warehouseCost = Sum(WareHouseOpen, data.WareHouseCosts)

12  # Cost of shops using warehouses
13  transpCost = Sum([Sum(varRow, costRow) for varRow, costRow in zip(
        ShopSupplied, data.SupplyCost)])

15  # Objective function
16  obj = warehouseCost + transpCost
17  model += Minimise(obj)

19  # Channel from store opening to store supply matrix
20  for col, store in zip(ShopSupplied.col, WareHouseOpen):
21      model += [var <= store for var in col]

23  # Make sure every shop is supplied by one warehouse
24  for row in ShopSupplied.row:
25      model += Sum(row) == 1

27  # Make sure that each warehouse does not exceed it's supply capacity
28  for col, cap in zip(ShopSupplied.col, data.Capacity):
29      model += Sum(col) <= cap

31  # Load the model with a named solver
32  solver = model.load("SCIP")

34  # Ask the solver to solve
35  solver.solve()

37  if solver.is_sat():
38      ... # print solution
39  elif solver.is_unsat():
40      print "Unsatisfiable"
```

Fig. 5. Model of the Warehouse Location Problem in Numberjack.

the Costas array problem, but when dealing with the optimisation problem of the Golomb ruler, it fails to scale. However, it does outperform the mixed integer programming solver which performs very poorly on these problems.

Table 2. Comparison between a mixed integer programming solver (SCIP) and a constraint programming solver (Mistral) on some instances of the Warehouse Location Problem.

Instance	SCIP			Mistral		
	Objective	Nodes	Time	Objective	Nodes	Time
cap44	1184690	1	0.84	1468957	10008044	>3600
cap63	1087190	14	1.82	1388391	10683754	>3600
cap71	957125	1	0.69	1297505	11029722	>3600
cap81	811324	1	0.65	1409091	3497095	>3600
cap131	954894	5	5.30	1457632	1281009	>3600

```
1   model = Model()

3   # N variables with domains 1..N representing the column of point in each row
4   seq = VarArray(N, 1, N)

6   # Points must be placed in distinct columns
7   model += AllDiff(seq)

9   # Each row of the triangular distance matrix contains no repeat distances
10  for i in range(N−2):
11      model += AllDiff([seq[j] − seq[j+i+1] for j in range(N−i−1)])
```

Fig. 6. Model of the Costas Array Problem in Numberjack.

```
1   model = Model()

3   # A vector of finite domain variables for the position of each mark
4   marks = VarArray(m, 2**(m−1))

6   # Pairwise distances are distinct
7   distance = [marks[i] − marks[j] for i in range(1, m) for j in range(i)]
8   model += AllDiff(distance)

10  # Symmetry breaking
11  model += marks[0] == 0
12  for i in range(1, m):
13      model += marks[i−1] < marks[i]

15  # Minimise the position of the last mark
16  model += Minimise(marks[−1])
```

Fig. 7. Model of the Golomb Ruler Problem in Numberjack.

Table 3. Performance of a constraint programming, satisfiability, and mixed integer programming solver on two arithmetic puzzles of increasing size. Values are CPU time in seconds, '-' represents a timeout, and 'M' a memory limit of 2 GB exceeded.

Instance	Mistral	MiniSat	SCIP
Costas (11)	0.0	0.0	27.0
Costas (12)	0.0	0.0	166.0
Costas (13)	0.0	0.0	286.0
Costas (14)	1.0	0.0	1065.0
Costas (15)	9.0	0.0	2564.0
Costas (16)	52.0	16.0	-
Costas (17)	562.0	163.0	-
Costas (18)	529.0	677.0	-
Golomb (6)	0.0	0.0	2.0
Golomb (7)	0.0	0.0	17.0
Golomb (8)	0.0	2.0	59.0
Golomb (9)	0.0	34.0	1778.0
Golomb (10)	3.0	M	-
Golomb (11)	133.0	M	-
Golomb (12)	3006.0	M	M

4 Systematic Search

Chronological backtracking search plays a central role in the solution process for combinatorial problems. Nodes in the search correspond to variables, and branches to assignments, thus the search explores the tree of possible partial solutions. Figure 8 illustrates a partial example of the search tree generated by backtracking search. Initially, from the *root node*, the variable X is branched on, taking one branch for each possible value in its domain.

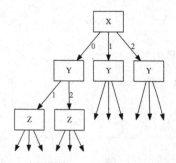

Fig. 8. An partial example of the search tree generated by backtracking search.

Modern constraint programming solvers typically perform *binary-branching* on the assignment or removal of a value from the domain. The process of *maintaining arc-consistency* (MAC) [48] during search has been shown to be highly effective. This consists of making the initial CSP arc-consistent before starting search, then again after every assignment and every backtrack. A *domain wipe-out* occurs when a variable has no values remaining in its domain. When this occurs search must backtrack and explore a different path. A solution has been found when all variables have been assigned a value in their domain which is globally consistent with the constraints.

Notably, if a bad decision is made early in the search, then the resulting sub-tree may be unsatisfiable. It may take exponential time for the search to prove that no solution exists in the sub-tree, a *refutation*, before backtracking to the bad decision node [32]. The *thrashing* phenomenon occurs when the current partial assignment cannot be extended to a solution but search continues backtracking on the remaining variables, trying all possible values when the real source of inconsistency is a bad decision higher up the tree.

To avoid such worst-case behaviour, a number of methods such as randomised restarting, back-jumping, and explanation-based search have been proposed. Nevertheless, an important decision to be made arises concerning what order the tree should be explored. These topics are discussed in the following sections.

4.1   Search Heuristics in Constraint Programming

Two closely-related decisions which are vital for success are the choice of variable to branch on and the subsequent value it will be assigned. These decisions have a dramatic affect on the size of the search tree that will be explored. Interestingly, an oracle proposing the value ordering could lead search directly to a solution without backtracking (if the problem is satisfiable), regardless of the variable ordering. In practice however, such an oracle is implausible so heuristic methods must be used.

The CSP community has devised a number of generic, problem independent heuristics for users to choose from. Options range from static heuristics such a selecting the variables in order of their domain size or degree of connectivity in the constraint-graph, to dynamic heuristics based on the activity of the solver during search such as weighted heuristics [11], and impact-based [44] to name a few.

To avoid bad decisions early in the search tree, the variable ordering heuristic, in general, follows a *fail-first* principle [28] whereby variables likely to lead to failure should be chosen first. Effort should be focused on difficult parts of the problem likely to lead to failure, which should ideally occur early in the search. Value ordering heuristics on the other hand try to select the most promising value, one most likely to lead to a solution [22].

Choosing an effective heuristic is a highly problem dependant task, often requiring intimate knowledge of the underlying technology, an undertaking often beyond the reach of many users. Automating such a task, simplifying the barrier to entry for users, has been proposed as one of the grand challenges for constraint programming [20]. One approach to this is to use a machine learning model to automatically select the heuristic based on instance specific features [12,23,35,39].

4.2 Restarting and Randomness

In practice, the search procedure will encounter many failures and have to backtrack. As mentioned previously, one risk occurs if a bad decision has been made early in the search process and proving that no solution exists in the sub-tree may take exponential time. One approach to avoiding such behaviour is to restart the search from the root node after a pre-defined limit on the number of failures has been reached [37].

To maintain the completeness of the search process, solvers adopt a restarting strategy whereby the failure limit eventually tends towards infinity. A restart strategy is defined by a sequence $\langle t_1, t_2, t_3, \ldots \rangle$ whereby each t_i specifies the limit on the number of failures for a particular run of the algorithm. Once the failure limit t_i is reached, the search is restarted from the root node with the new limit of t_{i+1}.

Two standard restart strategies are based on the *Luby* and *geometric* sequences. The Luby [37] sequence has the form $\langle 1, 1, 2, 1, 1, 2, 4, 1, 1, 2, 1, 1, 2, 4, \ldots \rangle$. In the context of Las Vegas algorithms [5] it is proven to be universally optimal, achieving a runtime that is only a logarithmic factor from an optimal restart strategy where the runtime distribution of the underlying algorithm is fully known, and no other universal strategy can do better by more than a constant factor [37]. Alternatively, the geometric [55] sequence increases the cutoff by a constant factor between each run.

Restarting is typically combined with randomisation in the variable and value heuristics to avoid repeatedly exploring the same search space. Such stochastic behaviour gives rise to solvers exhibiting a distribution of runtimes. In some cases, modelled by heavy- and fat-tailed distributions [26], possibly with infinite mean and variance. These distributions capture a non-negligible fraction of runs far to the right or left of the median, runs taking extremely long. *Rapid randomised restarting* [25,27] has been shown to eliminate heavy-tails to the right of the median and can even take advantage of heavy-tails to the left of the median.

5 Final Remarks

This chapter has presented an introduction to three areas of combinatorial optimisation, specifically constraint programming, satisfiability, and mixed integer programming. The contrasting approaches that each of these paradigms take to solving such problems is presented along with examples using Numberjack.

One of the underlying difficulties for new users of these technologies is in producing an effective solution. Some progress has been made to alleviate this burden, such as the Constraint Seeker [7] which identifies and ranks global constraints satisfying a given solution vector. The ModelSeeker [8] extends this to identify complete global constraint models satisfying a set of solutions. Conacq [13] interactively learns a constraint network by proposing partial solutions to the user. Automated Configuration tools help find good parameterisations of a solver, helping boost performance on problem classes [4,18,34]. Portfolio approaches unite the complimentary strengths of a collection of solvers [3,33,43,57], making decisions on an instance specific basis of which solver to be used.

References

1. Tourbier, Y., Oplobedu, A., Marcovitch, J., CHARME: un langage industriel de programmation par contraintes, illustr par une application chez Renault. In: Proceedings of the Ninth International Workshop on Expert Systems and their Applications, pp. 55–70 (1989)
2. Aggoun, A., Beldiceanu, N.: Extending CHIP in order to solve complex scheduling and placement problems. In: JFPL 1992, 1éres Journées Francophones de Programmation Logique, p. 51 (1992)
3. Amadini, R., Gabbrielli, M., Mauro, J.: A multicore tool for constraint solving. In: Proceedings of the Twenty-Fourth International Joint Conference on Artificial Intelligence, IJCAI 2015, pp. 232–238 (2015)
4. Ansótegui, C., Sellmann, M., Tierney, K.: A gender-based genetic algorithm for the automatic configuration of algorithms. In: Gent, I.P. (ed.) CP 2009. LNCS, vol. 5732, pp. 142–157. Springer, Heidelberg (2009). doi:10.1007/978-3-642-04244-7_14
5. Babai, L.: Monte-carlo algorithms in graph isomorphism testing. Technical report DMS 79–10, Université de Montréal (1979)
6. Beldiceanu, N., Carlsson, M., Rampon, J.-X.: Global constraint catalog. Technical report SICS-T 2005/08-SE (2005)
7. Beldiceanu, N., Simonis, H.: A constraint seeker: finding and ranking global constraints from examples. In: Lee, J. (ed.) CP 2011. LNCS, vol. 6876, pp. 12–26. Springer, Heidelberg (2011). doi:10.1007/978-3-642-23786-7_4
8. Beldiceanu, N., Simonis, H.: A model seeker: extracting global constraint models from positive examples. In: Milano, M. (ed.) CP 2012. LNCS, vol. 7514, pp. 141–157. Springer, Heidelberg (2012). doi:10.1007/978-3-642-33558-7_13
9. Bessiere, C.: Constraint propagation. In: Handbook of Constraint Programming, pp. 29–83 (2006)
10. Biere, A., Heule, M.J.H., van Maaren, H., Walsh, T. (eds.): Handbook of Satisfiability. Frontiers in Artificial Intelligence and Applications, vol. 185. IOS Press, Amsterdam (2009)
11. Boussemart, F., Hemery, F., Lecoutre, C., Sais, L.: Boosting systematic search by weighting constraints. In: Proceedings of the 16th Eureopean Conference on Artificial Intelligence, ECAI 2004, pp. 146–150 (2004)
12. Chu, G., Stuckey, P.J.: Learning value heuristics for constraint programming. In: Michel, L. (ed.) CPAIOR 2015. LNCS, vol. 9075, pp. 108–123. Springer, Heidelberg (2015). doi:10.1007/978-3-319-18008-3_8
13. Coletta, R., Bessiére, C., O'Sullivan, B., Freuder, E.C., O'Connell, S., Quinqueton, J.: Semi-automatic modeling by constraint acquisition. In: Rossi, F. (ed.) CP 2003. LNCS, vol. 2833, pp. 812–816. Springer, Heidelberg (2003). doi:10.1007/978-3-540-45193-8_58

14. Cook, S.A.: The complexity of theorem-proving procedures. In: Proceedings of the 3rd Annual ACM Symposium on Theory of Computing, pp. 151–158 (1971)
15. Costas, J.P.: A study of a class of detection waveforms having nearly ideal range - doppler ambiguity properties. Proc. IEEE **72**(8), 996–1009 (1984)
16. Dooms, G.: The CP(Graph) computation domain in constraint programming. Ph.D. thesis, Université catholique de Louvain, Faculté des sciences appliquées (2006)
17. Fages, J.-G.: Exploitation de structures de graphe en programmation par contraintes. (On the use of graphs within constraint-programming). Ph.D. thesis, École des mines de Nantes, France (2014)
18. Fitzgerald, T., Malitsky, Y., O'Sullivan, B., Tierney, K.: ReACT: real-time algorithm configuration through tournaments. In: Proceedings of the Seventh Annual Symposium on Combinatorial Search, SOCS 2014 (2014)
19. Freuder, E.C.: In pursuit of the holy grail. Constraints **2**(1), 57–61 (1997)
20. Freuder, E.C., O'Sullivan, B.: Grand challenges for constraint programming. Constraints **19**(2), 150–162 (2014)
21. Frisch, A.M., Harvey, W., Jefferson, C., Martínez-Hernández, B., Miguel, I.: Essence: a constraint language for specifying combinatorial problems. Constraints **13**(3), 268–306 (2008)
22. Geelen, P.A.: Dual viewpoint heuristics for binary constraint satisfaction problems. In: Proceedings of the 10th European Conference on Artificial Intelligence, ECAI 1992, pp. 31–35. Wiley (1992)
23. Gent, I.P., Jefferson, C., Kotthoff, L., Miguel, I., Moore, N.C.A., Nightingale, P., Petrie, K.E.: Learning when to use lazy learning in constraint solving. In: Proceedings of the 19th European Conference on Artificial Intelligence, ECAI 2010, pp. 873–878 (2010)
24. Gervet, C.: Set intervals in constraint-logic programming: definition and implementation of a language. Ph.D. thesis, Université de France-Compté (1995)
25. Gomes, C.P., Sabharwal, A.: Exploiting runtime variation in complete solvers. In: Handbook of Satisfiability, pp. 271–288 (2009)
26. Gomes, C.P., Selman, B., Crato, N., Kautz, H.: Heavy-tailed phenomena in satisfiability and constraint satisfaction problems. J. Autom. Reason. **24**(1–2), 67–100 (2000)
27. Gomes, C.P., Selman, B., Kautz, H.A.: Boosting combinatorial search through randomization. In: Proceedings of the 15th National Conference on Artificial Intelligence, AAAI 1998, pp. 431–437 (1998)
28. Haralick, R.M., Elliott, G.L.: Increasing tree search efficiency for constraint satisfaction problems. Artif. Intell. **14**(3), 263–313 (1980)
29. Hebrard, E., O'Mahony, E., O'Sullivan, B.: Constraint programming and combinatorial optimisation in numberjack. In: Lodi, A., Milano, M., Toth, P. (eds.) CPAIOR 2010. LNCS, vol. 6140, pp. 181–185. Springer, Heidelberg (2010). doi:10.1007/978-3-642-13520-0_22
30. Van Hentenryck, P., Carillon, J.-P., Generality versus specificity: an experience with AI and OR techniques. In: Proceedings of the 7th National Conference on Artificial Intelligence, AAAI 1988, pp. 660–664 (1988)
31. Hnich, B.: CSPLib problem 034: Warehouse location problem. http://www.csplib.org/Problems/prob034
32. Hulubei, T., O'Sullivan, B.: The impact of search heuristics on heavy-tailed behaviour. Constraints **11**(2–3), 159–178 (2006)

33. Hurley, B., Kotthoff, L., Malitsky, Y., O'Sullivan, B.: Proteus: a hierarchical portfolio of solvers and transformations. In: Simonis, H. (ed.) CPAIOR 2014. LNCS, vol. 8451, pp. 301–317. Springer, Heidelberg (2014). doi:10.1007/978-3-319-07046-9_22

34. Hutter, F., Hoos, H.H., Leyton-Brown, K.: Sequential model-based optimization for general algorithm configuration. In: Coello, C.A.C. (ed.) LION 2011. LNCS, vol. 6683, pp. 507–523. Springer, Heidelberg (2011). doi:10.1007/978-3-642-25566-3_40

35. Kotthoff, L., Gent, I.P., Miguel. I.: A preliminary evaluation of machine learning in algorithm selection for search problems. In: Proceedings of the 4th Annual Symposium on Combinatorial Search, SOCS 2011 (2011)

36. Laurière, J.-L.: A language and a program for stating and solving combinatorial problems. Artif. Intell. 10(1), 29–127 (1978)

37. Luby, M., Sinclair, A., Zuckerman, D.: Optimal speedup of las vegas algorithms. Inf. Process. Lett. 47(4), 173–180 (1993)

38. Mackworth, A.K.: Consistency in networks of relations. Artif. Intell. 8(1), 99–118 (1977)

39. Mehta, D., O'Sullivan, B., Kotthoff, L., Malitsky, Y.: Lazy branching for constraint satisfaction. In: Proceedings of the 25th International Conference on Tools with Artificial Intelligence, ICTAI 2013, pp. 1012–1019 (2013)

40. Mehta, D., O'Sullivan, B., Quesada, L., Ruffini, M., Payne, D.B., Doyle, L.: Designing resilient long-reach passive optical networks. In: Proceedings of the 23rd Conference on Innovative Applications of Artificial Intelligence, IAAI 2011 (2011)

41. Murty, K.G.: Linear Programming. Wiley, Hoboken (1983)

42. Nethercote, N., Stuckey, P.J., Becket, R., Brand, S., Duck, G.J., Tack, G.: MiniZinc: towards a standard CP modelling language. In: Bessière, C. (ed.) CP 2007. LNCS, vol. 4741, pp. 529–543. Springer, Heidelberg (2007). doi:10.1007/978-3-540-74970-7_38

43. O'Mahony, E., Hebrard, E., Holland, A., Nugent, C., O'Sullivan, B.: Using case-based reasoning in an algorithm portfolio for constraint solving. In: Irish Conference on Artificial Intelligence and Cognitive Science (2008)

44. Refalo, P.: Impact-based search strategies for constraint programming. In: Wallace, M. (ed.) CP 2004. LNCS, vol. 3258, pp. 557–571. Springer, Heidelberg (2004). doi:10.1007/978-3-540-30201-8_41

45. Régin, J.-C.: A filtering algorithm for constraints of difference in csps. In: Proceedings of the 12th National Conference on Artificial Intelligence, AAAI 1994, pp. 362–367 (1994)

46. Régin, J.-C.: Generalized arc consistency for global cardinality constraint. In: Proceedings of the 8th Innovative Applications of Artificial Intelligence Conference, IAAI 1996, pp. 209–215 (1996)

47. Rossi, F., van Beek, P., Walsh, T.: Handbook of Constraint Programming. Foundations of Artificial Intelligence. Elsevier, New York (2006)

48. Sabin, D., Freuder, E.C.: Contradicting conventional wisdom in constraint satisfaction. In: Proceedings of the 11th European Conference on Artificial Intelligence, ECAI 1994, pp. 125–129. Springer, Heidelberg (1994)

49. Simonin, G., Artigues, C., Hebrard, E., Lopez, P.: Scheduling scientific experiments on the rosetta/philae mission. In: Milano, M. (ed.) Principles and Practice of Constraint Programming. LNCS, vol. 7514, pp. 23–37. Springer, Heidelberg (2012)

50. Simonis, H.: Constraint applications in networks. In: Handbook of Constraint Programming, pp. 875–903 (2006)

51. Simonis, H.: Models for global constraint applications. Constraints 12(1), 63–92 (2007)

52. van Beek, P.: CSPLib problem 006: Golomb rulers. http://www.csplib.org/Problems/prob006
53. van Hoeve, W.-J., Katriel, I.: Global constraints. In: Handbook of Constraint Programming. Foundations of Artificial Intelligence, vol. 2, pp. 169–208. Elsevier (2006)
54. Wallace, M.: Practical applications of constraint programming. Constraints 1(1/2), 139–168 (1996)
55. Walsh, T.: Search in a small world. In: Proceedings of the 16th International Joint Conference on Artificial Intelligence, IJCAI 1999, pp. 1172–1177 (1999)
56. Wolsey, L.A.: Integer Programming. Wiley-Interscience, New York (1998)
57. Xu, L., Hutter, F., Hoos, H.H., Leyton-Brown, K.: SATzilla: portfolio-based algorithm selection for SAT. J. Artif. Intell. Res. **32**, 565–606 (2008)
58. Yip, Y.K.J.: The length-lex representation for constraint programming over sets. Ph.D. thesis, Brown University (2011)

Data Mining and Constraints: An Overview

Valerio Grossi$^{(\boxtimes)}$, Dino Pedreschi, and Franco Turini

Department of Computer Science, University of Pisa,
Largo B. Pontecorvo, 3, 56127 Pisa, Italy
{vgrossi,pedre}@di.unipi.it, turini@unipi.it

Abstract. This paper provides an overview of the current state-of-the-art on using constraints in knowledge discovery and data mining. The use of constraints requires mechanisms for defining and evaluating them during the knowledge extraction process. We give a structured account of three main groups of constraints based on the specific context in which they are defined and used. The aim is to provide a complete view on constraints as a building block of data mining methods.

1 Introduction

Data mining extracts synthetic models from datasets. Data are represented by collections of records characterizing data with respect to several dimensions. The use of constraints may be useful in the data mining process in at least three ways: *(i)* *filtering* and *organizing* the dataset before applying data mining methods; *(ii)* *improving* the performance of data mining algorithms by reducing the search space and focusing the search itself; and *(iii)* *reasoning* on the results of the mining step for sharpening them and presenting a more refined view of the extracted models.

The integration of constraints in data mining tasks has rapidly emerged as a challenging topic for the research community. A large number of ad-hoc extensions of mining algorithms use constraints for improving the quality of their results. The use of constraints requires a way for defining and satisfying them during the knowledge extraction process. This point is crucial both for the quality of the extracted data mining models, and for the scalability of the entire process. On the one hand, an analyst can define the knowledge extraction phase where a constraint must be satisfied. On the other hand, an optimizer is required to understand where a constraint must be satisfied inside the process flow, in an automatic way. Moreover, mining algorithms must be rewritten for satisfying constraints directly into model extraction.

The amount of data in our world has been exploding. This chapter ends offering the user a glimpse at the future by considering the emerging phenomenon of *big data*. With big data traditional analysis tools cannot be used because of the massive volume of data gathered by automated collection tools, there are already promising line researches addressing this issue.

Furthermore, this chapter represents a solid scientific basis for several advanced techniques developed inside the ICON project and outlined in this book.

© Springer International Publishing AG 2016

C. Bessiere et al. (Eds.): Data Mining and Constraint Programming, LNAI 10101, pp. 25–48, 2016.
DOI: 10.1007/978-3-319-50137-6_2

For example the reader can examine in depth the use of a constraint language for defining data mining tasks considering the Chapter *"Modeling Data Mining Problems in MiningZinc"*, or study clustering problems via constraints optimization reading the Chapter *"Partition-Based Clustering using Constraints Optimization"*.

For these aims, Sect. 2 provides an introduction to data mining and proposes several references useful to understand how the basic data mining concepts can be extended by using constraints. Section 3 reviews the use of constraints in data mining, introducing three different dimensions on which constraints can be classified. Finally, Sect. 4 draws some conclusions.

2   Data Mining

Today, data mining is both a technology that blends data analysis methods with sophisticated algorithms for processing large data sets, and an active research field that aims at developing new data analysis methods for novel forms of data. On the one hand, data mining tools are now part of mature data analysis systems and have been successfully applied to problems in various commercial and scientific domains. On the other hand, the increasing heterogeneity and complexity of new forms of data, such as those arriving from medicine, biology, the Web, Earth observation systems, call for new forms of patterns and models, together with new algorithms to discover such patterns and models efficiently.

Data mining is originally defined as the process of automatically discovering useful information in large data repositories. Traditionally, data mining is only a step of knowledge discovery in databases, the so-called KDD process for converting raw data into useful knowledge. The KDD process consists of a series of transformation steps: *data preprocessing*, which transforms the raw source data into an appropriate form for the subsequent analysis. *Actual data mining*, which transforms the prepared data into patterns or models, and *postprocessing of mined results*, which assesses validity and usefulness of the extracted patterns and models, and presents interesting knowledge to the final users - business analysts, scientists, planners, etc. – by using appropriate visual metaphors or integrating knowledge into decision support systems.

The three most popular data mining techniques are *predictive modelling*, *cluster analysis* and *association analysis*. In predictive modelling (Sect. 2.1), the goal is to develop classification models capable of predicting the value of a class label (or target variable) as a function of other variables (explanatory variables); the model is learnt from historical observations, where the class label of each sample is known: once constructed, a classification model is used to predict the class label of new samples whose class is unknown, as in forecasting whether a patient has a given disease based on the results of medical tests.

In association analysis, also called pattern discovery, the goal is precisely to discover patterns that describe strong correlations among features in the data or associations among features that occur frequently in the data (see Sect. 2.3). Often, the discovered patterns are presented in the form of association rules:

useful applications of association analysis include market basket analysis, i.e. the task of finding items that are frequently purchased together, based on point-of-sale data collected at cash registers.

Finally, in cluster analysis (Sect. 2.2), the goal is to partition a data set into groups of closely related data in such a way that the observations belonging to the same group, or cluster, are similar to each other, while the observations belonging to different clusters are not. Clustering can be used, for instance, to find segments of customers with a similar purchasing behaviour or categories of documents pertaining to related topics.

2.1 Predictive Modelling or Classification

Classification is one of the most popular approaches for mining useful information. The aim is to predict the behavior of new elements (*classification phase*), given a set of past and already classified instances. The process of classifying new data begins from a set of classified elements, and tries to extract some regularities from them (*training phase*) [WFH11, TSK06, HK12]. The model employs a set of input data called *training set* where the class label for each instance is provided. The process of classifying new data starts from a training set, and tries to extract some regularities from them. Classification is an example of *supervised learning*.

Based on the way learners actually subdivide the above-mentioned phases, they are categorized into two classes, namely *eager learners* or *lazy learners*. For example, decision trees or rule-based learners are examples of *eager* approaches. In this category, most of the computing resources are spent to extract a model, but once a model has been built, classifying a new object is a rather fast process.

By contrast, lazy learners, such as *nearest-neighbour classifiers* do not require an explicit model building phase, but classifying a test example can be very expensive, since the element to classify must be compared with all the samples in the training set. In the following, we provide a short description of the most popular classifiers available in the literature.

Decision Trees. The model has the form of a tree, where each node contains a test on an attribute, each branch from a node corresponds to a possible outcome of the test, and each leaf contains a predicted class label [Mor82]. Decision tree induction often uses a greedy top-down approach which recursively replaces leaves by test nodes, starting from the root. The attribute associated to each node is chosen through the comparison of all the available attributes, and the selection of the best one is based on some heuristic measures. Several impurity measures are available in the literature [Qui86, Qui93, BFOS84]. Typically, the measures developed are based on the degree of impurity of the child nodes. The lower is the value, the more skewed is the class distribution. The extraction procedure continues until a termination condition is satisfied.

The Hunt's algorithm represented in Algorithm 1 is the basis of several popular decision tree learners including ID3 [Qui86], CART [BFOS84], C4.5 [Qui93, Qui96] and EC4.5 [Rug02]. The cited approaches assume that all training

---

Algorithm 1. The Hunt's algorithm - *DecisionTree(TS, A)*

Require: Training set *TS*, an attribute set *A*
Ensure: Decision tree
1: **if** *stoppingCondition(TS, A)* = *true* **then**
2:    *leaf* ← *createLeaf(TS)* //given *TS* determines the class label to assign a leaf
      node
3:    **return** *leaf*
4: **else**
5:    *root* ← *createNode()*
6:    *root.testCondition* ← *findBestSplit(TS, A)*
7:    *TS_i* ← *splitData(root.testCondition)* //given the test condition splits *TS* in sub-
      sets
8:    **for each** *TS_i* **do**
9:       *root.child_i* ← *DecisionTree(TS_i, A)*
10:   **end for**
11: **end if**
12: **return** *root*

---

examples can be simultaneously stored in main memory, and thus have a limited number of examples from which they can learn. [LLS00] shows a comparison of complexity, training time and prediction accuracy of main memory classification algorithms, including decision trees. In several cases, training data can exceed the main memory capability. In order to avoid this limitation, disk-based decision tree learners, such as SLIQ [MAR96] and SPRINT [SAM96], assume the examples to be stored on disk, and are learned by repeatedly reading them in a sequence. More recently, new data structures and algorithms have been defined to tackle the classification problem in stream environments, also using decision trees [GT12, GS11].

Bayesian Approaches. In many situations, the relationship between the attributes and the class variable cannot be deterministic. This situation typically occurs in the presence of noisy data, or when external factors affecting classification, not included in our analysis, arises. Based on Bayes theorem, Bayesian classifiers are robust to isolate noisy points and irrelevant attributes.

A popular approach of Bayesian classification is naïve Bayes. This kind of classifier estimates the class-conditional probability, by assuming that the attributes are conditionally independent. To classify a record, the algorithm computes the posterior probability of a class value using Bayes theorem, and returns the class that maximizes this probability value. The way of computing class-conditional distribution varies in the presence of categorical or continuous attributes. In the first case, the conditional probability is estimated using the fraction of training samples with a specific class label considering an attribute value. By contrast, continuous attributes must be discretized, or a Gaussian distribution is typically chosen to compute the class-conditional probability.

Detailed discussions on Bayesian classifiers can be found in [DH73, Mic97, WK91]. An analysis of the accuracy of naïve Bayes classifiers without class

---

Algorithm 2. The k-nearest neighbour algorithm

Require: Training set TS, the number of nearest neighbour k
Ensure: Set of k nearest neighbours
 1: **for each** test example $z = (x',y')$ do **do**
 2:     $Distance(x', x) \leftarrow$ compute the distance between z and every training element
         $(x, y) \in TS$
 3:     $TS_s \leftarrow$ Select the k closest training example to z
 4:     $class \leftarrow FindClass(TS_s)$
 5:     **return** $class$
 6: **end for**

---

conditional independence hypothesis is available in [DP96], while [Jen96] provides a first overview of Bayesian networks.

Nearest Neighbour. This kind of classifier belongs to the family of lazy learners. In this case, every training example is viewed as a point in a multidimensional space, defined on the number of the available attributes.

As shown in Algorithm 2, given an element to classify, the call label is chosen based on the label of element neighbours selected by a proximity measure. In this case, specific training instances are employed to provide a prediction, without providing any model derived from data. Every training example is viewed as a point in a multidimensional space, defined on the number of the available attributes. In real applications only k points, that are closest to the element to classify are selected to decide the class label to return. The crucial aspect is to select the measures of proximity, that similarly to clustering are based on attribute types and special issues to solve. Due to its nature these models are rather sensible to noisy data and the prediction accuracy is highly influenced by the data preprocessing step and proximity measure.

With respect to decision trees, nearest-neighbor classifier provides a more flexible model representation. It produces arbitrarily-shaped boundaries, while decision trees are typically constrained to rectilinear decision boundaries [TSK06,HK12].

Support Vector Machine. This kind of approaches has its root in statistical learning theory. They have been successfully employed in many real applications, including handwritten digit recognition, and text categorization among others.

The main idea of this method is representing the decision boundary using a subset of training examples, known as support vectors. A support vector machine constructs a hyperplane (or set of hyperplanes) in a multi-dimensional space, which can be used for classification, regression, or other tasks. Essentially, given a set of possible hyperplanes (implicitly defined in the data), the classifier selects one hyperplane for representing its decision boundary, based on how well they are expected to perform on test examples. A support vector approach is typically described as linear or non-linear. The former involves a linear decision boundary to split the training objects into respective classes [ABR64]. Non-linear models try to compute a boundary for separating objects that cannot be represented by a linear model [BGV92]. The trick is to transform the data from its original space into a new space that can be divided by a linear bound. In the literature several approaches are available for learning a support vector model [CV95, Bur98,SC08].

2.2 Clustering

Clustering is the process of partitioning a set of data objects into subsets without any supervisory information such as data labels. Each subset is a cluster, such that objects in a cluster are similar to one another, yet dissimilar to objects in other clusters. The set of clusters resulting from a cluster analysis can be referred to as a clustering [WFH11,TSK06,HK12]. Clustering can lead to the discovery of previously unknown groups within the data. Examples of data objects include database records, graph nodes, a set of features describing individuals or images. Because there is no a priori knowledge about the class labels, clustering is also called unsupervised learning. Cluster analysis is used in a wide range of applications such as: business intelligence, image pattern recognition, web analysis, or biology.

The following general aspects are orthogonal characteristics in which clustering methods can be compared:

- **the partitioning criteria**: all the clusters are at the same level *vs.* partitioning data objects hierarchically, where clusters can be formed at different semantic levels.
- **separation of clusters**: methods partitioning data objects into mutually exclusive clusters *vs.* a data object may belong to more than one cluster.
- **similarity measure**: similarity measures play a fundamental role in the design of clustering methods. Some methods determine the similarity between two objects by the distance between them *vs.* the similarity may be defined by *connectivity* based on density or contiguity.
- **clustering space**: the entire given data space *vs.* subspace clustering.

The literature proposes several ways to compute and represent a cluster. The partition method is based on prototypes and is one of the most widely studied and applied approaches. In this case, every cluster is selected and represented by a prototype called *centroid* (e.g. *K-means* and *K-medoid*). Prototype-based techniques tend to consider the region only based on a distance value from a center. This approach typically provides clusters having globular shapes. Hierarchical-clustering is a method of cluster analysis which seeks to build a hierarchy of clusters. Also this kind of clustering is typically based on distance measures, but in this case, we permit clusters to have subclusters thus forming a tree. Each cluster i.e. a node in the tree, is the union of its subclusters, and the root of the tree is the cluster containing all the objects. The class of approaches for hierarchical clustering can be found under the agglomerative hierarchical clustering. BIRCH [ZRL96] is a famous example of hierarchical clustering algorithm.

Algorithm 3. The k-means algorithm

Require: Set of points P
Ensure: Set of k clusters
1: **repeat**
2:    Form k clusters by assigning each point $p_i \in P$ to the closest centroid
3:    *centroids* ← Recompute the centroid of each cluster
4: **until** *centroids* do not change

Density-based approaches work also with non-globular regions and they are designed for discovering dense areas surrounded by areas with low density (typically formed by noise or outliers). In this context a cluster consists of all density-connected objects, which can form a cluster of an arbitrary shape. DBSCAN [EKSX96] and its generalization OPTICS [ABKS99] are the most popular density based clustering methods. In several situations spectral and/or graph-based clustering are proposed for solving problems when the available information is encoded as a graph. If the data is represented as a graph, where the nodes are objects and the links represent connections among objects, then a cluster should be redefined as a connected component, i.e. a group of objects that are connected to one another, but that have no connection to objects outside the group. An important example of graph-based clusters are contiguity-based clusters, where two objects are connected only if they are within a specified distance of each other. This implies that each object in a contiguity-based cluster is closer to some other object in the cluster than to any point in a different cluster.

Finally, Fig. 1, taken from [HK12], summarizes the main characteristics related to the different clustering approaches considering the three main clustering methods proposed above. For each method, the figure highlights the specific features and the most well-known and basic algorithms widely studied in the literature. Finally, Fig. 1, taken from [HK12], summarizes the main characteristics related to the different clustering approaches considering the three main clustering methods proposed above. For each method, the figure highlights the specific features and the most well-known and basic algorithms widely studied in the literature.

Method	Specific Features	Algorithms
Partitioning methods	Distance based Discover mutual clusters of spherical shape Prototyped-based (mean or medoid) to represent centroid	K-means K-medoids
Hierarchical methods	Hierarchical decomposition May incorporate other techniques (e.g. microclustering) Cannot correct erroneous splits (or merges)	BIRCH
Density-based methods	Find arbitrary shaped clusters Based on concept of dense regions May filter out outliers	DBSCAN OPTICS

Fig. 1. Overview of clustering methods.

2.3 Pattern Discovery

Pattern analysis methods are fundamental in many application domains including market basket analysis, medicine, bioinformatics, web mining, network detection, DNA research. Unlike in predictive models, in pattern discovery the objective is to discover all patterns of interest. Here, we briefly recall the basic methods of pattern mining, including *frequent itemsets mining* (FIM), *association rule mining* (ARM) and *sequential patterns mining* (SPM). See [ZZ02,HCXY07,Sha09] for past surveys on ARM, and [ME10,CTG12] for surveys on SPM.

Let $I = \{i_1, \ldots, i_n\}$ be a set of distinct literals, called items. An itemset X is a subset of I. An itemset X has a support, $supp(X)$, in a transactional database D if $s\%$ of the transactions contains the itemset X in D. Given a user-defined minimum support $\bar{s}$, an itemset X such that $supp(X) \geq \bar{s}$ is called frequent itemset. The FIM problem can be stated as follows: given a transaction database D and a minimum support threshold $\bar{s}$, find all the frequent itemsets from the set of transactions w.r.t. $\bar{s}$.

A natural derivation of frequent itemsets is called association rule (AR), expressing an association between two itemsets. Given X and Y two itemsets, with $X \cap Y = \emptyset$, an AR is an expression of the form $X \Rightarrow Y$. X is called the body or antecedent, and Y is called the head or consequent of the rule. The support of an AR $X \Rightarrow Y$ is $supp(X \Rightarrow Y) = supp(X \cup Y)$. The confidence of an AR is $conf(X \Rightarrow Y) = \frac{supp(X \cup Y)}{supp(X)}$. Given a transaction database D, a minimum support threshold, $\bar{s}$, and a minimum confidence threshold, $\bar{c}$, the ARM problem is to find all the ARs from the set of transactions w.r.t. $\bar{s}$ and $\bar{c}$.

Finally, the concept of sequential pattern is introduced to capture typical behaviors over time, i.e. behaviors sufficiently repeated by individuals to be relevant for the decision maker. A sequence $S = < X_1 \ldots X_n >$ is an ordered list of itemsets. We say that S is a subsequence of another sequence $V = < Y_1 \ldots Y_m >$ with $n \leq m$, if there exist integers $1 \leq i_1 < \cdots < i_n \leq m$ such that $X_1 \subseteq Y_{i1}, \ldots, X_n \subseteq Y_{in}$. We denote with $X_i.time$ the timestamp of the itemset X_i and with $supp(S)$ the support of S, i.e. the number of tuples containing the sequence S. Given a sequence database and a minimum support threshold $\bar{s}$, the SPM problem is to find all the sequences from the set of transactions w.r.t. $\bar{\sigma}$. Sequential patterns are not the only form of patterns that can be mined. Consider for example the huge literature for gene mining [EZ13].

Different algorithms for FIM have been proposed in the literature [AS94, HPY00,SON95,Toi96,ZPOL97]. The most popular algorithm is Apriori [AS94]. The approach is outlined in Algorithm 4. It is based on a level-wise search process that makes multiple passes over the data. Initially, it computes the frequent itemsets of size 1. The core of the algorithm is then a cycle of passes each of them composed of two main phases: the candidate generation and the support counting. In the former phase, the set of all frequent k-itemsets, L_k, found in the pass k, is used to generate the candidate itemsets C_{k+1}. In the latter, data is scanned to determine the support of candidates. After the support counting, unfrequent itemsets are dropped, according to the *downward closure property*. Another algo-

---

Algorithm 4. The Apriori algorithm

Require: Set of transaction T
Ensure: Frequent itemsets
1: $k \leftarrow 1$
2: $F_k \leftarrow$ Find all frequent 1-itemsets
3: **repeat**
4:    $k \leftarrow k + 1$
5:    **for each** transaction $t \in T$ **do**
6:       Identify all candidates that belongs to t
7:       Compute support counting for each candidate C_t
8:    **end for**
9:    $F_k \leftarrow$ Extract the frequent k-itemsets
10: **until** $F_k = \emptyset$
11: **return** $\bigcup F_k$

---

rithm is the FP-Growth. It allows to reduce the number of transactions to be processed at each iteration via a *divide et impera* strategy [HPY00]. Basically, it divides the search space on a prefix base. After the first scan, the original problem can be divided into $|I|$ sub-problems, where I is the set of frequent singletons. Other algorithms based on the splitting of the input data into smaller datasets, are eclat [ZPQL97] and partition [SON95].

Sequential pattern mining methods can be classified into three classes: Apriori-based with an horizontal formatting methods; Apriori-based with a vertical formatting methods; projection-based pattern growth methods. The first class includes the GSP algorithm [SA96] and its derivations. The second class includes SPADE [Zak01]. The third class is based on the SPAM [AFGY02] and PrefixSpan algorithms [PHMA+04]. In particular, the latter works by means of a divide-and-conquer strategy with a single scan on the entire dataset. Each sequential pattern is treated as a prefix and mined recursively over the corresponding projected database.

Recently, mining frequent structural patterns from graph databases, e.g. web logs, citation networks, and social networks has become an important research problem with broad applications. Several efficient algorithms were proposed in the literature [WWZ+05, IWM00, YH02], ranging from mining graph patterns, with and without constraints, to mining closed graph patterns.

3   Using Constraints in Data Mining

The integration of constraints in data mining has rapidly emerged as a challenging topic for the research community. Many ad-hoc extensions of mining algorithms that use constraints for improving the quality of their results have been proposed for the different methods introduced along the Sect. 2. The definition and the integration of constraints allows the user to specify additional information on input data as well as requirements and expected properties of data mining models in output in a declarative way. For example, the extraction of association rules typically leads to a large quantity of useless rules.

An approach that extracts the rules by specifying the analyst's needs can speed up both the domain experts evaluation of the extracted rules and the extraction algorithm itself.

The literature proposes several works on using constraints in data mining tasks. Currently, every mining task has its own way for classifying constraints. A full view that binds mining tasks to the the objects on which constraints are defined, is still missing. For this reason, one of the aims of this chapter is to provide a general framework where a constraint can be classified. In this perspective, this section provides a description about the dimensions on which constraints can be classified. This view is based on the main characteristics that every kind of constraint proposes in its specific mining context.

We introduce the use of constraints considering three dimensions based on the characteristics that every kind of constraint presents in its specific context:

1. **Object Constraints**: considers which **objects** the constraints are applied to, namely *data*, *models* and *measures*. This kind of constraints is presented in Sect. 3.1.
2. **Hard &Soft Constraints**: considers the **type** of constraints: *hard* and *soft* constraints. Section 3.2 introduces this kind of constraints.
3. **Phase-defined Constraints**: considers the **phases** of the knowledge extraction process, in which the constraints are used, namely *pre*, *mining* and *post*. Section 3.3 overviews this class of constraints.

Before starting analysing the dimension dealing with the **objects** constraints, it is worth noting that the dimensions proposed above are not complementary or mutually exclusive, but they represent different perspectives on which we can classify constraints for data mining.

3.1   Object Constraints

We start by analyzing the dimension dealing with the **objects** constraints are applied to. Constraints can be defined on *data*, on the *mining model* and on *measures*. In particular, Sect. 3.1.1 overviews the constraints on data (or items), while Sect. 3.1.2 overviews the ones on mining models. Finally, Sect. 3.1.3 introduces the constraints defined on measures.

3.1.1   Constraints on Data

Referred to the literature also as constraints on *items*, this kind of object constraint involves specific data attributes. Data constraints require a complete knowledge about the data attributes and properties in order to define constraints on specific data features. Furthermore, they can involve some forms of background knowledge directly. Examples of constraints on data include the *must and cannot-link* in a clustering problem, or *consider only the items having a price higher than a given threshold* for pattern mining.

If we consider the *classification* task the literature in this field has explored constraints among *instances and classes*, and among different *classes* themselves.

This is principally due to the fact that a classifier is extracted from a training set specifically conceived on the requirements of the classification task. [HPRZ02] introduces a constrained classification task, where each example is labeled with a set of constraints relating multiple classes. Every constraint specifies the relative order of two classes and the goal is to learn a classifier consistent with these constraints. As reported in [PF08], in many applications explicit constraints among the labels can be easily discovered. For example, in the context of hierarchical classification, the presence of one label in the hierarchy often implies also the presence of all its ancestors. [TJHA05] proposes a constrained *support vector machine* approach. In this work, the authors consider cases where the prediction is a structured object or consists of multiple dependent constrained variables. An interesting approach is proposed in [DMM08] in case of a lack of labeled instances. In this case, the knowledge base is a set of labeled features, and the authors propose a method for training probabilistic models with labeled features (constrained from domain knowledge) from unlabeled instances. Labeled features are employed directly to constrain the model predictions on unlabeled instances.

Data constraints for *clustering* involves the concept of *instance-level* constraints. Well-established approaches on using data constraints for clustering problems focused on the introduction of instance-level constraints [WCRS01, WC00]. In this case a domain expert defines constraints that bind a pair of instances in the same cluster or that avoid that a pair of instances will be assigned to the same cluster. *(i) must-link* constraints enforce two instances to be placed in the same cluster, while *(ii) cannot-link* constraints enforce two instances to be in different clusters. Several properties are related to instance-level constraints [DR06]. Must-link constraints are symmetric, reflexive and transitive. The latter property enables a system to infer additional must-link constraints. On the contrary, cannot-links do not have the transitive property. Since *must* and *cannot*-link are relevant for a large amounts of works in the literature, where several types of constraints based on groups of instances have been defined in [DR05,DR09,DR07,DDV13], Chap. 1 in [BDW08] reports a detailed definition of the properties on which they are based.

In *pattern mining*, data constraints are introduced to specify patterns that include (or not) specific items. For example, when mining association rules out of a weblog, one might be interested in only rules having sport pages in the consequent, and not having shopping pages in the antecedent. In the case of sequential patterns, one might be interested to patterns that first visit finance, and then sport or books [PHW07]. There are two principal ways to express data constraints for pattern mining: *(i)* by means of a *concept hierarchy* (i.e. multi-level constraints) and *(ii)* *weighted pattern mining* emerges when considering a different semantic significance of the items.

Multi-level constraints enables the generalization of items at bottom level to higher levels of the hierarchy before applying the mining algorithm [SA95]. Methods to integrate multi-level constraints into mining algorithms are introduced in [HF99], in which frequent itemsets are generated one level at a time of the hierarchy. [SVA97] and [HLN99] can be seen as the first attempts to integrate multilevel mining directly into the Apriori. More recent works on generalized rule mining include [ST02] about exploiting the lattice of generalized itemsets, and [WH11], on using efficient data structures to retrieve item generalizations. [BCCG12] exploits schema constraints and the opportunistic confidence constraints to remove uninteresting rules.

Weighted pattern mining has been extensively proposed in *frequent itemset mining* and *association rule mining*, in discussing a new tree structure that is robust to database modifications [ATJ+12]; in pushing the weight constraint into pattern growth algorithms [YL05, TSWYng, YSRY12], or into level-wise methods [WYY00, TM03, LYC08]; in suggesting approximated weighted frequent pattern mining, as a fault tolerant factor [YR11].

3.1.2   Constraints on the Mining Model

This class of constraints defines specific requirements that an extracted model should satisfy. This kind of constraint does not involve background knowledge directly, but it requires a complete knowledge on the characteristics needed by the output model. For example, they include the extraction of association rules having a specific set of items in the body and in the head, or discovering clusters with a minimum number of elements.

Examples of model constraints for classification can be found in [NF07, NF10, NPS00]. [NPS00] proposes different kinds of constraints, related to the form of a decision tree, e.g. internal nodes should not have pure class distributions or rules about the class distribution. [NF10] defines a framework for determining which model constraints can be pushed into the pattern mining process, proposing an optimal classifier model. More precisely, [NF10] shows how several categories of constraints defined for frequent itemset mining, e.g. *monotonic*, *anti-monotonic* and *convertible*, can be applied in decision tree induction. It highlights the connection between constraints in pattern mining and constraints in decision tree extraction, developing a general framework for categorizing and managing decision tree mining constraints.

The algorithms K-means and K-medoid represent a basic approach for forcing clustering models to have specific properties [GMN+15]. In [BBD00, DBB08], the authors avoid empty clusters by adding k constraints to the clustering problem requiring that cluster h contains at least τ_h points. The solution proposed is equivalent to a minimum cost flow linear network optimization problem [Ber91]. Another approach for discovering balanced clusters can be found in [BG08, BG06]. In this case, the introduced constraint requires that the obtained clusters have a comparable size. The proposed method has three steps: *(i)* sampling; *(ii)* clustering of the sampled set; and *(iii)* populating and refining the clusters while satisfying the balancing constraints. Other methods for constraining the clustering approach to discover balanced clusters can be found in [SG03]. The authors propose the use of graph partition techniques or hierarchical approaches that encourage balanced results while progressively merging or splitting clusters [BK03, ZG03]. Many papers focus on metric learning driven by constraints. Distance measure learning and clustering with constraints in K-means were both considered in [BBM04b], and the result was extended to a Hidden Markov random field formulation in [BBM04a].

Pattern-model constraints are related to the form, or the structure of the entire pattern, as well as to relations among items. For example, one might wish to find patterns that include first visit of a sport page, then a shopping page, and finally a finance page. In this context, we are searching for meta-rules that are useful to specify the syntactic form of the patterns [FH95]. These constraints can be specified using either high-level user interfaces or declarative data mining query languages. Here, we briefly review the usage of regular expressions (RE) in sequential pattern mining. They are based on the typical RE operators, such as disjunction and Kleene closure, to constrain the set of items. Then, we deal with relaxation of constraints. There are several algorithms supporting RE constraints. SPIRIT [GRS99] is based on an evolution of the GSP algorithm. RE-Hackle represents RE by means of a tree structure [CMB03]. Prefix-growth extends the prefix-span approach with several kinds of constraints, among which RE are included [PHW07].

3.1.3   Constraints on Measures

Measures, e.g. *entropy* for classification, *support* and *confidence* for frequent itemsets and *euclidean distance* for clustering, play an important role in data mining, since they are related to the quality of the model extracted. This class of constraints specifies a requirement that the computation of a measure should respect. It involves both the knowledge about data and the knowledge about the characteristics of a model. For example, if we consider clustering people as moving objects, the trajectory implementing the shortest distance cannot cross a wall, or we can constraints a classifier to provide a minimum level of accuracy.

Starting from model constraints for classification, [YG04, VSKSvdH09] deal with the design of a classifier under constrained performance requirements. In particular, [VSKSvdH09] enables the user to define a desired classifier performance. The work provides a complete analysis when a classifier is constrained to a desired level of precision (defined as F-measure and/or to tp-/fp-rate related performance measures). The learned model is adjusted to achieve the desired performance, abstaining to classifying ambiguous examples in order to guarantee the required level of performance. Furthermore, [VSKSvdH09] studies the effect on an ROC curve when ambiguous instances are left unclassified. This is an example when a set of constraints defined on measures clearly influences also the learned model implicitly. Similarly in [YG04], an ensemble of neural networks is constrained by a given tp or fp-rate to ensure that the classification error for the most important class is within a desired limit. The final classifier is tuned by using a different structure (or architecture), employing different training samples, and training with a different subset of features for individual classifiers with respect to phase of employment. In most of the cases model constraints are used during the model construction phase.

Many papers focus on metric learning driven by constraints for clustering. Distance measure learning and clustering with constraints in K-means were both considered in [BBM04b], and the result was extended to a Hidden Markov random field formulation in [BBM04a]. In [SJ04], an SVM-like approach is employed to learn a weighted distance from relative constraints. The method learns a weighted euclidean distance from constraints by solving a convex optimization problem similar to SVMs to find the maximum margin weight vector. In this case, the approach integrates the input points with a set of training constraints that specify the distance requirements among points. Kumar and Kummamuru [KK08] proposed to learn an SVaD [KKA04] measure from relative comparisons. Relative comparisons were first employed in [SJ03] to learn distance measures using SVMs. The existing results on relative comparisons can be used to solve clustering problems with relative constraints (since each relative constraint is equivalent to two relative comparisons).

Besides those expressed on support and confidence, interestingness constraints specify thresholds on statistical measures of a pattern. We can find three kinds of interestingness measures. With *time constraints*, the user has the possibility of choosing not only the minimum support, but also time gaps and window size [SA96, PHW07, MPT09]. The former permits to constrain itemsets in a pattern to occur neither too close, nor too far w.r.t the time. Considering *recency, frequency and monetary constraints*, a model can be used to predict the behavior of a customer on the basis of history data, with the aim of analyzing how often and recently a customer purchases as well as how much he/she spends [BW95, WLW10]. Finally *aggregate constraints* are based on aggregates of items in a pattern, where the aggregate function can be sum, avg, max, min. See [ZZNS09] for a recent review on the various interestingness measures.

3.2   Hard and Soft Constraints

The use of constraints enables a mining method to explore only those solutions consistent with users expectations. Constraints may not always improve the reliability of the extracted model, e.g. data overfitting. Generally, it is not guaranteed that the use of constraints improves the reliability of the objective measures. Moreover in some cases constraints can be redundant, e.g. a constraint which does not affect the search solution space, and/or they can cause conflicts and introduce inconsistencies on final result.

For example, if we constrain two elements, say a and b, to be in the same cluster if their distance is lower than a given threshold t_1, and, at the same time, we require that a and b cannot be in the same cluster if their distance is greater than an additional threshold t_2, the satisfaction of these two constraints could not be solved by any cluster partitioning if t_2 is lower than t_1. Similarly, forcing a classifier to provide a desired performance can lead to find empty solutions since there is not a model extracted from the data that satisfies the required constraints, e.g. [VSKSvdH09] avoids this situation. The learned model is adjusted to achieve the desired performance by abstaining to classifying the most ambiguous example in order to guarantee the required level of performance.

Typically, these events happen when some sets of constraints work well but some others do not [Dav12]. This aspect requires the use of measures to evaluate how much a set of constraints is useful. Davidson et al. [DWB06, WBD06] introduce the concepts of *informativeness* and *coherence*. In the case of clustering, the authors define the informativeness as the amount of information in the constraint set that the algorithm cannot determine on its own. It is determined by the clustering algorithm's objective function (bias) and search preference. While given a distance matrix, the coherence measures the amount of agreement within the constraints themselves. The above definitions should be revised in the case of classification or pattern mining, but their relevance is already clear.

The above observations require that a user can define the way for computing the measure related to a constraint. Furthermore, the user expresses "how well" a constraint should be satisfied. Generally, the use of constraints does not necessarily guarantee the achievement of a solution. In order to control this effect it can be necessary to relax constraints. This leads to the need of offering the possibility of classifying constraints as either **hard** or **soft**, that is relaxable:

- **Hard constraint:** a constraint is called *hard* if a model that violates it is unacceptable. The use of only this class of constraints can involve the discovery of empty solutions. A hard-constrained algorithm halts when there does not exist a state that satisfies all the constraints, and it returns no results [OY12]. This situation is common when a large set of constraints is provided as input.
- **Soft constraint:** a constraint is called *soft* if even though a model that satisfies the constraint is preferable, a solution is acceptable anyway and especially when no any other (or better) solution is available [BMR97]. Typically, it is known that some constraints work well for finding the required solution, while others do not, and in some context where a result is needed in any case, it is important to select a set of useful constraints that should be considered as hard, while others can be treated as soft [DWB06].

This dimension is strictly related to the actual definition of a constraint and it should not be perceived as a rigid categorization. As explained above, there are some constraints that can be both hard and relaxed as soft based on the problem and the properties the solution requires.

3.3 Phase-Defined Constraints

Since a data mining task, or more generally a knowledge extraction process, is based on different iterated phases, constraints can be classified also with respect to where a knowledge extraction process can evaluate and satisfy the set of constraints defined by the user.

The *pre-processing* phase includes data cleaning, normalization, transformation, feature extraction and selection and its aim is to produce a set of data for the subsequent processing/mining step. [Pyl99] presents basic approaches for data pre-processing.

The *processing* step is the core phase where the actual knowledge extraction is performed. This is the mining phase where a model is extracted.

Finally, a *post-processing* step is required to verify if the model extracted by a data mining algorithm is valid and useful. If a model does not reach the desired standards, it is necessary to re-run the process and change parameters of the pre-processing and mining steps.

Given the above observations, techniques for constraint-driven mining can be roughly classified on the basis of the knowledge extraction phase in which they are satisfied:

- **Pre-processing constraints:** are satisfied during the pre-processing phase. They enable a restriction of the source data to the instances that can only generate patterns satisfying them.
- **Processing/Mining constraints:** are directly integrated into the mining algorithm used for extracting the model. The constraint evaluation in this case is embedded directly in the mining algorithms, enabling a reduction of the search space.
- **Post-processing constraints:** are satisfied either by filtering out patterns generated by the mining algorithm, or by highlighting only the relevant results given an interest measure provided by the user.

The phase of the knowledge extraction process where a constraint is satisfied is the last dimension we introduce. Also in this case, the above definition is useful to provide a complete picture about the use of constraints for data mining. Table 1 summarizes the main characteristics related to the different dimensions of constraints proposed in this chapter. The two main dimensions are the mining task and the kind of object where a constraint is applied. Furthermore, for each of the pairs the phase and the type of constraints are presented.

4   Conclusions: Towards New Frontiers of Data Mining

In this chapter, we presented an overview about the use of constraints in data mining. In particular, we have depicted a general multidimensional view for driving the reader into the world of constrained data mining. This chapter shows why the use of constraints is becoming an important and challenging task for the data mining community, since it requires a radical re-design of existing approaches in order to define and satisfy constraints during the whole knowledge extraction process.

Table 1. Main characteristics of the different classes of constraints

	Classification	Clustering	Pattern
Data	*phase*: pre, mining *type*: hard	*phase*: mining *type*: hard, soft	*phase*: pre, mining *type*: hard
Model	*phase*: mining, post *type*: soft	*phase*: mining *type*: soft, hard	*phase*: mining *type*: hard, soft
Measure	*phase*: mining, post *type*: hard, soft	*phase*: mining *type*: hard	*phase*: mining, post *type*: hard

Even though one of the aims of this chapter is to provide an introduction on the basic mining models and algorithms, it is worth stating that the basic concepts introduced along this overview are still valid also for advanced data mining analysis. We conclude this chapter considering the emerging phenomenon of *big data*. The final aim is to provide a set of features related to managing real data, in order to highlight that basic concepts introduced in the section of this chapter are actually the building blocks for real complex mining applications.

Often, traditional data analysis tools and techniques cannot be used because of the massive volume of data gathered by automated collection tools. The amount of data in our world has been exploding. Science gathers data at an ever-increasing rate across all scales and complexities of natural phenomena. New high-throughput scientific instruments, telescopes, satellites, accelerators, supercomputers, sensor networks and running simulations are generating massive amounts of scientific data. Companies capture trillions of bytes of information about their customers, suppliers, and operations. Smart sensing, including environment sensing, emergency sensing, people-centric sensing, smart health care, and new paradigms for communications, including email, mobile phone, social networks, blogs, Voip, are creating and communicating huge volumes of data. Sometimes, the non-traditional nature of the data implies that ordinary data analysis techniques are not applicable.

In this perspective, the challenge is particularly tough: which data mining tools are needed to master the complex dynamics of people in motion and construct concise and useful abstractions out of large volumes of mobility data is, by large, an unanswered question. Good news, hence, for researchers willing to engage in a highly interdisciplinary, highly risky and highly promising area, with a large potential impact on socially and economically relevant problems.

Big data requests a complete re-design of existing architectures and proposes new challenges on data management, privacy, and scalability among the other. Provide the *appropriate analytical* technology for distributed data mining and machine learning for big data, and a solid statistical framework adapting standard statistical data generation and analysis models to big data: once again, the sheer size and the complexity of big data call for novel analytical methods. At the same time, the kind of measures provided by the data and the population sample they describe cannot be easily modeled through standard statistical frameworks, which therefore need to be extended to capture the way the data are generated and collected.

The use of constrained-based tools, from the constraints programming to the solver, is finally under analysis from the researcher community. In this perspective, we are sure that the approaches developed along this book, generated from the experience inside the ICON project, not only represents a base for applying constrained methods to data mining but they are a first step for integrating a more versatile definition and formulation of mining approach as optimization problems by using constraint programming tools also considering the emerging phenomenon of big data.

References

[ABKS99] Ankerst, M., Breunig, M.M., Kriegel, H.-P., Sander, J.: Optics: ordering points to identify the clustering structure. In: Proceedings of the 1999 ACM SIGMOD International Conference on Management of Data, SIG-MOD 1999, pp. 49–60. ACM, New York, NY, USA (1999)

[ABR64] Aizerman, M.A., Braverman, E.A., Rozonoer, L.: Theoretical foundations of the potential function method in pattern recognition learning. Autom. Remote Control **25**, 821–837 (1964)

[AFGY02] Ayres, J., Flannick, J., Gehrke, J., Yiu, T.: Sequential pattern mining using a bitmap representation. In: Proceedings of the Eighth ACM SIGKDD International Conference on Knowledge Discovery and Data Mining (KDD), pp. 429–435 (2002)

[AS94] Agrawal, R., Srikant, R.: Fast algorithms for mining association rules. In: Proceedings of 20th International Conference on Very Large Data Bases (VLDB 1994), Santiago de Chile, Chile, 12–15 September, pp. 487–499 (1994)

[ATJ+12] Ahmed, C.F., Tanbeer, S.K., Jeong, B.-S., Lee, Y.-K., Choi, H.-J.: Single-pass incremental and interactive mining for weighted frequent patterns. Expert Syst. Appl. **39**(9), 7976–7994 (2012)

[BBD00] Bradley, P.S., Bennett, K.P., Demiriz, A.: Constrained k-means clustering. Technical report, MSR-TR-2000-65, Microsoft Research (2000)

[BBM04a] Basu, S., Bilenko, M., Mooney, R.J.: A probabilistic framework for semi-supervised clustering. In: Proceedings of the Tenth ACM SIGKDD International Conference on Knowledge Discovery and Data Mining (KDD), pp. 59–68 (2004)

[BBM04b] Bilenko, M., Basu, S., Mooney, R.J.: Integrating constraints and metric learning in semi-supervised clustering. In: Proceedings of the Twenty-First International Conference on Machine Learning, ICML 2004, p. 11. ACM, New York (2004)

[BCCG12] Baralis, E., Cagliero, L., Cerquitelli, T., Garza, P.: Generalized association rule mining with constraints. Inf. Sci. **194**, 68–84 (2012)

[BDW08] Basu, S., Davidson, I., Wagstaff, K.L.: Constrained Clustering: Advances in Algorithms, Theory, and Applications. Chapman and Hall/CRC, Boca Raton (2008)

[Ber91] Bertsekas, D.P.: Linear Network Optimization - Algorithms and Codes. MIT Press, Cambridge (1991)

[BFOS84] Breiman, L., Friedman, J., Olshen, R., Stone, C.: Classification and Regression Trees. Wadsworth International Group, Belmont (1984)

[BG06] Banerjee, A., Ghosh, J.: Scalable clustering algorithms with balancing constraints. Data Min. Knowl. Discov. **13**(3), 365–395 (2006)

[BG08] Banerjee, A., Ghosh, J.: Clustering with balancing constraints. Constrained Clustering: Advances in Algorithms. Theory, and Applications, pp. 171–200. Chapman and Hall/CRC, Boca Raton (2008)

[BGV92] Boser, B.E., Guyon, I.M., Vapnik, V.N.: A training algorithm for optimal margin classifiers. In Proceedings of the Fifth Annual Workshop on Computational Learning Theory, COLT 1992, pp. 144–152. ACM, New York (1992)

[BK03] Barbará, D., Kamath, C. (eds.): Proceedings of the Third SIAM International Conference on Data Mining, 1–3 May 2003. SIAM, San Francisco (2003)

[BMR97] Bistarelli, S., Montanari, U., Rossi, F.: Semiring-based constraint solving and optimization. J. ACM **44**(2), 201–236 (1997)

[Bur98] Burges, C.J.C.: A tutorial on support vector machines for pattern recognition. Data Min. Knowl. Discov. **2**(2), 121–167 (1998)

[BW95] Bult, J.R., Wansbeek, T.J.: Optimal selection for direct mail. Mark. Sci. **14**(4), 378–394 (1995)

[CMB03] Capelle, M., Masson, C., Boulicaut, J.F.: Mining frequent sequential patterns under regular expressions: a highly adaptive strategy for pushing constraints. In: Proceedings of the Third SIAM International Conference on Data Mining, pp. 316–320 (2003)

[CTG12] Chand, C., Thakkar, A., Ganatra, A.: Sequential pattern mining: survey and current research challenges. Int. J. Soft Comput. Eng. (IJSCE) **2**(1), 2231–2307 (2012)

[CV95] Cortes, C., Vapnik, V.: Support-vector networks. Mach. Learn. **20**(3), 273–297 (1995)

[Dav12] Davidson, I.: Two approaches to understanding when constraints help clustering. In: The 18th ACM SIGKDD International Conference on Knowledge Discovery and Data Mining (KDD), pp. 1312–1320 (2012)

[DBB08] Demiriz, A., Bennett, K.P., Bradley, P.S.: Using assignment constraints to avoid empty clusters in k-means clustering. Constrained Clustering: Advances in Algorithms. Theory, and Applications, pp. 201–220. Chapman and Hall/CRC, Boca Raton (2008)

[DDV13] Dao, T.-B.-H., Duong, K.-C., Vrain, C.: A declarative framework for constrained clustering. In: Blockeel, H., Kersting, K., Nijssen, S., Železný, F. (eds.) ECML PKDD 2013. LNCS (LNAI), vol. 8190, pp. 419–434. Springer, Heidelberg (2013). doi:10.1007/978-3-642-40994-3_27

[DH73] Duda, R.O., Hart, P.E.: Pattern Classification and Scene Analysis. Wiley, New York (1973)

[DMM08] Druck, G., Mann, G.S., McCallum, A.: Learning from labeled features using generalized expectation criteria. In: Proceedings of the 31st Annual International ACM SIGIR Conference on Research and Development in Information Retrieval (SIGIR), pp. 595–602 (2008)

[DP96] Domingos, P., Pazzani, M.J.: Beyond independence: conditions for the optimality of the simple Bayesian classifier. In: Proceedings of the 13th International Conference on Machine Learning (ICML 1996), Bari, Italy, pp. 148–156 (1996)

[DR05] Davidson, I., Ravi, S.S.: Clustering with constraints: feasibility issues and the k-means algorithm. In: Proceedings of the SIAM International Conference on Data Mining (SDM) (2005)

[DR06] Davidson, I., Ravi, S.S.: Identifying and generating easy sets of constraints for clustering. In: Proceedings of the Twenty-First National Conference on Artificial Intelligence and the Eighteenth Innovative Applications of Artificial Intelligence Conference (AAAI), pp. 336–341 (2006)

[DR07] Davidson, I., Ravi, S.S.: The complexity of non-hierarchical clustering with instance and cluster level constraints. Data Min. Knowl. Discov. **14**(1), 25–61 (2007)

[DR09] Davidson, I., Ravi, S.S.: Using instance-level constraints in agglomerative hierarchical clustering: theoretical and empirical results. Data Min. Knowl. Discov. **18**(2), 257–282 (2009)

[DWB06] Davidson, I., Wagstaff, K.L., Basu, S.: Measuring constraint-set utility for partitional clustering algorithms. In: Fürnkranz, J., Scheffer, T., Spiliopoulou, M. (eds.) PKDD 2006. LNCS (LNAI), vol. 4213, pp. 115–126. Springer, Heidelberg (2006). doi:10.1007/11871637_15

[EKSX96] Ester, M., Kriegel, H.-P., Sander, J., Xiaowei, X.: A density-based algorithm for discovering clusters in large spatial databases with noise. In: Proceedings of the Second International Conference on Knowledge Discovery and Data Mining (KDD), pp. 226–231 (1996)

[EZ13] Elloumi, M., Zomaya, A.Y.: Biological Knowledge Discovery Handbook: Preprocessing, Mining and Postprocessing of Biological Data, 1st edn. Wiley, New York (2013)

[FH95] Yongjian, F., Han, J.: Meta-rule-guided mining of association rules in relational databases. In: Proceedings of the Post-Conference Workshops on Integration of Knowledge Discovery in Databases with Deductive and Object-Oriented Databases (KDOOD/TDOOD), pp. 39–46 (1995)

[GMN+15] Grossi, V., Monreale, A., Nanni, M., Pedreschi, D., Turini, F.: Clustering formulation using constraint optimization. In: Bianculli, D., Calinescu, R., Rumpe, B. (eds.) SEFM 2015. LNCS, vol. 9509, pp. 93–107. Springer, Heidelberg (2015). doi:10.1007/978-3-662-49224-6_9

[GRS99] Garofalakis, M.N., Rastogi, R., Shim, K.: SPIRIT: Sequential pattern mining with regular expression constraints. In: Proceedings of 25th International Conference on Very Large Data Bases (VLDB), pp. 223–234 (1999)

[GS11] Grossi, V., Sperduti, A.: Kernel-based selective ensemble learning for streams of trees. In: Walsh, T. (ed.) IJCAI 2011, Proceedings of the 22nd International Joint Conference on Artificial Intelligence, Barcelona, Catalonia, Spain, 16–22 July 2011, pp. 1281–1287. IJCAI/AAAI (2011)

[GT12] Grossi, V., Turini, F.: Stream mining: a novel architecture for ensemble-based classification. Knowl. Inf. Syst. **30**(2), 247–281 (2012)

[HCXY07] Han, J., Cheng, H., Xin, D., Yan, X.: Frequent pattern mining: current status and future directions. Data Min. Knowl. Discov. **15**(1), 55–86 (2007)

[HF99] Han, J., Fu, Y.: Mining multiple-level association rules in large databases. IEEE Trans. Knowl. Data Eng. **11**(5), 798–805 (1999)

[HK12] Han, J., Kamber, M.: Data Mining: Concepts and Techniques, 2nd edn. Morgan Kaufmann, San Francisco (2012)

[HLN99] Han, J., Lakshmanan, L.V.S., Ng, R.T.: Constraint-based multidimensional data mining. IEEE Comput. **32**(8), 46–50 (1999)

[HPRZ02] Har-Peled, S., Roth, D., Zimak, D.: Constraint classification: a new approach to multiclass classification. In: Proceedings of the 13th International Conference Algorithmic Learning Theory (ALT), pp. 365–379 (2002)

[HPY00] Han, J., Pei, J., Yin, Y.: Mining frequent patterns without candidate generation. In: Proceedings of the 2000 ACM SIGMOD International Conference on Management of Data, Dallas, Texas, USA, 16–18 May, pp. 1–12 (2000)

[IWM00] Inokuchi, A., Washio, T., Motoda, H.: An apriori-based algorithm for mining frequent substructures from graph data. In: Zighed, D.A., Komorowski, J., Żytkow, J. (eds.) PKDD 2000. LNCS (LNAI), vol. 1910, pp. 13–23. Springer, Heidelberg (2000). doi:10.1007/3-540-45372-5_2

[Jen96] Jensen, F.V.: An introduction to Bayesian networks. Springer, New York (1996)

[KK08]   Kumar, N., Kummamuru, K.: Semisupervised clustering with metric learn-
          ing using relative comparisons. IEEE Trans. Knowl. Data Eng. **20**(4),
          496–503 (2008)
[KKA04]   Kummamuru, K., Krishnapuram, R., Agrawal, R.: Learning spatially vari-
          ant dissimilarity (SVaD) measures. In: Proceedings of the Tenth ACM
          SIGKDD International Conference on Knowledge Discovery and Data
          Mining (KDD), pp. 611–616 (2004)
[LLS00]   Lin, T.S., Loh, W.Y., Shib, Y.S.: A comparison of prediction accuracy,
          complexity, and training time of thirty-tree old and new classification algo-
          rithms. Mach. Learn. **40**(3), 203–228 (2000)
[LYC08]   Li, Y.-C., Yeh, J.-S., Chang, C.-C.: Isolated items discarding strategy for
          discovering high utility itemsets. Data Knowl. Eng. **64**(1), 198–217 (2008)
[MAR96]   Mehta, M., Agrawal, R., Rissanen, J.: SLIQ: A fast scalable classifier for
          data mining. In: Proceedings of 5th International Conference on Extending
          Database Technology (EBDT 1996), Avignon, France, pp. 18–32 (1996)
[ME10]   Mabroukeh, N.R., Ezeife, C.I.: A taxonomy of sequential pattern mining
          algorithms. ACM Comput. Surv. **43**(1), 3: 1–3: 41 (2010)
[Mic97]   Michell, T.: Machine Learning. McGraw Hill, New York (1997)
[Mor82]   Moret, B.M.E.: Decision trees and diagrams. Comput. Surv. **14**(4), 593–
          623 (1982)
[MPT09]   Masseglia, F., Poncelet, P., Teisseire, M.: Efficient mining of sequential
          patterns with time constraints: reducing the combinations. Expert Syst.
          Appl. **36**(2), 2677–2690 (2009)
[NF07]   Nijssen, S., Fromont, É.: Mining optimal decision trees from itemset lat-
          tices. In: Proceedings of the 13th ACM SIGKDD International Conference
          on Knowledge Discovery and Data Mining (KDD), pp. 530–539 (2007)
[NF10]   Nijssen, S., Fromont, E.: Optimal constraint-based decision tree induction
          from itemset lattices. Data Min. Knowl. Discov. Fromont. **21**(1), 9–51
          (2010)
[NPS00]   Niyogi, P., Pierrot, J.-B., Siohan, O.: Multiple classifiers by constrained
          minimization. In: Proceedings of the Acoustics, Speech, and Signal
          Processing of 2000 IEEE International Conference on ICASSP 2000, vol.
          06, pp. 3462–3465. IEEE Computer Society, Washington, DC (2000)
[OY12]   Okabe, M., Yamada, S.: Clustering by learning constraints priorities.
          In: Proceedings of the 12th International Conference on Data Mining
          (ICDM2012), pp. 1050–1055 (2012)
[PF08]   Park, S.H., Furnkranz, J.: Multi-label classification with label constraints.
          Technical report, Knowledge Engineering Group, TU Darmstadt (2008)
[PHMA+04] Pei, J., Han, J., Mortazavi-Asl, B., Wang, J., Pinto, H., Chen, Q., Dayal,
          U., Hsu, M.: Mining sequential patterns by pattern-growth: the prefixspan
          approach. IEEE Trans. Knowl. Data Eng. **16**(11), 1424–1440 (2004)
[PHW07]   Pei, J., Han, J., Wang, W.: Constraint-based sequential pattern mining:
          the pattern growth methods. J. Intell. Inf. Syst. **28**(2), 133–160 (2007)
[Pyl99]   Pyle, D.: Data Preparation for Data Mining. Morgan Kaufmann Publish-
          ers Inc., San Francisco (1999)
[Qui86]   Quinlan, J.R.: Induction of decision trees. Mach. Learn. **1**, 81–106 (1986)
[Qui93]   Quinlan, J.R.: C4.5 Programs for Machine Learning. Wadsworth Interna-
          tional Group, Belmont (1993)
[Qui96]   Quinlan, J.R.: Improved use of continuous attributes in C4.5. J. Artif.
          Intell. Res. **4**, 77–90 (1996)

[Rug02] Ruggieri, S.: Efficient C4.5. IEEE Trans. Knowl. Data Eng. **14**(2), 438–444 (2002)

[SA95] Srikant, R., Agrawal, R.: Mining generalized association rules. In: Proceedings of the 21st Conference on Very Large Data Bases (VLDB), pp. 407–419 (1995)

[SA96] Srikant, R., Agrawal, R.: Mining sequential patterns: generalizations and performance improvements. In: Proceedings of the 5th International Conference on Extending Database Technology (EDBT), pp. 3–17 (1996)

[SAM96] Shafer, J., Agrawal, R., Mehta, M.: Sprint: a scalable parallel classifier for data mining. In: Proceedings of 1996 International Conference on Very Large Data Bases (VLDB 1996), Bombay, India, pp. 544–555 (1996)

[SC08] Steinwart, I., Christmann, A.: Support Vector Machines, 1st edn. Springer Publishing Company, Incorporated, Heidelberg (2008)

[SG03] Strehl, A., Ghosh, J.: Relationship-based clustering and visualization for high-dimensional data mining. INFORMS J. Comput. **15**(2), 208–230 (2003)

[Sha09] Shankar, S.: Utility sentient frequent itemset mining and-association rule mining: a literature survey and comparative study. Int. J. Soft Comput. Appl. **4**, 81–95 (2009)

[SJ03] Schultz, M., Joachims, T.: Learning a distance metric from relative comparisons. In: Proceedings of Conference Advances in Neural Information Processing Systems (NIPS) (2003)

[SJ04] Schultz, M., Joachims, T.: Learning a distance metric from relative comparisons. In: NIPS, MIT Press (2004)

[SON95] Savasere, A., Omiecinski, E., Navathe, S.B.: An efficient algorithm for mining association rules in large databases. In: Proceedings of the 21st International Conference on Very Large Data Bases (VLDB), Zurich, Switzerland, 11–15 September 1995, pp. 432–444 (1995)

[ST02] Sriphaew, K., Theeramunkong, T.: A new method for finding generalized frequent itemsets in generalized association rule mining. In: Proceedings of the 7th IEEE Symposium on Computers and Communications (ISCC), pp. 1040–1045 (2002)

[SVA97] Srikant, R., Quoc, V., Agrawal, R.: Mining association rules with item constraints. In: Proceedings of the Third International Conference on Knowledge Discovery and Data Mining (KDD), pp. 67–73 (1997)

[TJHA05] Tsochantaridis, I., Joachims, T., Hofmann, T., Altun, Y.: Large margin methods for structured and interdependent output variables. J. Mach. Learn. Res. **6**, 1453–1484 (2005)

[TM03] Tao, F., Murtagh, F.: Weighted association rule mining using weighted support and significance framework. In: Proceedings of the 9th ACM SIGKDD International Conference on Knowledge Discovery and Data Mining, pp. 661–666 (2003)

[Toi96] Toivonen, H.: Sampling large databases for association rules. In: Proceedings of the 22nd International Conference on Very Large Data Bases (VLDB), Mumbai (Bombay), India, 3–6 September, pp. 134–145 (1996)

[TSK06] Tan, P.N., Steinbach, M., Kumar, V.: Introduction to Data Mining. Addison Wesley, Boston (2006)

[TSWYng] Tseng, V.S., Shie, B.-E., Wu, C.-W., Philip, S.: Efficient algorithms for mining high utility itemsets from transactional databases. IEEE Transactions on Knowledge and Data Engineering, forthcoming

[VSKSvdH09] Vanderlooy, S., Sprinkhuizen-Kuyper, I.G., Smirnov, E.N., Jaap van den Herik, H.: The ROC isometrics approach to construct reliable classifiers. Intell. Data Anal. **13**(1), 3–37 (2009)

[WBD06] Wagstaff, K., Basu, S., Davidson, I.: When is constrained clustering beneficial, and why? In: Proceedings of The Twenty-First National Conference on Artificial Intelligence and the Eighteenth Innovative Applications of Artificial Intelligence Conference (AAAI) (2006)

[WC00] Wagstaff, K., Cardie, C.: Clustering with instance-level constraints. In: Proceedings of the Seventeenth National Conference on Artificial Intelligence and Twelfth Conference on on Innovative Applications of Artificial Intelligence (AAAI/IAAI), p. 1097 (2000)

[WCRS01] Wagstaff, K., Cardie, C., Rogers, S., Schrödl, S.: Constrained k-means clustering with background knowledge. In: Proceedings of the Eighteenth International Conference on Machine Learning, ICML 2001, pp. 577–584. Morgan Kaufmann Publishers Inc., San Francisco (2001)

[WFH11] Witten, I.H., Frank, E., Hall, M.: Data Mining, Pratical Machine Learning Tools and Techiniques, 3rd edn. Morgan Kaufmann, San Francisco (2011)

[WH11] Wu, C.-M., Huang, Y.-F.: Generalized association rule mining using an efficient data structure. Expert Syst. Appl. **38**(6), 7277–7290 (2011)

[WK91] Weiss, S.M., Kulikowski, C.A.: Computer Systems That Learn: Classification and Prediction Methods from Statistics, Neural Nets. Machine Learning and Expert Systems. Morgan Kaufmann, San Francisco (1991)

[WLW10] Wei, J.-T., Lin, S.-Y., Hsin-Hung, W.: A review of the application of RFM model. Afr. J. Bus. Manag. **4**(19), 4199–4206 (2010)

[WWZ+05] Wang, W., Wang, C., Zhu, Y., Shi, B., Pei, J., Yan, X., Han, J.: Graphminer: a structural pattern-mining system for large disk-based graph databases and its applications. In: zcan, F. (ed.) SIGMOD Conference, pp. 879–881. ACM (2005)

[WYY00] Wang, W., Yang, J., Philip, S.: Efficient mining of weighted association rules (WAR). In: Proceedings of the 6th ACM SIGKDD International Conference on Knowledge Discovery and Data Mining, pp. 270–274 (2000)

[YG04] Yan, W., Goebel, K.F.: Designing classifier ensembles with constrained performance requirements. In: Proceedings of SPIE Defense and Security Symposium, Multisensor Multisource Information Fusion: Architectures, Algorithms, and Applications 2004, pp. 78–87 (2004)

[YH02] Yan, X., Han, J.: gSpan: Graph-based substructure pattern mining. In: Proceedings of the 2002 IEEE International Conference on Data Mining, ICDM 2002, p. 721. IEEE Computer Society, Washington, DC, USA (2002)

[YL05] Yun, U., Leggett, J.J.: WFIM: weighted frequent itemset mining with a weight range and a minimum weight. In: Proceeding of the 2005 SIAM International Data Mining Conference, Newport Beach, CA, pp. 636–640 (2005)

[YR11] Yun, U., HoRyu, K.: Approximate weighted frequent pattern mining with/without noisy environments. Knowl.-Based Syst. **24**(1), 73–82 (2011)

[YSRY12] Yun, U., Shin, H., Ho Ryu, K., Yoon, E.: An efficient mining algorithm for maximal weighted frequent patterns in transactional databases. Knowl.-Based Syst. **33**, 53–64 (2012)

[Zak01] Zaki, M.J.: SPADE: an efficient algorithm for mining frequent sequences. Mach. Learn. **42**(1/2), 31–60 (2001)

[ZG03]   Zhong, S., Ghosh, J.: Scalable, balanced model-based clustering. In: Proceedings of the Third SIAM International Conference on Data Mining, San Francisco (SDM) (2003)

[ZPOL97] Zaki, M.J., Parthasarathy, S., Ogihara, M., Li, W.: New algorithms for fast discovery of association rules. In: Proceedings of the Third International Conference on Knowledge Discovery and Data Mining (KDD 1997), Newport Beach, California, USA, 14–17 August, pp. 283–286 (1997)

[ZRL96]   Zhang, T., Ramakrishnan, R., Livny, M.: Birch: an efficient data clustering method for very large databases. SIGMOD Rec. **25**(2), 103–114 (1996)

[ZZ02]   Zhang, C., Zhang, S.: Association Rule Mining: Models and Algorithms. LNCS, vol. 2307. Springer, Heidelberg (2002)

[ZZNS09] Zhang, Y., Zhang, L., Nie, G., Shi, Y.: A survey of interestingness measures for association rules. In: Proceedings of the Second International Conference on Business Intelligence and Financial Engineering (BIFE), pp. 460–463 (2009)

Learning to Model

New Approaches to Constraint Acquisition

Christian Bessiere[1(✉)], Abderrazak Daoudi[1,2], Emmanuel Hebrard[3],
George Katsirelos[4], Nadjib Lazaar[1], Younes Mechqrane[2], Nina Narodytska[5],
Claude-Guy Quimper[6], and Toby Walsh[7]

[1] CNRS, University of Montpellier, Montpellier, France
bessiere@lirmm.fr
[2] University Mohammed V of Rabat, Rabat, Morocco
[3] LAAS-CNRS, Toulouse, France
[4] MIAT, INRA, Toulouse, France
[5] Samsung Research America, Mountain View, USA
[6] Université Laval, Quebec City, Canada
[7] NICTA, UNSW, Sydney, Australia

Abstract. In this chapter we present the recent results on constraint
acquisition obtained by the Coconut team and their collaborators. In a
first part we show how to learn constraint networks by asking the user
partial queries. That is, we ask the user to classify assignments to subsets
of the variables as positive or negative. We provide an algorithm, called
QUACQ, that, given a negative example, finds a constraint of the target
network in a number of queries logarithmic in the size of the example. In a
second part, we show that using some background knowledge may improve
the acquisition process a lot. We introduce the concept of generalization
query based on an aggregation of variables into types. We propose a generalization algorithm together with several strategies that we incorporate
in QUACQ. Finally we evaluate our algorithms on some benchmarks.

1 Introduction

A major bottleneck in the use of constraint solvers is modelling. How does the
user write down the constraints of a problem? Several techniques have been
proposed to tackle this bottleneck. For example, the matchmaker agent [13]
interactively asks the user to provide one of the constraints of the target problem each time the system proposes an incorrect solution. In *Conacq.1* [4,5], the
user provides examples of solutions and non-solutions. Based on these examples,
the system learns a set of constraints that correctly classifies all examples given
so far. This is a form of *passive* learning. In [16], a system based on inductive
logic programming uses background knowledge on the structure of the problem to learn a representation of the problem correctly classifying the examples.
A last passive learner is *ModelSeeker* [3]. Positive examples are provided by

---

Sections 3 and 4 of this paper describe material published in [9], Sect. 5 describes
material published in [8], and Sect. 6 describes results coming from both of these
two papers.

© Springer International Publishing AG 2016
C. Bessiere et al. (Eds.): Data Mining and Constraint Programming, LNAI 10101, pp. 51–76, 2016.
DOI: 10.1007/978-3-319-50137-6_3

the user to the system, which arranges each of them as a matrix and identifies constraints in the global constraints catalog [2] that are satisfied by particular subsets of variables in all the examples. Such particular subsets are for instance rows, columns, diagonals, etc. An efficient ranking technique combined with a representation of solutions as matrices allows *ModelSeeker* to find quickly a good model when a problem has an underlying matrix structure.

By contrast, in an *active learner* like *Conacq.2* [6], the system proposes examples to the user to classify as solutions or non solutions. Such questions are called *membership queries* [1]. Such active learning has several advantages. It can decrease the number of examples necessary to converge to the target set of constraints. Another advantage is that the user needs not be a human. It might be a previous system developed to solve the problem. For instance, the Normind company has hired a constraint programming specialist to transform their expert system for detecting failures in electric circuits in Airbus airplanes into a constraint model in order to make it more efficient and easier to maintain. As another example, active learning is used to build a constraint model that encodes non-atomic actions of a robot (e.g., catch a ball) by asking queries of the simulator of the robot in [18]. Such active learning introduces two computational challenges. First, how does the system generate a useful query? Second, how many queries are needed for the system to converge to the target set of constraints? It has been shown that the number of membership queries required to converge to the target set of constraints can be exponentially large [7].

In this chapter, we propose QUACQ (for QuickAcquisition), an active learner that asks the user to classify *partial* queries. Given a negative example, QUACQ is able to learn a constraint of the target constraint network in a number of queries logarithmic in the number of variables. In fact, we identify information theoretic lower bounds on the complexity of learning constraint networks which show that QUACQ is optimal on some simple languages. However, even that good theoretical bounds can be hard to put in practice. For instance, QUACQ requires the user to classify more than 8000 examples to get the complete Sudoku model. We then propose a new technique to make constraint acquisition more efficient in practice by using variable types. In real problems, variables often represent components of the problem that can be classified in various types. For instance, in a school time-tabling problem, variables can represent teachers, students, rooms, courses, or time-slots. Such types are often known by the user. To deal with types of variables, we introduce a new kind of query, namely, *generalization query*. We expect the user to be able to decide if a learned constraint can be generalized to other scopes of variables of the same type as those in the learned constraint. We propose an algorithm, GENACQ for *generalized acquisition*, that asks such generalization queries each time a new constraint is learned. We propose several strategies and heuristics to select the good candidate generalization query. We plugged our generalization functionality into the QUACQ constraint acquisition system, leading to the G-QUACQ algorithm. We experimentally evaluate the benefits of our algorithms on several benchmark problems. The results show the polynomial behavior of QuAcq. They also show that G-QuAcq dramatically improves the basic QuAcq algorithm in terms of number of queries.

One application for QuAcq and G-QuAcq would be to learn a general purpose model. In constraint programming, a distinction is made between model and data. For example, in a sudoku puzzle, the model contains generic constraints like each subsquare contains a permutation of the numbers. The data, on the other hand, gives the pre-filled squares for a specific puzzle. As a second example, in a time-tabling problem, the model specifies generic constraints like no teacher can teach multiple classes at the same time. The data, on the other hand, specifies particular room sizes, and teacher availability for a particular time-tabling problem instance. The cost of learning the model can then be amortized over the lifetime of the model. Another advantage of this approach is that it provides less of a burden on the user. First, it often converges quicker than other methods. Second, partial queries will be easier to answer than complete queries. Third, as opposed to existing techniques, the user does not need to give positive examples. This might be useful if the problem has not yet been solved, so there are no examples of past solutions.

The rest of the paper is organized as follows. Section 2 gives the necessary definitions to understand the technical presentation. Section 3 presents QuAcq, the algorithm that learns constraint networks by asking partial queries. In Sect. 4, we show how QuAcq behaves on some simple languages. Section 5 describes the generalization algorithm G-QuAcq. In Sect. 5.5, several strategies are presented to make G-QuAcq more efficient. Section 6 presents the experimental results we obtained when comparing G-QuAcq to the basic QuAcq and when comparing the different strategies in G-QuAcq. Section 7 concludes the paper and gives some directions for future research.

2   Background

The learner and the user need to share some common knowledge to communicate. We suppose this common knowledge, called the *vocabulary*, is a (finite) set of n variables X and a domain $D = \{D(X_1), \ldots, D(X_n)\}$, where $D(X_i) \subset \mathbb{Z}$ is the finite set of values for X_i. Given a sequence of variables $S \subseteq X$, a *constraint* is a pair (c, S) (also written c_S), where c is a relation over $\mathbb{Z}$ specifying which sequences of $|S|$ values are allowed for the variables S. S is called the *scope* of c_S. A *constraint network* is a set C of constraints on the vocabulary (X, D). An assignment e_Y on a set of variables $Y \subseteq X$ *is rejected by* a constraint c_S if $S \subseteq Y$ and the projection $e_Y[S]$ of e on the variables in S is not in c. An assignment on X is a *solution* of C iff it is not rejected by any constraint in C. We write $sol(C)$ for the set of solutions of C, and $C[Y]$ for the set of constraints from C whose scope is included in Y.

In addition to the vocabulary, the learner owns a *language* Γ of bounded arity relations from which it can build constraints on specified sets of variables. To simplify the descriptions, we only consider languages closed by conjunction, that is, languages Γ such that if relations c_1 and c_2 belong to Γ, then relation

$c_1 \cap c_2$ also belongs to Γ. Adapting terms from machine learning, the *constraint basis*, denoted by B, is a set of constraints built from the constraint language Γ on the vocabulary (X, D) from which the learner builds the constraint network. Formally speaking, $B = \{c_S \mid (S \subseteq X) \wedge (c \in \Gamma) \wedge arity(c) = |S|)\}$.

A *concept* is a Boolean function over $D^X = \Pi_{X_i \in X} D(X_i)$, that is, a map that assigns to each $e \in D^X$ a value in $\{0, 1\}$. A *target concept* is a concept f_T that returns 1 for e if and only if e is a solution of the problem the user has in mind. e is called a positive or negative *example* depending on whether $f_T(e) = 1$ or $f_T(e) = 0$. A *membership query* $ASK(e)$ takes as input a *complete* assignment e in D^X and asks the user to classify it. The answer to $ASK(e)$ is "yes" if and only if $f_T(e) = 1$.

To be able to use *partial queries*, we have an extra condition on the capabilities of the user. Even if she is not able to articulate the constraints of her problem, she is able to decide if partial assignments of X violate some requirements or not. More formally, we consider that the user has in mind her problem in the form of a target constraint network. A *target constraint network* is a network C_T such that $sol(C_T) = \{e \in D^X \mid f_T(e) = 1\}$. A *partial query* $ASK(e_Y)$, with $Y \subseteq X$, is a classification question asked of the user, where e_Y is a *partial* assignment in $D^Y = \Pi_{X_i \in Y} D(X_i)$. A set of constraints C *accepts* a partial query e_Y if and only if there does not exist any constraint c_S in C rejecting $e_Y[S]$. The answer to $ASK(e_Y)$ is "yes" if and only if C_T accepts e_Y. It is important to observe that "$ASK(e_Y) = yes$" does not mean that e_Y extends to a solution of C_T, which would put an NP-complete problem on the shoulders of the user. For any assignment e_Y on Y, $\kappa_B(e_Y)$ denotes the set of all constraints in B rejecting e_Y. A classified assignment e_Y is called positive or negative *example* depending on whether $ASK(e_Y)$ is "yes" or "no".

We now define *convergence*, which is the constraint acquisition problem we are interested in. We are given a set E of (partial) examples labelled by the user 0 or 1. We say that a constraint network C agrees with E if C accepts all examples labelled 1 in E and does not accept those labelled 0. The learning process has *converged* on the network $C_L \subseteq B$ if C_L agrees with E and for every other network $C' \subseteq B$ agreeing with E, we have $sol(C') = sol(C_L)$. If there does not exist any $C_L \subseteq B$ such that C_L agrees with E, we say that we have *collapsed*. This happens when $C_T \not\subseteq B$.

Finally, we define types of variables to be used to generalize constraints. A *type* T_i is a subset of variables defined by the user as having a common property. A variable X_i is of type T_i iff $X_i \in T_i$. A scope $S = (X_1, \ldots, X_k)$ of variables *belongs* to a sequence of types $T = (T_1, \ldots, T_k)$ (denoted by $S \in T$) if and only if $X_i \in T_i$ for all $i \in 1..k$. Consider $T = (T_1, \ldots, T_k)$ and $T' = (T'_1, \ldots, T'_k)$ two sequences of types. We say that T' *covers* T (denoted by $T \sqsubseteq T'$) iff $T_i \subseteq T'_i$ for all $i \in 1..k$. A relation c *holds* on a sequence of types T if and only if $c_S \in C_T$ for all $S \in T$. A sequence of types T is *maximal* with respect to a relation c if and only if c holds on T and there does not exist T' covering T on which c holds.

Algorithm 1. QUACQ: Acquiring a constraint network C_T with partial queries

---

1 $C_L \leftarrow \varnothing$;
2 **while** *true* **do**
3   **if** $sol(C_L) = \varnothing$ **then return** "collapse";
4   choose e in D^X accepted by C_L and rejected by B ;
5   **if** $e = nil$ **then return** "convergence on C_L";
6   **if** $ASK(e) = yes$ **then** $B \leftarrow B \setminus \kappa_B(e)$;
7   **else**
8          $S \leftarrow \mathtt{FindScope}(e, \varnothing, X, \mathbf{false})$;
9          $c_S \leftarrow \mathtt{FindC}(S)$;
10   **if** $c_S = nil$ **then return** "collapse";
11   **else** $C_L \leftarrow C_L \cup \{c_S\}$;

---

3 Constraint Acquisition with Partial Queries

We propose QUACQ, a novel active learning algorithm. QUACQ takes as input a basis B on a vocabulary (X, D). It asks partial queries of the user until it has converged on a constraint network C_L equivalent to the target network C_T, or collapses. When a query is answered *yes*, constraints rejecting it are removed from B. When a query is answered *no*, QUACQ enters a loop (functions FindScope and FindC) that will end by the addition of a constraint to C_L.

3.1 Description of QUACQ

QUACQ (see Algorithm 1) initializes the network C_L it will learn to the empty set (line 1). If C_L is unsatisfiable (line 3), the space of possible networks collapses because there does not exist any subset of the given basis B that is able to correctly classify the examples the user has already been asked. In line 4, QUACQ computes a complete assignment e satisfying C_L but violating at least one constraint from B. (Observe that for this task, the constraint solver needs to be able to express the negation of the constraints in B. This is not a problem as we have only bounded arity constraints in B.) If such an example does not exist (line 5), then all constraints in B are implied by C_L, and we have converged. If we have not converged, we propose the example e to the user, who will answer by *yes* or *no*. If the answer is *yes*, we can remove from B the set $\kappa_B(e)$ of all constraints in B that reject e (line 6). If the answer is *no*, we are sure that e violates at least one constraint of the target network C_T. We then call the function FindScope to discover the scope S of one of these violated constraints (line 8). FindC will return a constraint of C_T whose scope is in S (line 9). If no constraint is returned (line 10), this is again a condition for collapsing as we could not find in B a constraint rejecting one of the negative examples. Otherwise, the constraint returned by FindC is added to the learned network C_L (line 11).

Algorithm 2. Function `FindScope`: returns the scope of a constraint in C_T

---

1 **function** $FindScope($**in** e: example, R, Y: scopes, ask_query: Boolean$)$:
   scope;
2 **begin**
3   **if** ask_query **then**
4   **if** $ASK(e[R]) = yes$ **then** $B \leftarrow B \setminus \kappa_B(e[R])$;
5   **else return** $\varnothing$;
6   **if** $|Y| = 1$ **then return** Y;
7   split Y into $< Y_1, Y_2 >$ such that $|Y_1| = \lceil |Y|/2 \rceil$;
8     $S_1 \leftarrow$ FindScope$(e, R \cup Y_1, Y_2, \mathbf{true})$;
9     $S_2 \leftarrow$ FindScope$(e, R \cup S_1, Y_1, (S_1 \neq \varnothing))$;
10   **return** $S_1 \cup S_2$;

The recursive function `FindScope` (see Algorithm 2) takes as parameters an example e, two sets R and Y of variables, and a Boolean ask_query. An invariant of `FindScope` is that e violates at least one constraint whose scope is a subset of $R \cup Y$. A second invariant is that `FindScope` always returns a subset of Y that is also the subset of the scope of a constraint violated by e. When `FindScope` is called with $ask\_query = \mathbf{false}$, we already know that R does not contain the scope of a constraint that rejects e (line 3). If $ask\_qery = \mathbf{true}$ we ask the user whether $e[R]$ is positive or not (line 4). If yes, we can remove all the constraints that reject $e[R]$ from the basis, otherwise we return the empty set because we have no guarantee that any variable of Y belongs to a scope (line 5). We reach line 6 only in case $e[R]$ does not violate any constraint. We know that $e[R \cup Y]$ violates a constraint. Hence, if Y is a singleton, the variable it contains necessarily belongs to the scope of a constraint that violates $e[R \cup Y]$. The function returns Y. If none of the return conditions are satisfied, the set Y is split in two balanced parts (line 7) and we apply a technique similar to QuickXplain [15] to elucidate the variables of a constraint violating $e[R \cup Y]$ in a logarithmic number of steps (lines 8–10). The rationale of lines 8 and 9 is to quickly find sets R and Y so that $e[R]$ is positive and $e[R \cup Y]$ is negative. If Y is a singleton this ensures that the variable in Y belongs to the scope of a constraint we are looking for. This variable is returned and forced to be in R in all subsequent calls to `FindScope`.

The function `FindC` (see Algorithm 3) takes as parameter Y, the scope on which `FindScope` has found that there is a constraint from the target network C_T. `FindC` first removes from B all constraints with scope Y that are implied by C_L because there is no need to learn them (line 3).[1] The set Δ is initialized to all candidate constraints (line 4). In line 6, an example e is chosen in such a way that Δ contains both constraints rejecting e and constraints satisfying e. If no such example exists (line 7), this means that all constraints in Δ are equivalent wrt $C_L[Y]$. Any of them is returned except if Δ is empty (lines 8–9). If a suitable

---

[1] This operation could proactively be done in QuAcq, just after line 11, but we preferred the lazy mode as this is a computationally expensive operation.

Algorithm 3. Function FindC: returns a constraint of C_T with scope Y

---

1 **function** *FindC*(**in** Y: scope): **constraints**;
2 **begin**
3     $B \leftarrow B \setminus \{c_Y \mid C_L \models c_Y\}$;
4     $\Delta \leftarrow \{c_Y \mid (c_Y \in B[Y])\}$;
5   **while** *true* **do**
6   choose e in $sol(C_L[Y])$ such that $\exists c_Y, c'_Y \in \Delta, e \models c_Y$ and $e \not\models c'_Y$;
7   **if** $e = nil$ **then**
8   **if** $\Delta = \varnothing$ **then return** *nil*;
9   **else** pick c_Y in Δ; **return** c;
10   **if** $ASK(e) = yes$ **then**
11           $B \leftarrow B \setminus \kappa_B(e)$; $\Delta \leftarrow \Delta \setminus \kappa_B(e)$;
12   **else**
13   **if** $\exists c_S \in \kappa_B(e) \mid S \subsetneq Y$ **then**
14   **return** FindC(FindScope($e, \varnothing, S, \textbf{false}$));
15   **else** $\Delta \leftarrow \Delta \cap \kappa_B(e)$;

---

example was found, it is proposed to the user for classification (line 10). If classified positive, all constraints rejecting it are removed from B and Δ (line 11). Otherwise we test whether the example e does not violate constraints with scope strictly included in Y (line 13). If yes, we recursively call FindScope and FindC to find that smaller arity constraint before the one having scope Y (line 14). If no, we remove from Δ all constraints accepting that example e (line 15) and we continue the loop of line 5.

3.2 Example

We illustrate the behavior of QUACQ on a simple example. Consider the set of variables $X_1, \ldots, X_5$ with domains $\{1..5\}$, a language $\Gamma = \{=, \neq\}$, a basis $B = \{=_{ij}, \neq_{ij} \mid i, j \in 1..5, i < j\}$, and a target network $C_T = \{=_{15}, \neq_{34}\}$. Suppose the first example generated in line 4 of QUACQ is $e_1 = (1, 1, 1, 1, 1)$.

Table 1. The example

Call	R	Y	ASK	Return
0	$\varnothing$	X_1, X_2, X_3, X_4, X_5	×	X_3, X_4
1	X_1, X_2, X_3	X_4, X_5	Yes	X_4
1.1	X_1, X_2, X_3, X_4	X_5	No	$\varnothing$
1.2	X_1, X_2, X_3	X_4	×	X_4
2	X_4	X_1, X_2, X_3	Yes	X_3
2.1	X_4, X_1, X_2	X_3	Yes	X_3
2.2	X_4, X_3	X_1, X_2	No	$\varnothing$

The trace of the execution of FindScope($e_1, \varnothing, X_1 \ldots X_5,$ **false**) is in Table 1. Each line corresponds to a call to FindScope. Queries are always on the variables in R. '$\times$' in the column ASK means that the previous call returned $\varnothing$, so the question is skipped. The initial call (call-0 in Table 1) does not ask the question because the initial call to FindScope has the Boolean ask_query set to **false**. Y is split in two sets $Y_1 = \{X_1, X_2, X_3\}$ and $Y_2 = \{X_4, X_5\}$ and the recursive call-1 is performed with $R = Y_1$ and $Y = Y_2$. As $e_1[R]$ is classified as positive, line 4 of FindScope removes the constraints $\neq_{12}, \neq_{13}$ and $\neq_{23}$ from B. A new split of Y leads to the call-1.1 with $R = \{X_1, X_2, X_3, X_4\}$ and $Y = \{X_5\}$. As $e_1[R]$ is negative, the empty set is returned in line 5. Call-1.2 (line 9) is performed with $R = \{X_1, X_2, X_3\}$, $Y = \{X_4\}$ and $ask\_query =$ **true**. It merely detects that Y is a singleton and thus returns $\{X_4\}$. Call-1 finishes by returning $\{X_4\}$ one level above in the recursivity (line 10). Call-2 classifies $e_1[X_4]$ as positive and goes down to call-2.1 with $R = \{X_4, X_1, X_2\}$ and $Y = \{X_3\}$. In call-2.1, $e_1[R]$ is classified positive. FindScope thus removes constraints $\neq_{14}$ and $\neq_{24}$ from B and returns the singleton $\{X_3\}$. In call-2.2, $e_1[R]$ is classified as negative with $R = \{X_4, X_3\}$ and $Y = \{X_1, X_2\}$. This proves that $\{X_3, X_4\}$ is the scope of a constraint rejecting e_1. Empty set is returned by call-2.2. As a result, call-2 returns $\{X_3\}$, and call-0 returns $\{X_3, X_4\}$. Once the scope (X_3, X_4) is returned, FindC requires a single example to return $\neq_{34}$ and prune $=_{34}$ from B. Suppose the next example generated by QuAcq is $e_2 = (1, 2, 3, 4, 5)$. FindScope will find the scope (X_1, X_5) and FindC will return $=_{15}$ in a way similar to the processing of e_1. The constraints $=_{12}, =_{13}, =_{14}, =_{23}, =_{24}$ are removed from B by a partial positive query on $X_1, \ldots, X_4$ and $\neq_{15}$ by FindC. Finally, examples $e_3 = (1, 1, 1, 2, 1)$ and $e_4 = (3, 2, 2, 3, 3)$, both positive, will prune $\neq_{25}, \neq_{35}, =_{45}$ and $=_{25}, =_{35}, \neq_{45}$ from B respectively, leading to convergence.

3.3   Analysis

We analyse the complexity of QuAcq in terms of the number of queries it can ask of the user. Queries are proposed to the user in lines 6 of QuAcq, 4 of FindScope and 10 of FindC.

Proposition 1. *Given a basis B built from a language Γ, a target network C_T, a scope Y, FindC uses $O(|\Gamma|)$ queries to return a constraint c_Y from C_T if it exists.*

Proof. Each time FindC asks a query, whatever the answer of the user, the size of Δ strictly decreases. Thus the total number of queries asked in FindC is bounded above by $|\Delta|$, which itself, by construction in line 4, is bounded above by the number of constraints from Γ of arity $|Y|$.

Proposition 2. *Given a basis B, a target network C_T, an example $e \in D^X \setminus sol(C_T)$, FindScope uses $O(|S| \cdot \log|X|)$ queries to return the scope S of one of the constraints of C_T violated by e.*

Proof. FindScope is a recursive algorithm that asks at most one query per call (line 4). Hence, the number of queries is bounded above by the number of nodes of the tree of recursive calls to FindScope. We will show that a leaf node is either on a branch that leads to the elucidation of a variable in the scope S that will be returned, or is a child of a node of such a branch. When a branch does not lead to the elucidation of a variable in the scope S that will be returned, that branch necessarily only leads to leaves that correspond to calls to FindScope that returned the empty set. The only way for a leaf call to FindScope to return the empty set is to have received a *no* answer to its query (line 5). Let R_{child}, Y_{child} be the values of the parameters R and Y for a leaf call with a *no* answer, and R_{parent}, Y_{parent} be the values of the parameters R and Y for its parent call in the recursive tree. From the *no* answer to the query $ASK(e[R_{child}])$, we know that $S \subseteq R_{child}$ but $S \nsubseteq R_{parent}$ because the parent call received a *yes* answer. Consider first the case where the leaf is the left child of the parent node. By construction, $R_{parent} \subsetneq R_{child} \subsetneq R_{parent} \cup Y_{parent}$. As a result, Y_{parent} intersects S, and the parent node is on a branch that leads to the elucidation of a variable in S. Consider now the case where the leaf is the right child of the parent node. As we are on a leaf, if the *ask_query* Boolean is false, we have necessarily exited from FindScope through line 6, which means that this node is the end of a branch leading to a variable in S. Thus, we are guaranteed that the *ask_query* Boolean is true, which means that the left child of the parent node returned a non empty set and that the parent node is on a branch to a leaf that elucidates a variable in S.

We have proved that every leaf is either on a branch that elucidates a variable in S or is a child of a node on such a branch. Hence the number of nodes in the tree is at most twice the number of nodes in branches that lead to the elucidation of a variable from S. Branches can be at most $\log |X|$ long. Therefore the total number of queries FindScope asks is at most $2 \cdot |S| \cdot \log |X|$, which is in $O(|S| \cdot \log |X|)$.

Theorem 1. *Given a basis B built from a language Γ of bounded arity constraints, and a target network C_T, QuAcq uses $O(|C_T| \cdot (\log |X| + |\Gamma|))$ queries to find the target network or to collapse and $O(|B|)$ queries to prove convergence.*

Proof. Each time line 6 of QuAcq classifies an example as negative, the scope of a constraint c_S from C_T is found in at most $|S| \cdot \log |X|$ queries (Proposition 2). As Γ only contains constraints of bounded arity, either $|S|$ is bounded and c_S is found in $O(|\Gamma|)$ or we collapse (Proposition 1). Hence, the number of queries necessary for finding C_T or collapsing is in $O(|C_T| \cdot (\log |X| + |\Gamma|))$. Convergence is obtained once B is wiped out thanks to the examples that are classified positive in line 6 of QuAcq. Each of these examples necessarily leads to at least one constraint removal from B because of the way the example is built in line 4. This gives a total in $O(|B|)$.

4   Learning Simple Languages

In order to gain a theoretical insight into the "efficiency" of QuAcq, we look at some simple languages, and analyze the number of queries required to learn networks on these languages. In some cases, we show that QuAcq will learn problems of a given language with an asymptotically optimal number of queries. However, for some other languages, a suboptimal number of queries can be necessary in the worst case. Our analysis assumes that when generating a complete example in line 4 of QuAcq, the solution of C_L *maximizing* the number of violated constraints in the basis B is chosen.

4.1   Languages for Which QuAcq is Optimal

Theorem 2. QuAcq *learns Boolean networks on the language* $\{=, \neq\}$ *in an asymptotically optimal number of queries.*

Proof (Sketch). First, we give a lower bound on the number of queries required to learn a constraint network in this language. Consider the restriction to equalities only. In an instance of this language, all variables of a connected component must be equal. This is isomorphic to the set of partitions of n objects, whose size is given by *Bell's Number*:

$$C(n+1) = \begin{cases} 1 & \text{if } n = 0 \\ \sum_{i=1}^{n} \binom{n}{i} C(n-i) & \text{if } n > 0 \end{cases} \tag{1}$$

By an information theoretic argument, at least $\log C(n)$ queries are required to learn such a problem. This entails a lower bound of $\Omega(n \log n)$ since $\log C(n) \in \Omega(n \log n)$ (see [12] for the proof). The language $\{=, \neq\}$ is richer and thus requires at least as many queries.

Second, we consider the query submitted to the user in line 6 of QuAcq and count how many times it can receive the answer *yes* and *no*. The key observation is that an instance of this language contains at most $O(n)$ non-redundant constraints. For each *no* answer in line 6 of QuAcq, a new constraint will eventually be added to C_L. Only non-redundant constraints are discovered in this way because the query must satisfy C_L. It follows that at most $O(n)$ such queries are answered *no*, each one entailing $O(\log n)$ more queries through the procedure FindScope.

Now we bound the number of *yes* answers in line 6 of QuAcq. The same observation on the structure of this language is useful here as well. We show in the complete proof that a query maximizing the number of violations of constraints in the basis B while satisfying the constraints in C_L violates at least $\lceil |B|/2 \rceil$ constraints in B. Thus, each query answered *yes* at least halves the number of constraints in B. It follows that the query submitted in line 6 of QuAcq cannot receive more than $O(\log n)$ *yes* answers. The total number of queries is therefore bounded by $O(n \log n)$.

The same argument holds for simpler languages ($\{=\}$ and $\{\neq\}$ on Boolean domains). Moreover, this is still true for $\{=\}$ on arbitrary domains.

Corollary 1. QuAcq *can learn constraint networks with unbounded domains on the language* {=} *in an asymptotically optimal number of queries.*

4.2   Languages for Which QuAcq Is Not Optimal

First, we show that a Boolean constraint network on the language {<} can be learnt with $O(n)$ queries. Then, we show that QuAcq requires $\Omega(n \log n)$ queries.

Theorem 3. *Boolean constraint networks on the language* {<} *can be learned in $O(n)$ queries.*

Proof. Observe that in order to describe such a problem, the variables can be partionned into three sets, one for variables that must take the value 0 (i.e., on the left side of a < constraint), a second for variables that must take the value 1 (i.e., on the right side of a < constraint), and the third for unconstrained variables. In the first phase, we greedily partition variables into three sets, L, R, U initially empty and standing respectively for *Left*, *Right* and *Unknown*. During this phase, we have three invariants:

1. There is no $x, y \in U$ such that $x < y$ belongs to the target network
2. $x \in L$ iff there exists $y \in U$ and a constraint $x < y$ in the target network
3. $x \in R$ iff there exists $y \in U$ and a constraint $y < x$ in the target network

We go through all variables of the problem, one at a time. Let x be the last variable picked. We query the user with an assignment where x, as well as all variables in U are set to 0, and all variables in R are set to 1 (variables in L are left unassigned). If the answer is *yes*, then there is no constraints between x and any variable in $y \in U$, hence we add x to the set of undecided variables U without breaking any invariant. Otherwise we know that x is either involved in a constraint $y < x$ with $y \in U$, or a constraint $x < y$ with $y \in U$. In order to decide which way is correct, we make a second query, where the value of x is flipped to 1 and all other variables are left unchanged. If this second query receives a *yes* answer, then the former hypothesis is true and we add x to R, otherwise, we add it to L. Here again, the invariants still hold.

At the end of the first phase, we therefore know that variables in U have no constraints between them. However, they might be involved in constraints with variables in L or in R. In the second phase, we go over each undecided variable $x \in U$, and query the user with an assignment where all variables in L are set to 0, all variables in R are set to 1 and x is set to 0. If the answer is *no*, we conclude that there is a constraint $y < x$ with $y \in L$ and therefore x is added to R (and removed from U). Otherwise, we ask the same query, but with the value of x flipped to 1. If the answer is *no*, there must exists $y \in R$ such that $x < y$ belongs to the network, hence x is added to R (and removed from U). Last, if both queries get the answer *yes*, we conclude that x is not constrained. When every variable has been examined in this way, variables remaining in U are not constrained.

Theorem 4. QUACQ *does not learn Boolean networks on the language* $\{<\}$ *with a minimal number of queries.*

Proof. By Theorem 3, we know that these networks can be learned in $O(n)$ queries. Such networks can contain up to $n-1$ non redundant constraints. QUACQ learns constraints one at a time, and each call to FindScope takes $\Omega(\log n)$ queries. Therefore, QUACQ requires $\Omega(n \log n)$ queries.

5   Constraint Acquisition with Generalization Queries

In this section we present GENACQ, a *generalized acquisition* algorithm, The idea behind this algorithm is, given a constraint c_S learned on S, to generalize this constraint to sequences of types T covering S by asking generalization queries $AskGen(T, c)$. A generalization query $AskGen(T, c)$ is answered *yes* by the user if and only if for every sequence S of variables covered by T the relation c holds on S in the target constraint network C_T.

5.1   Description of GENACQ

The algorithm GENACQ (see Algorithm 4) takes as input a target constraint c_S that has already been learned and a set $NonTarget$ of constraints that are known not to belong to the target network. GENACQ returns the set of all sequences of scopes that are maximal with respect to the relation c. GENACQ uses the global data structure $NegativeQ$, which is a set of pairs (T, c) for which we know that c does not hold on all sequences of variables covered by T. c_S and $NonTarget$ can come from any constraint acquisition mechanism or as background knowledge. $NegativeQ$ is built incrementally by each call to GENACQ. GENACQ also uses the set $Table$ as local data structure. $Table$ will contain all sequences of types that are candidates for generalizing c_S.

In line 2, GENACQ initializes the set $Table$ to all possible sequences T of types that contain the scope S of the constraint c_S. In line 3, GENACQ initializes the set G to the sequence S. G will contain the output of GENACQ, that is, the set of maximal sequences from $Table$ on which c holds. The counter $\#NoAnswers$ counts the number of consecutive times generalization queries have been classified negative by the user. It is initialized to zero (line 4). $\#NoAnswers$ is not used in the basic version of GENACQ but it will be used in the version with cutoffs. (In other words, the basic version uses cutoffNo $= +\infty$ in line 8).

The first loop in GENACQ (line 5) eliminates from $Table$ all sequences T for which we already know the answer to the query $AskGen(T, c)$. In line 6, GENACQ eliminates from $Table$ all sequences T such that a relation c' entailed by c is already known not to hold on a sequence T' covered by T (i.e., (T', c') is in $NegativeQ$). We can remove such sequences because the absence of c' on some scope in T' implies the absence of c on some scope in T (see Lemma 1). In line 7, GENACQ eliminates from $Table$ all sequences T such that we know from $NonTarget$ that there exists a scope S' in T such that $c_{S'}$ does not belong to C_T.

Algorithm 4.   GENACQ: returns the maximum generalizations of a constraint c_S

---

1 **function** GENACQ(**in** c_S: constraint, $NonTarget$: set): **Generalizations**;
2 $Table \leftarrow \{T \mid S \in T\} \setminus \{S\}$;
3 $G \leftarrow \{S\}$;
4 $\#NoAnswers \leftarrow 0$;
5 **foreach** $T \in Table$ **do**
6   **if** $\exists (T', c') \in NegativeQ \mid c \subseteq c' \wedge T' \sqsubseteq T$ **then** $Table \leftarrow Table \setminus \{T\}$;
7   **if** $\exists c_{S'} \in NonTarget \mid S' \in T$ **then** $Table \leftarrow Table \setminus \{T\}$;
8 **while** $Table \neq \varnothing \ \wedge \ \#NoAnswers < $ cutoffNo **do**
9   pick T in $Table$;
10   **if** $AskGen(T, c) = yes$ **then**
11 $G \leftarrow G \cup \{T\} \setminus \{T' \in G \mid T' \sqsubseteq T\}$;
12 $Table \leftarrow Table \setminus \{T' \in Table \mid T' \sqsubseteq T\}$;
13 $\#NoAnswers \leftarrow 0$;
14   **else**
15 $Table \leftarrow Table \setminus \{T' \in Table \mid T \sqsubseteq T'\}$;
16 $NegativeQ \leftarrow NegativeQ \cup \{(T, c)\}$;
17 $\#NoAnswers + +$;
18 **return** G;

In the main loop of GENACQ (line 8), we pick a sequence T from $Table$ at each iteration and we ask a generalization query to the user (line 10). If the user says *yes*, T is a sequence on which c holds. We put T in G and remove from G all sequences covered by T, so as to keep only the maximal ones (line 11). We also remove from $Table$ all sequences T' covered by T (line 12) to avoid asking redundant questions later. If the user says *no*, we remove from $Table$ all sequences T' that cover T (line 15) because we know they are no longer candidate for generalization of c and we store in $NegativeQ$ the fact that (T, c) has been answered *no*. The loop finishes when $Table$ is empty and we return G (line 18).

5.2   Completeness and Complexity

We analyze the completeness and complexity of GENACQ in terms of number of generalization queries it ask of the user.

Lemma 1. *If $AskGen(T, c) = no$ then for any (T', c') such that $T \sqsubseteq T'$ and $c' \subseteq c$, we have $AskGen(T', c') = no$.*

Proof. Assume that $AskGen(T, c) = no$. Hence, there exists a sequence $S \in T$ such that $c_S \notin C_T$. As $T \sqsubseteq T'$ we have $S \in T'$ and then we know that $c_S \notin C_T$. As $c' \subseteq c$, we also have $c'_S \notin C_T$. As a result, $AskGen(T', c') = no$.

Lemma 2. *If $AskGen(T, c) = yes$ then for any T' such that $T' \sqsubseteq T$, we have $AskGen(T', c) = yes$.*

Proof. Assume that $AskGen(T, c) = yes$. As $T' \sqsubseteq T$, for all $S \in T'$ we have $S \in T$ and then we know that $c_S \in C_T$. As a result, $AskGen(T', c) = yes$.

Proposition 3 (Completeness). *When called with constraint c_S as input, the algorithm* GENACQ *returns all maximal sequences of types covering S on which the relation c holds.*

Proof. All sequences covering S are put in *Table*. A sequence in *Table* is either asked for generalization or removed from *Table* in lines 6, 7, 12, or 15. We know from Lemma 1 that a sequence removed in line 6, 7, or 15 would necessarily lead to a *no* answer. We know from Lemma 2 that a sequence removed in line 12 is subsumed and less general than another one just added to G.

Proposition 4. *Given a learned constraint c_S and its associated Table,* GENACQ *uses $O(|Table|)$ generalization queries to return all maximal sequences of types covering S on which the relation c holds.*

Proof. For each query on $T \in Table$ asked by GENACQ, the size of *Table* strictly decreases regardless of the answer. As a result, the total number of queries is bounded above by $|Table|$.

5.3 Illustrative Example

Let us take the Lewis Carroll's Zebra problem to illustrate our generalization approach. The problem is to find where the zebra lives, given five houses of five different colors, owned by five men of five different nationalities, having five different drinks, cigarets, and pets. (See for instance [10] for a complete description of the Zebra problem.) The Zebra problem has a single solution. The target network is formulated using 25 variables, partitioned in 5 types of 5 variables each. The ith variable of a given type represents the number of the house where the ith element of the given type is located. The types are *color*, *nationality*, *drink*, *cigaret*, *pet*, and the trivial type X of all variables. There is a clique of $\neq$ constraints on all pairs of variables of the same non trivial type and 14 additional constraints given in the description of the problem.

Figure 1 shows the variables of the Zebra problem and their types. In this example, the constraint $X_2 \neq X_5$ has been learned between the two color variables X_2 and X_5. This constraint is given as input of the GENACQ algorithm. GENACQ computes the *Table* of all sequences of types covering the scope (X_2, X_5). $Table = \{(X_2, color), (X_2, X), (color, X_5), (color, color), (color, X), (X, X_5), (X, color), (X, X)\}$. Suppose we pick $T = (X, X_5)$ at line 9 of GENACQ. According to the user's answer (*no* in this case), the *Table* is reduced to $Table = \{(X_2, color), (X_2, X), (color, X_5), (color, color), (color, X)\}$. As next iteration, let us pick $T = (color, color)$. The user will answer *yes* because there is indeed a clique of $\neq$ on the *color* variables. Hence, $(color, color)$ is added to G and the *Table* is reduced to $Table = \{(X_2, X), (color, X)\}$. If we pick (X_2, X), the user answers *no* and we reduce the *Table* to the empty set and return $G = \{(color, color)\}$, which means that the constraint $X_2 \neq X_5$ can be generalized to all pairs of variables in the sequence $(color, color)$, that is, $(X_i \neq X_j) \in C_T$ for all $(X_i, X_j) \in (color, color)$.

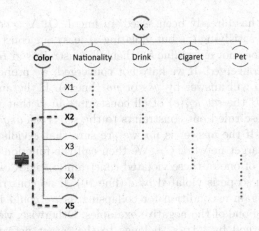

Fig. 1. Variables and types for the Zebra problem.

5.4   Using Generalization in QuAcq

GENACQ is a generic technique that can be plugged into any constraint acquisition system. In this section we present G-QUACQ, a constraint acquisition algorithm obtained by plugging GENACQ into QUACQ, G-QUACQ is presented in Algorithm 5.

Algorithm 5. G-QUACQ = QUACQ + GENACQ

---

1 $C_L \leftarrow \varnothing$, $NonTarget \leftarrow \varnothing$;
2 **while** *true* **do**
3   **if** $sol(C_L) = \varnothing$ **then** **return** "collapse";
4   choose e in D^X accepted by C_L and rejected by B;
5   **if** $e = nil$ **then** **return** "convergence on C_L";
6   **if** $Ask(e) = yes$ **then**
7          $B \leftarrow B \setminus \kappa_B(e)$;
8          $NonTarget \leftarrow NonTarget \cup \kappa_B(e)$;
9   **else**
10          $c_S \leftarrow \text{FindC}(e, \text{FindScope}(e, \varnothing, X, \textbf{false}))$;
11   **if** $c_S = nil$ **then return** "collapse";
12   **else**
13              $G \leftarrow \text{GENACQ}(c_S, NonTarget)$;
14   **foreach** $T \in G$ **do** $C_L \leftarrow C_L \cup \{c_{S'} \mid S' \in T\}$;

G-QUACQ has a structure very similar to QUACQ. It initializes the set $NonTarget$ and the network C_L it will learn to the empty set (line 1). If C_L is unsatisfiable (line 3), the space of possible networks collapses because there does not exist any subset of the given basis B that is able to correctly classify the examples the user has already been asked. In line 4, QuAcq computes a complete assignment e satisfying C_L but violating at least one constraint from B. If such an example does not exist (line 5), then all constraints in B are implied by C_L, and we have converged. If we have not converged, we propose the example e to the user, who will answer by *yes* or *no* (line 6). If the answer is *yes*, we can remove from B the set $\kappa_B(e)$ of all constraints in B that reject e (line 7) and we add all these ruled out constraints to the set *NonTarget* to be used in GenAcq (line 8). If the answer is *no*, we are sure that e violates at least one constraint of the target network C_T. We then call the function FindScope to discover the scope of one of these violated constraints. FindC will select which one with the given scope is violated by e (line 10). If no constraint is returned (line 11), this is again a condition for collapsing as we could not find in B a constraint rejecting one of the negative examples. Otherwise, we know that the constraint c_S returned by FindC belongs to the target network C_T. This is here that the algorithm differs from QuAcq as we call GenAcq to find all the maximal sequences of types covering S on which c holds. They are returned in G (line 13). Then, for every sequence of variables S' belonging to one of these sequences in G, we add the constraint $c_{S'}$ to the learned network C_L (line 14).

5.5 Strategies

GenAcq learns the maximal sequences of types on which a constraint can be generalized. The order in which sequences are picked from *Table* in line 9 of Algorithm 4 is not specified by the algorithm. As shown on the following example, different orderings can lead more or less quickly to the good (maximal) sequences on which a relation c holds. Let us come back to our example on the Zebra problem (Sect. 5.3). In the way we developed the example, we needed only 3 generalization queries to empty the set *Table* and converge on the maximal sequence $(color, color)$ on which $\neq$ holds:

1. $AskGen((X, X_5), \neq) = no$
2. $AskGen((color, color), \neq) = yes$
3. $AskGen((X_2, X), \neq) = no$

Using another ordering, GenAcq needs 8 generalization queries:

1. $AskGen((X, X), \neq) = no$
2. $AskGen((X, color), \neq) = no$
3. $AskGen((color, X), \neq) = no$
4. $AskGen((X, X_5), \neq) = no$
5. $AskGen((X_2, X), \neq) = no$
6. $AskGen((X_2, color), \neq) = yes$
7. $AskGen((color, X_5), \neq) = yes$
8   $AskGen((color, color), \neq) = yes$

If we want to reduce the number of generalization queries, we may wonder which strategy to use. In this section we propose two techniques. The first idea is to pick sequences in the set $Table$ following an order given by a heuristic that will try to minimize the number of queries. The second idea is to put a cutoff on the number of consecutive negative queries we accept to face, leading to a non complete generalization startegy: the output of GENACQ will no longer be guaranteed to be the *maximal* sequences.

Query Selection Heuristics. We propose some query selection heuristics to decide which sequence to pick next from $Table$. We first propose *optimistic* heuristics, which try to take the best from positive answers:

- **max_CST:** This heuristic selects a sequence T maximizing the number of possible constraints c_S in the basis such that S is in T and c is the relation we try to generalize. The intuition is that if the user answers *yes*, the generalization will be maximal in terms of number of constraints.
- **max_VAR:** This heuristic selects a sequence T involving a maximum number of variables, that is, maximizing $|\bigcup_{S \in T} S|$. The intuition is that if the user answers *yes*, the generalization will involve many variables.

Dually, we propose *pessimistic* heuristics, which try to take the best from negative answers:

- **min_CST:** This heuristic selects a sequence T minimizing the number of possible constraints c_S in the basis such that S is in T and c is the relation we try to generalize. The intuition is to maximize the chances to receive a *yes* answer. If, despite this, the user answers *no*, a great number of sequences are removed from $Table$ (see Lemma 1).
- **min_VAR:** This heuristic selects a sequence T involving a minimum number of variables, that is, minimizing $|\bigcup_{S \in T} S|$. The intuition is to maximize the chances of a *yes* answer while focusing on smaller sets of variables than min_CST. Again, a *no* answer leads to a great number of sequences removed from $Table$.

As a baseline for comparison, we define a random selector.

- **random:** It picks randomly a sequence T in $Table$.

Using Cutoffs. The idea here is to exit GENACQ before having proved the maximality of the sequences returned. We put a threshold cutoffNo on the number of consecutive negative answers to avoid using queries to check unpromising sequences. The hope is that GENACQ will return near-maximal sequences of types despite not proving maximality. This cutoff strategy is implemented by setting the variable cutoffNo to a predefined value. In lines 13 and 17 of GENACQ, a counter of consecutive negative answers is respectively reset and incremented depending on the answer from the user. In line 8, that counter is compared to cutoffNo to decide to exit or not.

6 Experimental Evaluation

We made some experiments to evaluate the behavior of active learning with partial queries (QUACQ) and to test the impact of using generalization (GENACQ). GENACQ was plugged in QUACQ, leading to G-QUACQ. All the experiments were done on an Intel Xeon E5462 @ 2.80 GHz with 16 Gb of RAM.

We first present the benchmark problems we used for our experiments. Then, we report the results of several experiments. The first experiment presents the performance of QUACQ to learn our benchmark problems. The second one compares the performance of G-QUACQ to the basic QUACQ. The third reports experiments evaluating the different strategies we proposed (query selection heuristics and cutoffs) on G-QUACQ. Finally, we evaluate the performance of G-QUACQ when our knowledge of the types of variables is incomplete.

6.1 Benchmark Problems

Problems Without Types

Random. We generated binary random target networks with 50 variables, domains of size 10, and m binary constraints. The binary constraints are selected from the language $\Gamma = \{\geq, \leq, <, >, \neq, =\}$. QUACQ is initialized with the basis B containing the complete graph of 7350 binary constraints taken from Γ. For densities $m = 12$ (under-constrained) and $m = 122$ (phase transition) we launched QUACQ on 100 instances and report averages.

Golomb Rulers (prob006 in [14]). A Golomb ruler is a set of m marks to put on a ruler so that the distances between marks are all distinct. This is encoded as a target network with m variables corresponding to the m marks, and constraints of varying arity. We learned the target network encoding the 8 mark ruler. We initialized QUACQ and G-QUACQ with a basis of 55,484 constraints taken from a language with 24 basic arithmetic and distance constraints with unary, binary, ternary and quaternary scopes.

Problems with Types

Zebra Problem. Lewis Carroll's zebra problem has a single solution. The target network is formulated using 25 variables of domain size of 5 with 5 cliques of $\neq$ constraints and 11 additional constraints given in the description of the problem. We initialized QUACQ and G-QUACQ with a basis of 6,850 constraints taken from a language with 24 basic arithmetic and distance constraints with unary, binary, ternary and quaternary scopes. In G-QUACQ, the variables are given as the 5 types of 5 variables that naturally occur from the problem description (color, nationality, pet, cigaret, drink).

Sudoku. The Sudoku logic puzzle is a 9×9 grid pre-filled with some numbers. It must be completed in such a way that all the rows, all the columns and the 9 non overlapping 3×3 squares contain the numbers 1 to 9. The target network of the Sudoku has 81 variables with domains of size 9 and 810 binary $\neq$ constraints on rows, columns and squares. QUACQ and G-QUACQ are initialized with a basis B of 6,480 binary constraints from the language $\Gamma = \{=, \neq\}$. In this problem, the types are the 9 rows, 9 columns and 9 squares, of 9 variables each.

Latin Square. The Latin square problem consists of an $n \times n$ table in which each element occurs once in every row and column. For this problem, we use 25 variables with domains of size 5 and 100 binary $\neq$ constraints on rows and columns. Rows and columns are the types of variables (10 types). QUACQ and G-QUACQ are initialized with a basis of constraints based on the language $\Gamma = \{=, \neq\}$.

Radio Link Frequency Assignment Problem. The RLFAP problem is to provide communication channels from limited spectral resources [11]. The constraint model of the instance we selected has 25 variables with domains of size 25 and 125 binary constraints. We have five stations of five terminals (transmitters/receivers), which form five types. We initialized QUACQ and G-QUACQ with a basis of 1,800 binary constraints taken from a language of 6 arithmetic and distance constraints.

Purdey [17]. Like Zebra, this problem has a single solution. Four families have stopped by Purdeys general store, each to buy a different item and paying differently. Under a set of additional constraints given in the description, the problem is how can we match family with the item they bought and how they paid for it. The target network of Purdey has 12 variables with domains of size 4 and 30 binary constraints. Here we have three types of variables, which are *family*, *bought* and *paid*, each of them contains four variables. We initialized QUACQ and G-QUACQ with a basis of constraints based on the language $\Gamma = \{=, \neq\}$.

6.2 QUACQ Evaluation

To ensure rapid converge, we want a query answered *yes* to prune B as much as possible. This is best achieved when the query generated in line 4 of QUACQ is an assignment violating a large number of constraints in B. We implemented the *max* heuristic to generate a solution of C_L that violates a maximum number of constraints from B. However, this heuristic can be time consuming as it solves an optimization problem. We then added a cutoff of 1 or 10 seconds to the solver using *max*, which gives the two heuristics *max-1* and *max-10* respectively. We also implemented a cheaper heuristic that we call *sol*. It simply solves C_L and stops at its first solution violating at least one constraint from B.

Our first experiment was to compare *max-1* and *max-10* on large problems. We observed that the performance when using *max-1* is not significantly worse in number of queries than when using *max-10*. For instance, on the rand_50_10_122, $\#Ask = 1074$ for *max-1* and $\#Ask = 1005$ for *max-10*. The average time for

Table 2. Results of QuAcq learning until convergence.

		$|C_L|$	$\#Ask$	$\#Ask_c$	$\overline{Ask}$	Time
rand_50_10_12	max-1	12	196	34	24.04	0.23
	sol	12	286	133	33.22	0.09
rand_50_10_122	max-1	86	1074	94	13.90	0.14
	sol	83	1062	120	15.64	0.06
Golomb-8	max-1	83	438	83	5.12	0.85
	sol	127	645	132	5.34	0.46
Zebra	max-1	60	764	61	10.99	0.29
	sol	60	752	61	11.17	0.06
Sudoku 9×9	max-1	810	8645	821	20.58	0.16
	sol	810	9593	815	20.84	0.06

generating a query is 0.14 s for *max-1* and 0.86 s for *max-10* with a maximum of 1 and 10 s respectively. We then chose not to report results for *max-10*.

Table 2 reports the results obtained with QuAcq to reach convergence using the heuristics *max-1* and *sol*. We report the size $|C_L|$ of the learned network (which can be smaller than the target network due to redundant constraints), the total number $\#Ask$ of queries, the number $\#Ask_c$ of complete queries (i.e., generated in line 4 of QuAcq), the average size $\overline{Ask}$ of all queries, and the average time needed to compute a query (in seconds). A first observation is that *max-1* generally requires less queries than *sol* to reach convergence. This is especially true for rand_50_10_12, which is very sparse, and Golomb-8, which contains many redundant constraints. If we have a closer look, these differences are mainly explained by the fact that *max-1* requires significantly less complete positive queries than *sol* to totally prune B and prove convergence (22 complete positive queries for *max-1* and 121 for *sol* on rand_50_10_12). But in general, *sol* is not as bad as we could have expected. The reason is that, except on very sparse networks, the number of constraints from B violated 'by chance' with *sol* is large enough. The second observation is that when the network contains a lot of redundancies, *max-1* converges on a smaller network than *sol*. We observed this on Golomb-8, and other problems not reported here. The third observation is that the average size of queries is always significantly smaller than the number of variables in the problem. A last observation is that *sol* is very fast for all its queries (see the time column). We can expect it to be usable on even larger problems.

As a second experiment we evaluated the effect of the size of the basis on the number of queries. On the zebra problem we initialized QuAcq with bases of different sizes and stored the number of queries for each run. Figure 2 shows that when $|B|$ grows, the number of queries follows a logarithmic scale. This is very good news as it means that learning problems with expressive bases will scale well.

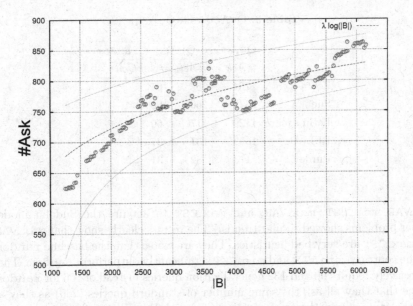

Fig. 2. QuAcq behavior on different basis sizes for Zebra

QuAcq has two main advantages over learning with membership queries, as in Conacq. One is the small average size of queries $\overline{Ask}$, which are probably easier to answer by the user. The second advantage is the time to generate queries. *Conacq.2* needs to find examples that violate exactly one constraint of the basis to make progress towards convergence. This can be expensive to compute, preventing the use of *Conacq.2* on large problems. QuAcq, on the other hand, can use cheap heuristics like *max-1* and *sol* to generate queries.

6.3   Using Generalization Queries

For all our experiments we report, the total number *#Ask* of standard queries asked by the basic QuAcq, the total number *#AskGen* of generalization queries, and the numbers *#no* and *#yes* of negative and positive generalization queries, respectively, where *#AskGen = #no + #yes*. The time overhead of using G-QuAcq rather than QuAcq is not reported. Computing a generalization query takes a few milliseconds.

Our first experiment compares QuAcq and G-QuAcq in its baseline version, G-QuAcq + random, on our benchmark problems. Table 3 reports the results. We observe that the number of queries asked by G-QuAcq is dramatically reduced compared to QuAcq. This is especially true on problems with many types involving many variables, such as Sudoku or Latin square. G-QuAcq acquires the Sudoku with 260 standard queries plus 166 generalization queries, when QuAcq acquires it in 8645 standard queries.

Let us now focus on the behavior of our different heuristics in G-QuAcq. The upper part of Table 4 reports the results obtained with G-QuAcq using

Table 3. QuAcq vs g-QuAcq.

	QuAcq	g-QuAcq + random	
	#Ask	#Ask	#AskGen
Zebra	764	257	67
Sudoku	8645	260	166
Latin square	1129	117	60
RFLAP	1653	151	37
Purdey	173	82	31

min_VAR, min_CST, max_VAR, and max_CST to acquire the Sudoku model. (Other problems showed similar trends.) The results clearly show that max_VAR, and max_CST are very bad heuristics. They are worse than the baseline random. On the contrary, min_VAR and min_CST significantly outperform random. They respectively require 90 and 132 generalization queries instead of 166 for random. Notice that they all ask the same number of standard queries (260) as they all find the same maximal sets of sequences for each learned constraint.

The lower part of Table 4 we compare the behavior of our two best heuristics (min_VAR and min_CST) when combined with the cutoff strategy. We tried all values of the cutoff from 1 to 3. A first observation is that min_VAR remains the best whatever the value of the cutoff is. Interestingly, even with a cutoff equal to 1, min_VAR requires the same number of standard queries as the versions of g-QuAcq without cutoff. This means that using min_VAR as selection heuristic in *Table*, g-QuAcq is able to return the maximal sequences despite being stopped after the first negative generalization answer. We also observe that the number of generalization queries with min_VAR decreases when the cutoff

Table 4. g-QuAcq with heuristics and cutoff strategy on Sudoku.

	Cutoff	#Ask	#AskGen	#yes	#no
Random	+∞	260	166	42	124
min_VAR			90	21	69
min_CST			132	63	69
max_VAR			263	63	200
max_CST			247	21	226
min_VAR	3	260	75	21	54
	2		57	21	36
	1		**39**	21	18
min_CST	3	626	238	112	126
	2	679	231	132	99
	1	837	213	153	60

becomes smaller (from 90 to 39 when the cutoff goes from $+\infty$ to 1). By looking at the last two columns we see that this is the number #no of negative answers which decreases. The good performance of min_VAR + cutoff = 1 can thus be explained by the fact that min_VAR selects first queries that cover a minimum number of variables, which increases the chances to have a *yes* answer. Finally, we observe that the heuristic min_CST does not have the same nice characteristics as min_VAR. The smaller the cutoff, the more standard queries are needed, not compensating for the saving in number of generalization queries (from 260 to 837 standard queries for min_CST when the cutoff goes from $+\infty$ to 1). This means that with min_CST, when the cutoff becomes too small, GENACQ does not return the maximal sequences of types where the learned constraint holds.

Table 5. G-QUACQ with random, min_VAR, and cutoff = 1 on Zebra, Latin square, RLFAP, and Purdey.

		#Ask	#AskGen	#yes	#no
Zebra	Random	257	67	10	57
	min_VAR		48	5	43
	min_VAR + cutoff = 1		**23**	5	18
L.Square	Random	117	60	16	44
	min_VAR		34	10	24
	min_VAR + cutoff = 1		**20**	10	10
RLFAP	Random	151	37	16	21
	min_VAR		41	14	27
	min_VAR + cutoff = 1		**22**	14	8
Purdey	Random	82	31	5	26
	min_VAR		24	3	21
	min_VAR + cutoff = 1		**12**	3	9

In Table 5, we report the performance of G-QUACQ with random, min_VAR and min_VAR + cutoff = 1 on all the other problems. We see that min_VAR + cutoff = 1 significantly improve the performance of G-QUACQ on all problems. As in the case of Sudoku, we observe that min_VAR + cutoff = 1 does not lead to an increase in the number of standard queries. This means that on all these problems min_VAR + cutoff = 1 always returns the maximal sequences while asking less generalization queries with negative answers.

From these experiments we see that G-QUACQ with min_VAR + cutoff = 1 leads to tremendous savings in number of queries compared to QUACQ: $257 + 23$ instead of 764 on Zebra, $260 + 39$ instead of 8645 on Sudoku, $117 + 20$ instead of 1129 on Latin square, $151 + 22$ instead of 1653 on RLFAP, $82 + 12$ instead of 173 on Purdey.

In our last experiment, we show the effect on the performance of G-QUACQ of a lack of knowledge on some variable types. We took again our 5 benchmark

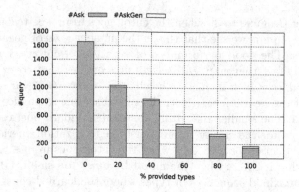

Fig. 3. G-QuAcq on RLFAP when the percentage of provided types increases.

problems in which we have varied the amount of types known by the algorithm. This simulates a situation where the user does not know that some variables are from the same type. For instance, in Sudoku, the user could not have noticed that variables are arranged in columns. Figure 3 shows the number of standard queries and generalization queries asked by G-QuAcq with min_VAR + cutoff = 1 to learn the RLFAP model when fed with an increasingly more accurate knowledge of types. We observe that as soon as a small percentage of types is known

Table 6. G-QuAcq when the percentage of provided types increases.

	% of types	#Ask	#AskGen
Zebra	0	764	0
	20	619	12
	40	529	20
	60	417	27
	80	332	40
	100	257	48
Sudoku 9 × 9	0	8645	0
	33	3583	232
	66	610	60
	100	260	39
Latin square	0	1129	0
	50	469	49
	100	117	20
Purdey	0	173	0
	33	111	8
	66	100	10
	100	82	12

(20%), G-QuAcq reduces drastically its number of queries. Table 6 gives the same information for all other problems.

7 Conclusion

We have proposed QuAcq, an algorithm that learns constraint networks by asking the user to classify partial assignments as positive or negative. Each time it receives a negative example, the algorithm converges on a constraint of the target network in a logarithmic number of queries. We have shown that QuAcq is optimal on certain constraint languages. Asking the user to classify partial assignments allows to converge on the target constraint network in a polynomial number of queries. Furthermore, as opposed to existing techniques, the user does not need to provide positive examples to converge. This last feature can be very useful when the problem has not been previously solved. We have also proposed GenAcq, a technique to make constraint acquisition more efficient in practice by using information on the types of components the variables in the problem represent. We have introduced generalization queries. They are asked to the user to generalize a constraint to other scopes of variables of the same type where this constraint possibly applies. GenAcq can be called to generalize each new constraint that is learned. We have proposed several heuristics and strategies to select the good candidate generalization query. We have plugged GenAcq into the QuAcq constraint acquisition system, leading to the G-QuAcq version. Our experimental evaluation shows that generating good queries in QuAcq is not computationally difficult and that when the basis increases in size, the increase in number of queries follows a logarithmic shape. These results are promising for the use of QuAcq on real problems. However, problems with dense constraint networks require a number of queries that could be too large for a human user. We have then evaluated the benefit of generalization queries, with and without complete knowledge on the types of variables. The results show that G-QuAcq dramatically improves the basic QuAcq algorithm in terms of number of queries.

References

1. Angluin, D.: Queries and concept learning. Mach. Learn. **2**(4), 319–342 (1987)
2. Beldiceanu, N., Carlsson, M., Rampon, J.: Global constraint catalog. Technical report, T2005: 08, Swedish Institute of Computer Science, Kista, Sweden, May 2005
3. Beldiceanu, N., Simonis, H.: A model seeker: extracting global constraint models from positive examples. In: Milano, M. (ed.) CP 2012. LNCS, vol. 7514, pp. 141–157. Springer, Heidelberg (2012)
4. Bessiere, C., Coletta, R., Freuder, E.C., O'Sullivan, B.: Leveraging the learning power of examples in automated constraint acquisition. In: Wallace, M. (ed.) CP 2004. LNCS, vol. 3258, pp. 123–137. Springer, Heidelberg (2004). doi:10.1007/978-3-540-30201-8_12

5. Bessiere, C., Coletta, R., Koriche, F., O'Sullivan, B.: A SAT-based version space algorithm for acquiring constraint satisfaction problems. In: Gama, J., Camacho, R., Brazdil, P.B., Jorge, A.M., Torgo, L. (eds.) ECML 2005. LNCS (LNAI), vol. 3720, pp. 23–34. Springer, Heidelberg (2005). doi:10.1007/11564096_8
6. Bessiere, C., Coletta, R., O'Sullivan, B., Paulin, M.: Query-driven constraint acquisition. In: Proceedings of the Twentieth International Joint Conference on Artificial Intelligence (IJCAI 2007), Hyderabad, India, pp. 44–49 (2007)
7. Bessiere, C., Koriche, F., Lazaar, N., O'Sullivan, B.: Constraint acquisition. Artif. Intell. (in press)
8. Bessiere, C., Coletta, R., Daoudi, A., Lazaar, N., Mechqrane, Y., Bouyakhf, E.: Boosting constraint acquisition via generalization queries. In: Proceedings of the 21st European Conference on Artificial Intelligence. Frontiers in Artificial Intelligence and Applications, vol. 263, pp. 99–104. IOS Press, Prague (2014)
9. Bessiere, C., Coletta, R., Hebrard, E., Katsirelos, G., Lazaar, N., Narodytska, N., Quimper, C., Walsh, T.: Constraint acquisition via partial queries. In: Proceedings of the 23rd International Joint Conference on Artificial Intelligence, pp. 475–481. IJCAI/AAAI, Beijing (2013)
10. Bessiere, C., Cordier, M.: Arc-consistency and arc-consistency again. In: Proceedings of the 11th National Conference on Artificial Intelligence, pp. 108–113. AAAI Press/The MIT Press, Washington, D.C. (1993)
11. Cabon, B., de Givry, S., Lobjois, L., Schiex, T., Warners, J.P.: Radio link frequency assignment. Constraints 4(1), 79–89 (1999)
12. De Bruijn, N.: Asymptotic Methods in Analysis. Dover Books on Mathematics. Dover Publications, New York (1970)
13. Freuder, E.C., Wallace, R.J.: Suggestion strategies for constraint-based matchmaker agents. In: Maher, M., Puget, J.-F. (eds.) CP 1998. LNCS, vol. 1520, pp. 192–204. Springer, Heidelberg (1998). doi:10.1007/3-540-49481-2_15
14. Gent, I., Walsh, T.: CSPLib: a benchmark library for constraints. http://www.csplib.org/ (1999)
15. Junker, U.: QUICKXPLAIN: preferred explanations and relaxations for over-constrained problems. In: Proceedings of the Nineteenth National Conference on Artificial Intelligence (AAAI 2004), San Jose, CA, pp. 167–172 (2004).
16. Lallouet, A., Lopez, M., Martin, L., Vrain, C.: On learning constraint problems. In: Proceedings of the 22nd IEEE International Conference on Tools for Artificial Intelligence (IEEE-ICTAI 2010), Arras, France, pp. 45–52 (2010)
17. Mason, J.: Purdey's general store. Dell Mag. 54, 10 (1997)
18. Paulin, M., Bessiere, C., Sallantin, J.: Automatic design of robot behaviors through constraint network acquisition. In: Proceedings of the 20th IEEE International Conference on Tools for Artificial Intelligence (IEEE-ICTAI 2008), Dayton, OH, pp. 275–282 (2008)

ModelSeeker: Extracting Global Constraint Models from Positive Examples

Nicolas Beldiceanu[1] and Helmut Simonis[2(✉)]

[1] TASC Team (INRIA/CNRS), Mines Nantes, 44307 Nantes, France
Nicolas.Beldiceanu@mines-nantes.fr
[2] Insight Centre for Data Analytics, Department of Computer Science, University College Cork, Cork, Ireland
helmut.simonis@insight-centre.org

Abstract. We describe a system which generates finite domain constraint models from positive example solutions, for highly structured problems. The system is based on the global constraint catalog, providing the library of constraints that can be used in modeling, and the Constraint Seeker tool, which finds a ranked list of matching constraints given one or more sample call patterns.

We have tested the modeler with 230 examples, ranging from 4 to 6,500 variables, using between 1 and 7,000 samples. These examples come from a variety of domains, including puzzles, sports-scheduling, packing & placement, and design theory. When comparing against manually specified "canonical" models for the examples, we achieve a hit rate of 50%, processing the complete benchmark set in less than one hour on a laptop. Surprisingly, in many cases the system finds usable candidate lists even when working with a single, positive example.

1 Introduction

In this chapter[1] we present the *Model Seeker* system which generates constraint models from example solutions. We focus on problems with a regular structure (this encompasses *matrix models* [18]), whose models can be compactly represented as a small set of conjunctions of identical constraints. We exploit this structure in our learning algorithm to focus the search for the strongest (i.e. most restrictive) possible model.

In our system we use global constraints from the global constraint catalog [2] mainly as modeling constructs, and not as a source of filtering algorithms. The global constraints are the primitives from which our models are created, each capturing some particular aspect of the overall problem. Using existing work on global constraints for mixed integer programming [24] or constraint based local search [20], our results are not only applicable for finite domain constraint programming, but can potentially reach a wider audience.

The second author is supported by EU FET grant ICON (project number 284715).
[1] This chapter is an extended version of reference [8].

© Springer International Publishing AG 2016
C. Bessiere et al. (Eds.): Data Mining and Constraint Programming, LNAI 10101, pp. 77–95, 2016.
DOI: 10.1007/978-3-319-50137-6_4

The input format we have chosen consists of a flat vector of integer values, allowing for different representations of the same problem. We do not force the user to adapt his input to any particular technology, but rather aim to be able to handle examples taken from a variety of existing sources.

In our method we extensively use meta-data about the constraints in the catalog, which describe their properties and their connection. We have added a number of new, useful information classes during our work, which prove to be instrumental in recognizing the structure of different models.

The main contribution of this chapter is the presentation of the implemented Model Seeker tool, which can deal with a variety of problem types at a practical scale. The examples we have studied use up to 6,500 variables, and deal with up to 7,000 samples, even though the majority of the problems are restricted to few, and often unique solution samples. We currently only work with positive examples, which seems to provide enough information to achieve quite accurate models of problems. As a side-effect of our work we also have strengthened the constraint description in the constraint catalog with new categories of meta-data, in particular to show implications between different constraints.

Our chapter is structured in the following way: We first introduce a running example, that we will use to explain the core features of our system. In Sect. 2, we describe the basic workflow in our system, also detailing the types of meta-data that are used in its different components. We present an overview of our evaluation in Sect. 3, which is followed by a discussion of related work (Sect. 4), before finishing with limitations and possible future work in Sect. 5. For space reasons we can only give an overview of the learning algorithm and the obtained results. A full description can be found in a companion technical report at http:// 4c.ucc.ie/~hsimonis/modelling/report.pdf.

1.1   A Running Example

As a running example we use the 2010/2011 season schedule of the Bundesliga, the German soccer championship. The problem representation is based on the format in http://www.weltfussball.de/alle_spiele/bundesliga-2010-2011/, replacing team names with numbers from 1 to 18. The schedule is given as a set of games on each day of the season. Table 1 shows days 1, 2, 3, 18 and 19 of the schedule. Each line shows all games of one day; on the first day, team 1 (at home) is playing against team 2 (away), team 3 (at home) plays team 4, etc. The second half of the season (days 18–34) repeats the games of the first half, exchanging the home and away teams, on day 18, for example, team 18 (at home) plays team 17, team 2 (at home) plays team 1, and so on. Overall, each team plays each other twice, once at home, and once away in a double round-robin scheme.

As input data we receive the flat vector of numbers, we will reconstruct the matrix as part of our analysis. Note that for most sports scheduling problems we will have access to only one example solution, the published schedule for a given year, schedules from different years encode different teams and constraints, and are thus incomparable.

Table 1. Bundesliga running example: input data

1	2	3	4	5	6	7	8	9	10	11	12	13	14	15	16	17	18
8	1	14	11	4	7	2	15	12	13	6	9	10	3	18	5	16	17
3	14	17	2	13	6	5	12	9	16	11	18	1	4	15	8	7	10
								...									
18	17	2	1	4	3	6	5	10	9	16	15	14	13	12	11	8	7
13	12	11	14	17	16	15	2	9	6	1	8	7	4	5	18	3	10
								...									

2   Workflow

We will now describe how we proceed from the given positive examples to a candidate list of constraints modeling the problem. The workflow is described in Fig. 1. Data are shown in green, software modules in blue/bold outline, and specific global constraint catalog meta-data are shown in yellow/italics. We first give a brief overview of the modules, and then discuss each step in more detail.

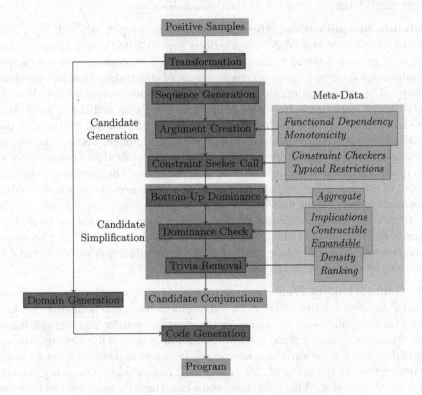

Fig. 1. Workflow in the Model Seeker (Color figure online)

Transformation. In a first step, we try to convert the input samples to other, more appropriate representations. This might involve replacing a 0/1 format with finite domain values, or converting different graph representations into the successor variable form used by the global constraints in the catalog. For some transformations, we keep both the original and the transformed representation for further analysis, for others we replace the original sample with the transformed data.

Candidate Generation. The second step (Sequence Generation) tries to group the variables of the sample into regular subsets, for example interpreting the input vector as a matrix and creating subsequences for all rows and all columns. In the Argument Creation step, we create call patterns for constraints from the subsequences. We can try each subsequence on its own, or combine pairs of subsequences or use all subsequences together in a single collection. We also try to add additional arguments based on functional dependencies and monotonic arguments of constraints, described as meta-data in the global constraint catalog. For each of these generated call patterns, we then call the Constraint Seeker to find matching constraints, which satisfy all subsequences of all samples. For this we enforce the *Typical Restrictions*, meta-data in the catalog, which describe how a constraint is typically used in a constraint program. Only the highest ranking candidates are retained for further analysis.

Candidate Simplification. After the seeker calls, we potentially have a very large list of possible candidate conjunctions (up to 2,000 conjunctions in our examples), we now have to reduce this set as much as possible. We first apply a Dominance Check to remove all conjunctions of constraints that we can show are implied by other conjunctions of constraints in our candidate list. Instead of showing the implication from first principles, we rely on additional meta-data in the catalog, which describe implications between constraints, but we also use conditional implications which only hold if certain argument restrictions apply, and expandible and contractible [27] properties, which state that a constraint still holds if we add or remove some of its decision variables. The dominance check is the core of our modeling system, helping to remove irrelevant constraints from the candidate list. In the last step before our final candidate list output the system removes trivial constraints and simplifies some constraint pattern. This also performs a ranking of the candidates based on the constraint and sequence generator used, trying to place the most relevant conjunction of constraints at the top of the candidate list.

Code Generation. As a side effect of the initial transformation, we also create potential domains for the variables of our problem. In the default case, we just use the range of all values occurring in the samples, but for some graph-based transformations a more refined domain definition is used. Given the candidate list and domains for the variables, we can easily generate code for a model using the constraints. At the moment, we produce SICStus Prolog code using the call format of the catalog. The generated code can then be used to validate the samples given, or to find solutions for the model that has been found.

After this brief overview, we will now discuss the different process steps in more detail.

2.1 Transformation

In finite domain programming, there are implicit conventions on how to express models leading to effective solution generation. In our system, we can not assume that the user is aware of these conventions, nor that the sample solutions are already provided in the "correct" form. We therefore try to apply a series of (currently 12) input transformations, that convert between different possible representations of problems, and that retain the form which matches the format expected by the constraint catalog. In each case, some pre- and post-conditions must be satisfied for the transformation to be considered valid. We now give some examples.

Converting 0/1 Samples. If a solution is given using only 0/1 values, there may be a way of re-interpreting the sample as a permutation with finite domain variables. If we consider the 0/1 values as an $n \times n$ matrix (x_{ij}) where each row and each column contains a single one, we can transform this into a vector v_i of n finite domain values based on the equivalence

$$\forall_{1 \le i \le n} \forall_{1 \le j \le n} : \quad x_{ij} = 1 \Rightarrow v_i = j$$

This transformation is the equivalent of a channeling constraint between 0/1 and finite domain variables, described for example in [23].

Using Successor Notation for Graph Constraints. Most graph constraints in the catalog use a *successor* notation to describe circuits, cycles or paths, i.e. the value of a node indicates the next node in the final assignment. But this is not the only way of describing graphs. In the original Knight's Tour formulation [37], the value of a node is the position in the path, i.e. the first node has value one, the second value two, and so on. We have defined transformations which map between these (and several other) formats, while checking that the resulting graph can be compactly described.

Table 2. Bundesliga running example: transformed problem

2	−1	4	−3	6	−5	8	−7	10	−9	12	−11	14	−13	16	−15	18	−17
−8	15	−10	7	−18	9	−4	1	−6	3	−14	13	−12	11	−2	17	−16	5
4	−17	14	−1	12	−13	10	−15	16	−7	18	−5	6	−3	8	−9	2	−11
...																	
−2	1	−4	3	−6	5	−8	7	−10	9	−12	11	−14	13	−16	15	−18	17
8	−15	10	−7	18	−9	4	−1	6	−3	14	−13	12	−11	2	−17	16	−5
...																	

Using Schreuder Tables. Another transformation is linked to sports scheduling problems. In many cases, users like to give the schedule as a list of fixtures, listing which games will be played on each day. The first team is the home team, the second the away team. For constraint models, the format of Schreuder tables [34], as shown in Table 2 for our running example, can lead to more compact models [21,22,29,36]. For each time-point t over q rounds and for an even number of teams n, they can be obtained from the fixture representation as follows:

$$\forall_{1 \leq t \leq q(n-1)} \forall_{1 \leq i \leq \lfloor n/2 \rfloor} : \quad x_{2i-1,t} = k, x_{2i,t} = l \Rightarrow v_{k,t} = l, v_{l,t} = -k$$

2.2 Sequence Generator

After the input transformation, we have to consider possible, regular substructures which group the samples into subsequences. For space reasons again, we only give some examples of the sequence generators used in our running example, the full list (containing 21 generators) with their formal definition can be found in the technical report, some were already described in [4].

vector(n). This is the most basic sequence generator of treating all elements of the sample as a single sequence of size n.

scheme(n, r, c, a, b). By far the most common sequence generator treats a sample of length n as an $r \times c$ matrix, and creates non-overlapping blocks of size $a \times b$, creating n/ab sequences of size ab. The number of such partitions depends on the number of factors of n, as $n = rc$. For our running example (Sect. 1.1) with 612 values, we have to consider the matrices 2×306, 3×204, 4×153, 6×102, 9×68, 17×36, 18×34, 34×18, 36×17, 68×9, 102×6, 153×4, 204×3 and 306×2. Some of the blocks created from these matrices lead to the same partition of the variables, only one representative is kept.

repart(n, r, c, a, b). This sequence generator also treats the sample of size n as a $r \times c$ matrix, and considers blocks of size $a \times b$. But it groups elements in the same position from each block, creating $a \times b$ sequences of size $n/(ab)$.

For the running example, a total of 68 subsequence collections are generated. Note that the subsequences often, but not always, have the same size. We also provide an API where the user can provide his own sequence generators, this can be helpful to deal with known, but irregular structure in the problem.

2.3 Argument Creation

In the next step of the operation, we convert the generated subsequences into call patterns for the Constraint Seeker [3]. In order to consider more of the constraints in the catalog, we have to provide different argument signatures by organizing the subsequences in different ways, and by adding arguments.

Single, Pairs and Collection. In the first part we decide how we want to use the subsequences. Consider we have k subsequences, each of length m. If we use each subsequence on its own, we create k call patterns with a single argument, each a collection of m variables. This corresponds to the argument pattern used by `alldifferent` constraints, for example. We can also consider pairs of subsequences, creating a call patterns with two arguments, for $k-1$ calls to a predicate like `lex_greater`. Finally we can use all subsequences as a single collection of collections, which creates one call with a collection of k collections of m elements each. This would match a constraint like `lex_alldifferent`. We generate all these potential calls in parallel, and perform the steps described in the following two paragraphs.

Value Projection. For some problems (like our transformed, running example), a projection from the original domain to a smaller domain can lead to a more compact model. If, for example, some of the values in the sample are positive, and others are negative, we can try a projection using the absolute value or the sign of the numbers, in addition to the original values.

Adding Arguments. Many global constraints use additional arguments besides the main decision variables. If we do not generate these arguments in the call pattern, we can not find such constraints with the Constraint Seeker. But just enumerating all possible values for these additional arguments would lead to a combinatorial explosion. Fortunately, we can compute values for these arguments in case of functional dependencies and monotonic arguments. This is similar to the argument generation discussed for the `gcc` constraint discussed in [12].

2.4   Constraint Seeker

The Constraint Seeker [3] will find a ranked list of global constraints that satisfy a collection of positive and negative sample calls, using the available constraint checkers of the catalog. We use this seeker as a black-box for all call patterns with all additional argument values and value projections defined in the previous section.

Using Multiple, Positive Samples. The seeker first checks that the call signature matches the constraint, then tries to evaluate the constraint on the samples. In our case, these are the call patterns prepared in the previous step for all subsequences of all positive examples given. In our modeling system we currently do not consider negative examples. They would require a slightly different treatment, as a negative example can be rejected by just one constraint, while all positive examples must be accepted by all constraints found.

Typical Restrictions. In addition to the restrictions that must hold for the constraint to be applied, in our modeling tool we also check for the *typical* restrictions that are specified in the catalog. The `alldifferent` constraint for example can be called with an empty collection, but a typical use would have more than two variables in the collection. The typical constraints are expressed using the same language as the formal restrictions of the catalog, checking their validity thus does not require any additional code.

Selecting Top-Ranked Elements. The Constraint Seeker returns a ranked list of candidates, this ranking is a combination of structural properties (functional dependencies or monotonic arguments), implications between constraints, estimated solution density and estimated popularity of the constraint described in [3]. In our system we only use the top ranked element that satisfies all subsequences of all samples. This reduces the number of candidates to be considered, while at the same time it does not seem to exclude constraints that are required for the highly structured problems considered.

For our running example, we perform 1,099 calls to the Constraint Seeker, which performs 82,458 constraint checks, and which results in 589 possible candidate conjunctions. We now face the task of reducing this candidate list as much as possible, keeping only interesting conjunctions.

2.5 Bottom-Up Dominance

Some constraints like sum or gcc have the *aggregate* property, one can combine multiple such constraints over disjoint variable sets by adding the right hand sides or summing the counter values. As an example, we can combine

$$x_1 + x_2 = 5 \wedge x_3 + x_4 = 2 \Rightarrow x_1 + x_2 + x_3 + x_4 = 7$$

We want to remove aggregated constraints of this type, as they are implied by conjunctions of smaller constraints. We perform a bottom-up saturation of combining constraints with the aggregate property up to a limited size, and remove any candidate conjunctions where all constraints are dominated.

2.6 Dominance Check

The dominance check compares all conjunction candidates against each other (worst case quadratic number of comparisons), and marks dominated entries. Note that dominated entries may be used to dominate other entries, and thus can not be removed immediately. We use a number of meta-data fields to check for dominance.

Implications. In our final candidate list, we are interested in only the strongest, most restrictive constraints, all constraints that are implied by other candidate constraints can be excluded. Note that this will sometimes lead to overly restrictive solutions, especially if only a few samples are given.

Checking if some conjunction is implied by some other conjunction for a particular set of input values is a complex problem, a general solution would require sophisticated theorem proving tools like those used in [15] for a restricted problem domain. We do not attempt such a generic solution, but instead rely on meta-data in the catalog linking the constraints. That meta-data is useful also for understanding the relations between constraints, and thus serves multiple purposes. This syntactic implication check is easy to implement, but only can be used if both constraints have the same arguments.

Conditional Implications. For some constraints additional implications exist, but only if certain restrictions apply. The cycle constraint for example implies the circuit constraint, but only if the NCYCLE argument is equal to one.

For conditional implications the arguments do not have to be the same, but the main decision variables used must match.

Contractibility and Expandability. Other useful properties are contractibility [27] and expandibility. A constraint (like `alldifferent`) is *contractible* if it still holds if we remove some of its decision variables. This allows us to dominate large conjunctions of constraints with few variables with small conjunctions of constraints with many variables. Due to the way we systematically generate all subsequence collections, this is often possible. In a similar way, some constraints like `atleast` are *expandible*, they still hold if we add decision variables. We can again use this property to dominate some conjunctions of constraints. Details and possible extensions have been described in [4].

Hand-coded Domination Rules. Some dominance rules are currently hand-crafted in the program, if the required meta-data have not yet been formalized in the catalog description. Such examples can be an important source of requirements for the catalog itself, enhancing the expressive power of the constraint descriptions.

2.7   Trivia Removal

Even after the dominance check, we can still have candidate explanations which are valid and not dominated, but which are not useful for modeling. In the trivia removal section, we eliminate or replace most of these based on sets of rules.

Functional Dependencies on Single Samples. In Sect. 2.3 we have described how we can add some arguments to a call pattern for functional dependencies. In the case of pure functional dependencies [6], we have to worry about pattern consisting of a single subsequence with a single sample. In that case, the constraint does not filter any pattern, as for each pattern the correct value can be selected. We therefore remove such candidates.

Constraint Simplification. At this point we can also try to simplify some constraints that have particular structure. A typical example are `lex_chain` constraints on a subsequence, where already the first entries of the collections are ordered in strictly increasing order. We can therefore replace the `lex_chain` constraint on the subsequences with a `strictly_increasing` constraint on the first elements of the collections, using a special `first` sequence generator. These constraints often occur as symmetry-breaking restrictions in models, which we find if all the samples given respect the symmetry breaking.

Uninteresting Constraints. Even with the typical restrictions in the Constraint Seeker, we often find candidates (like `not_all_equal`) which are not very interesting for defining models. As a final safe-guard, we use a black-list to remove some combinations of constraints and sequence generators that should not be included in our models.

2.8   Candidate List for Bundesliga Schedule

Table 3 shows the list of the candidate conjunctions generated for our transformed example problem. Entries in green match a manually defined model, ten other candidates are also proposed. The arguments of constraints in the Constraint Conjunction column indicate any additional parameters, the *n indicates how many constraints form the conjunction. The value projections absolute_value and sign convert each element of the input data, id denotes the identity projection.

Table 3. Constraint conjunctions for problem Bundesliga

-	Sequence Generator	Projection	Constraint Conjunction
1	scheme(612,34,18,34,1)	id	alldifferent*18
2	scheme(612,34,18,2,2)	id	alldifferent*153
3	scheme(612,34,18,1,18)	id	alldifferent*34
4	scheme(612,34,18,1,18)	absolute_value	symmetric_alldifferent([1..18])*34
5	scheme(612,34,18,17,1)	absolute_value	alldifferent*36
6	repart(612,34,18,34,9)	id	sum_ctr(0)*306
7	repart(612,34,18,34,9)	id	twin*1
8	repart(612,34,18,34,9)	id	elements([i,-i])*1
9	first(9,[1,3,5,7,9,11,13,15,17])	id	strictly_increasing*1
10	vector(612)	id	global_cardinality([-18.. -1-17,0-0,1..18-17])*1
11	repart(612,34,18,34,9)	id	sum_powers5_ctr(0)*306
12	repart(612,34,18,34,9)	id	sum_cubes_ctr(0)*306
13	repart(612,34,18,34,3)	sign	global_cardinality([-1-3,0-0,1-3])*102
14	scheme(612,34,18,34,1)	sign	global_cardinality([-1-17,0-0,1-17])*18
15	repart(612,34,18,17,9)	sign	global_cardinality([-1-2,0-0,1-2])*153
16	repart(612,34,18,2,9)	sign	global_cardinality([-1-17,0-0,1-17])*18
17	scheme(612,34,18,1,18)	sign	global_cardinality([-1-9,0-0,1-9])*34
18	repart(612,34,18,34,9)	sign	sum_ctr(0)*306
19	repart(612,34,18,34,9)	sign	twin*1
20	repart(612,34,18,34,9)	absolute_value	twin*1
21	repart(612,34,18,34,9)	sign	elements([i,-i])*1
22	scheme(612,34,18,34,1)	sign	among_seq(3,[-1])*18
23	repart(612,34,18,34,9)	absolute_value	elements([i,i])*1
24	first(9,[1,3,5,7,9,11,13,15,17])	absolute_value	strictly_increasing*1
25	first(6,[1,4,7,10,13,16])	absolute_value	strictly_increasing*1
26	scheme(612,34,18,34,1)	absolute_value	nvalue(17)*18

Some of the constraints mentioned are perhaps unfamiliar, we provide a short definition. The constraint symmetric_alldifferent($[x_1,..,x_n]$) [2, p. 1854] in line 4 states that

$$\forall_{1 \leq i \leq n}: \quad x_i \in [1,n]; x_i = j \iff x_j = i$$

It expresses the constraint that if team A plays team B on some day, then team B will play team A. The constraints twin($[\langle x_1, y_1 \rangle, ..., \langle x_n, y_n \rangle]$) [2, p. 1896] in lines 7, 19 and 20 state that

$$\forall_{1 \leq i \leq n} : \quad (x_i = u \land y_i = v) \Rightarrow (\forall_{1 \leq j \leq n} : \; x_j = u \iff y_j = v)$$

These constraints express the fact that the tournament is played in two symmetric half-seasons, with home and away games swapped. Note that constraints 8, 21 and 23 also express this condition, but using an `elements` constraint, pairing positive and negative numbers. The `alldifferent` constraint in line 1 expresses that no repeat games occur in the season, while that of line 5 states that all teams play on each day. The `strictly_increasing` constraint in line 9 results from the simplification of a symmetry breaking `lex_chain` constraint. The `gcc` in line 14 states that each team plays 17 home (positive value) and 17 away (negative value) games. Finally, the `among_seq` constraint in line 22 states that no team has more than two consecutive away games.

2.9   Domain Creation

By default, the domains of the variables in our generated models are the interval between the smallest and largest value occurring in the samples. Based on the transformation used, we can use more restricted domains for graph models like graph partitioning and domination [19], where the domain of each variable/node specifies the initial graph.

2.10   Code Generation

The code generation builds flat models for the given instances. The programs consist of five parts, we first define all variables with their domains, then state all restrictions due to fixed values as assignments, state any projections used to simplify the variables, then build the constraints in the catalog syntax, and finally call a generic value assignment routine to search for a solution. We can use the generated model as a test to check if it accepts the given samples, or to generate new solutions for the problem. Many puzzles have a unique solution, we can count solutions of our program to see if the generated model is restrictive enough to capture this property.

It would be straightforward to generate the code for other systems than SICStus Prolog, provided that the catalog constraints are supported. A version generating FlatZinc [28] or XCSP [33] would be especially attractive to benefit from the variety of backend solvers which support these formats.

3   Evaluation

Table 4 shows summary results for selected problems of our evaluation set. The problems range from sports scheduling (ACC Basketball Scheduling, csplib11; Bundesliga; DEL2011 (German ice hockey league); Scottish Premier League (soccer); Rabodirect Pro 12 (rugby)), to scheduling (Job-shop 10×10 [14]) and placement (Duijvestijn, csplib9; Conway $5 \times 5 \times 5$ [9]; Costas Array [16]), design theory (BIBD, csplib28; Kirkman [17]; Orthogonal Latin Squares [13]), event scheduling (Social Golfer, csplib10; Progressive Party, csplib13) and puzzles. Details of

Table 4. Selected example results

Name	Transformation Id	Instance Size	Nr Samples	Nr Sequences	Nr Seeker Calls	Constraint Checks	Nr Relevant	Nr Non Dom	NrSpecified	Nr Models	Nr Missing	Hit Rate	Time [s]
12 Amazons All	-	12	156	36	127	35596	55	9	5	8	0	62.50	2,64
8 Queens All	-	8	92	20	100	12077	34	7	3	5	0	60.00	0.83
ACC Basketball Schedule	-	162	1	109	1786	29117	772	263	23	36	7	n/a	9.17
BIBD (8,14,7,4,3)	-	112	92	151	626	92461	232	112	4	15	0	26.67	26.01
Bundesliga	18	612	1	68	1099	52933	589	169	16	26	0	61.54	51.44
Conway 5x5x5 Packing	-	102	1	60	184	2619	78	35	1	2	0	50.00	0.42
Costas Array 12	-	12	48	36	121	14820	42	2	2	2	0	100.00	1.01
DEL 2011	-	728	1	235	1334	66999	555	173	3	8	0	37.50	54.23
De Jaenisch Tour	-	64	1	83	568	10130	283	58	2	13	0	15.38	1.66
De Jaenisch Tour	7	64	1	36	219	12952	113	67	1	1	0	100.00	0.46
Dominating Knights 8	9	64	2	36	141	11021	51	42	1	1	0	100.00	0.31
Duijvestijn 21	-	84	1	111	504	11625	240	102	1	12	0	8.33	1.77
Euler Knight Cube 4x4x4	7	64	1	36	208	12759	97	58	1	1	0	100.00	0.42
Job Shop 10x10 (10 sol)	-	400	10	326	1521	80589	582	130	2	2	0	100.00	40.27
Kirkman Wikipedia	-	105	1	40	179	2634	89	39	3	5	0	60.00	1.21
Leaper Tour 18x18	7	324	1	60	298	105955	140	61	1	1	0	100.00	2.18
Magic Square All	-	16	7040	33	176	1068574	57	5	4	4	0	100.00	115.07
Magic Square Duerer	-	16	1	33	212	2074	115	44	9	15	0	60.00	0.25
Orthogonal Latin Squares 10	-	200	1	147	910	15441	443	118	3	8	0	37.50	6.04
Progressive Party	-	174	1	45	171	4279	61	31	4	3	1	n/a	0.70
Rabodirect Pro12	18	264	1	66	1041	46898	539	155	8	13	0	61.54	7.91
Scottish Premier League	18	396	1	68	992	58959	459	157	9	12	0	75.00	14.18
Social Golfer	-	288	1	528	2813	69681	1221	256	5	36	0	13.89	61.93
Sudoku 81x81	-	6561	1	91	657	101075	334	68	3	5	0	60.00	244.16

these problems can be found in the technical report mentioned before. Smaller problems are solved within seconds, even the largest require less than 5 min on a single core of a MacBook Pro (2.2 GHz) with 8 Gb of memory.

The columns denote: *Transformation Id*: the number of the transformation applied (if any), *Instance Size*: the number of values in the solution, i.e. the number of variables in the model, *Nr Samples*: the number of solutions given as input, *Nr Sequences*: the number of sequence sets generated, *Nr Seeker Calls*: the number of times the Constraint Seeker is called, *Constraint Checks*: the number of calls to constraint checkers inside the seeker, *Nr Relevant*: the number of initial candidate conjunctions found by the Constraint Seeker, *Nr Non Dom*: the number of non-dominated candidates remaining after the dominance checkers, *Nr Specified*: the number of conjunctions specified in the manual, "canonical" model, *Nr Models*: the number of conjunctions given as output of the Model Seeker, *Nr Missing*: how many of the manually defined conjunctions were not found by the system, *Hit Rate*: the percentage rate of Nr Specified to Nr Models, a value of 100% indicates that exactly the candidates of the canonical model were found, and *Time*: the execution time in seconds.

For two of the problems, we only find part of the complete model. The Progressive party problem [35] requires a bin-packing constraint that we currently do not recognize, as it relies on additional data for the boat sizes, while the ACC basketball problem contains several constraints which apply only for specific parts of the schedule, and which can not be learned from a single solution. Also note that for the De Jaenisch problem [31], we show results with and without a transformation. This problem combines a "near" magic square, found without transformation, with an Euler Knight tour, using transformation 7.

For our full evaluation, we have used 230 examples from various sources. For 10 of the examples no reasonable model was generated, either because we did not have the right sequence generator, or we are currently missing the global constraint required to express the problem. For a further 37 problems, only part of the model was found. This is typically caused by some constraint requiring additional data, not currently given as input, or by an over-specialization of the output, where the Model Seeker finds a more restrictive constraint than the one specified manually. Overall, we considered 73 constraints in the Constraint Seeker, and selected 53 different global constraints as potential solutions. This is only a fraction of the 380 constraint in the catalog, many of the missing constraints have more complex argument signatures or use finite sets, which are currently not available in SICStus Prolog.

Figure 2 shows the number of candidates found for all examples studied as a function of the instance size, split between single samples and multiple samples. Note that the plot uses a log-log scale. The results indicate that even with a single sample, the number of candidate conjunctions found is quite limited, this drops further if multiple samples are used.

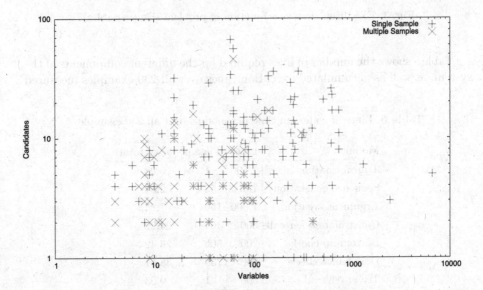

Fig. 2. Candidates as a function of problem size (variables)

Another view of all the results is shown in Fig. 3. It shows the relationship between number of variables and execution time, again grouped by problems with a single sample and problems for which multiple samples were provided. While no formal complexity analysis has been attempted, as several subproblems are expressed as constraint problems, results seem to indicate a low-polynomial link between problem size and execution time. The non-linear least square fit for the single sample problems is $8.5e^{-2}x^{0.90}$, and for multiple samples $6.1e^{-3}x^{1.45}$.

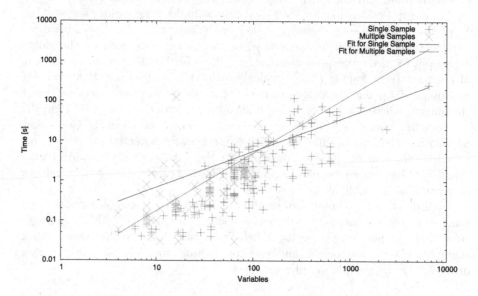

Fig. 3. Execution time as a function of problem size (variables)

Table 5 shows the number of lines required for the different components of the system, as well as accumulated execution times over all 230 examples measured

Table 5. Lines of code/run time per module over all 230 examples

Module	Lines	Time [s]	% of total
Transformation	1,500	1	0.03
Sequence generation	1,000	53	2.81
Argument creation	1,000	150	7.84
Constraint seeker call	300	464	24.22
Bottom-up check	200	506	26.42
Dominance check	800	739	38.61
Trivia removal	500	1	0.03
Glue/IO/test	2,000	-	-

for these components. The programming effort is fairly evenly split amongst the different components, while the two dominance checkers require nearly two-thirds of the total execution time, with the constraint seeker using another quarter of the time. The system is written in SICStus Prolog 4.2, and uses the Constraint Seeker [3] with an additional 6,500 lines of code, and the global constraint catalog meta-data description of 60,000 lines of Prolog.

4   Related Work

Our approach of searching for conjunctions of identical constraints generalizes the idea of *matrix models* [18], which are an often-used pattern in constraint modeling.

The method proposed is a special, restricted case of *constraint acquisition* [30], which is the process of finding a constraint network from a training set of positive and negative examples. The learning process operates on a library of allowed constraints, and a resulting solution is a conjunction of constraints from that library, each constraint ranging over a subset of the variables.

The most successful of these systems is the CONACQ system [10], which proposes the use of version space learning to generate models interactively with the user, relying on an underlying SAT model to perform the learning. This is shown to work for binary constraints, but the method breaks down for global constraints over an unlimited number of variables.

In [11], the authors study the problem of determining argument values for global constraints like the gcc from example solutions, in the context of timetabling problems. This is similar to the argument creation we describe in Sect. 2.3.

The more recent work of [25] considers the use of inductive logic programming for finding models for problems given as a set of logic formulas. This can be powerful to find generic, size-independent models for a problem, but again, it is unclear how to deal with a library of given global constraints, which may not have a simple description as logic formulas.

Our dominance check based on meta-data is related to the work described in [15], where they use a theorem prover to find certain implications between constraints for a restricted domain. This does not rely on meta-data provided in the system, but instead would require a very powerful theorem prover to work for a collection of constraints for problems of the size considered here.

Common to all these results is that they have not been evaluated on a large variety of problems, that they consider only a limited number of potential constraints, and that problem sizes have been quite small.

5   Limitations and Future Work

We are currently only considering some 70 constraints in the global constraint catalog in our seeker calls. Many of the missing constraints require additional information (cost matrix, lookup tables) which have to be provided as additional input data to the system. For some problems, such additional data, like a precedence graph, may also express implicit, less regular sequence generators, which define for which variables a constraint should be stated. Extending our input format to allow for such data would drastically increase both the number of constraints that can be considered, as well as the range of application problems that can be modelled.

Most other constraint acquisition systems use both positive and negative examples. The negative examples are used to interactively differentiate between competing models of the system. We currently only use positive examples, but given recent results on global constraint reification [6], we could extend our system to include this functionality.

A first application of the ModelSeeker to an application problem is described in [7]. It discusses the use of the ModelSeeker to learn models of electrical power plants as part of a Unit Commitment Model of the French electricity company EDF. We learn constraints on the power output over consecutive time periods from positive samples by using specific time-series constraints [5].

If we want to provide the functionality we have presented here to end-users, we will have to consider issues of usability and interactivity, allowing the user to filter and change constraint candidates, as well as being able to suggest custom sequence generators tailored to a specific problem. An interactive tool that helps programers to refactor their MiniZinc constraint programs with global constraints was described in [26]. It tries to replace code fragments by corresponding global constraints, generating sample solutions as required, and is supported by a web-based interface.

Ultimately, we are looking for a modeling tool which can analyze samples of different sizes, and generate a generic, size independent model. Building on top of our existing framework, this would require to express both the sequence generator parameters and any additional arguments for constraints in terms of a variable problem size, to produce more compact, iterative code instead of the flat models currently generated. A first step towards this goal was discussed in [32], where we learn coefficients of simple polynomials that characterize generic models of constraint problems.

A very different approach to generating constraint models semi-automatically is used in the Conjure system [1]. Instead of using sample solutions, the input consists of a naive, high-level model of a problem. This model is then transformed into a more effective constraint model by recognizing structure and symmetries. It is an interesting open question to unterstand whether sample solutions or high-level models are more readily available for practical problems.

6   Conclusion

Exploiting the idea that many highly structured combinatorial problems can be described by a small set of conjunctions of identical global constraints, this chapter described the ModelSeeker system for extracting global constraint models from positive sample solutions. It relies on a detailled description of the constraints in terms of meta-data in the global constraint catalog. The system provides promising results on a variety of problems even when working from a very limited number of only positive examples.

Acknowledgement. The help of Hakan Kjellerstrand in finding example problems is gratefully acknowledged.

References

1. Akgun, O., Frisch, A.M., Gent, I.P., Hussain, B.S., Jefferson, C., Kotthoff, L., Miguel, I., Nightingale, P.: Automated symmetry breaking and model selection in CONJURE. In: Schulte, C. (ed.) CP 2013. LNCS, vol. 8124, pp. 107–116. Springer, Heidelberg (2013). doi:10.1007/978-3-642-40627-0_11
2. Beldiceanu, N., Carlsson, M., Rampon, J.: Global constraint catalog, 2nd edn. (revision a). Technical report T2012:03, SICS (2012)
3. Beldiceanu, N., Simonis, H.: A constraint seeker: finding and ranking global constraints from examples. In: Lee, J. (ed.) CP 2011. LNCS, vol. 6876, pp. 12–26. Springer, Heidelberg (2011). doi:10.1007/978-3-642-23786-7_4
4. Beldiceanu, N., Simonis, H.: Using the global constraint seeker for learning structured constraint models: a first attempt. In: The 10th International Workshop on Constraint Modelling and Reformulation (ModRef 2011), Perugia, Italy, pp. 20–34, September 2011
5. Beldiceanu, N., Carlsson, M., Douence, R., Simonis, H.: Using finite transducers for describing and synthesising structural time-series constraints. Constraints **21**(1), 22–40 (2016)
6. Beldiceanu, N., Carlsson, M., Flener, P., Pearson, J.: On the reification of global constraints. Constraints **18**(1), 1–6 (2013)
7. Beldiceanu, N., Ifrim, G., Lenoir, A., Simonis, H.: Describing and generating solutions for the EDF unit commitment problem with the ModelSeeker. In: Schulte, C. (ed.) Principles and Practice of Constraint Programming. LNCS, vol. 8124, pp. 733–748. Springer, Heidelberg (2013). doi:10.1007/978-3-642-40627-0_54
8. Beldiceanu, N., Simonis, H.: A model seeker: extracting global constraint models from positive examples. In: Milano, M. (ed.) CP 2012. Lecture Notes in Computer Science, vol. 7514, pp. 141–157. Springer, Heidelberg (2012). doi:10.1007/978-3-642-33558-7_13
9. Berlekamp, E.R., Conway, J.H., Guy, R.K.: Winning Ways for Your Mathematical Plays, vol. 4, 2nd edn. A K Peters/CRC Press, Natick (2004)
10. Bessière, C., Coletta, R., Freuder, E.C., O'Sullivan, B.: Leveraging the learning power of examples in automated constraint acquisition. In: Wallace, M. (ed.) CP 2004. LNCS, vol. 3258, pp. 123–137. Springer, Heidelberg (2004). doi:10.1007/978-3-540-30201-8_12
11. Bessiere, C., Coletta, R., Koriche, F., O'Sullivan, B.: A SAT-based version space algorithm for acquiring constraint satisfaction problems. In: Gama, J., Camacho, R., Brazdil, P.B., Jorge, A.M., Torgo, L. (eds.) ECML 2005. LNCS (LNAI), vol. 3720, pp. 23–34. Springer, Heidelberg (2005). doi:10.1007/11564096_8
12. Bessiere, C., Coletta, R., Petit, T.: Acquiring parameters of implied global constraints. In: Beek, P. (ed.) CP 2005. LNCS, vol. 3709, pp. 747–751. Springer, Heidelberg (2005). doi:10.1007/11564751_57

13. Bose, R.C., Shrikhande, S.S., Parker, E.T.: Further results on the construction of mutually orthogonal latin squares and the falsity of Euler's conjecture. Can. J. Math. **12**, 189–203 (1960)
14. Carlier, J., Pinson, E.: An algorithm for solving the job shop problem. Manag. Sci. **35**, 164–176 (1989)
15. Charnley, J., Colton, S., Miguel, I.: Automatic generation of implied constraints. In: Brewka, G., Coradeschi, S., Perini, A., Traverso, P. (eds.) ECAI. Frontiers in Artificial Intelligence and Applications, vol. 141, pp. 73–77. IOS Press, Amsterdam (2006)
16. Drakakis, K.: A review of Costas arrays. J. Appl. Math. **2006**, 1–32 (2006)
17. Dudeney, H.E.: Amusements in Mathematics. Dover, New York (1917)
18. Flener, P., Frisch, A., Hnich, B., Kiziltan, Z., Miguel, I., Walsh, T.: Matrix modelling. Technical report 2001–023, Department of Information Technology, Uppsala University, September 2001
19. Haynes, T.W., Hedetniemi, S.T., Slater, P.J.: Fundamentals of Domination in Graphs. Monographs and Textbooks in Pure and Applied Mathematics. Marcel Dekker, New York (1998)
20. Van Hentenryck, P., Michel, L.: Constraint-Based Local Search. MIT Press, Boston (2005)
21. Henz, M.: Scheduling a major college basketball conference - revisited. Oper. Res. **49**, 163–168 (2001)
22. Henz, M., Müller, T., Thiel, S.: Global constraints for round robin tournament scheduling. Eur. J. Oper. Res. **153**(1), 92–101 (2004)
23. Hernández, B.M.: The systematic generation of channelled models in constraint satisfaction. PhD thesis, University of York, York, YO10 5DD, UK, Department of Computer Science (2007)
24. Hooker, J.N.: Integrated Methods for Optimization. Springer Science + Business Media LLC, New York (2007)
25. Lallouet, A., Lopez, M., Martin, L., Vrain, C.: On learning constraint problems. In: ICTAI, vol. 1, pp. 45–52. IEEE Computer Society (2010)
26. Leo, K., Mears, C., Tack, G., Garcia de la Banda, M.: Globalizing constraint models. In: Schulte, C. (ed.) CP 2013. LNCS, vol. 8124, pp. 432–447. Springer, Heidelberg (2013). doi:10.1007/978-3-642-40627-0_34
27. Maher, M.J.: Open constraints in a boundable world. In: van Hoeve, W.-J., Hooker, J.N. (eds.) CPAIOR 2009. LNCS, vol. 5547, pp. 163–177. Springer, Heidelberg (2009)
28. Marriott, K., Nethercote, N., Rafeh, R., Stuckey, P.J., de la Banda, M.G., Wallace, M.: The design of the Zinc modelling language. Constraints **13**(3), 229–267 (2008)
29. Nemhauser, G., Trick, M.: Scheduling a major college basketball conference. Oper. Res. **46**, 1–8 (1998)
30. O'Sullivan, B.: Automated modelling and solving in constraint programming. In: Fox, M., Poole, D. (eds.) AAAI, pp. 1493–1497. AAAI Press, Palo Alto (2010)
31. Petkovic, M.S.: Famous Puzzles of Great Mathematicians. American Mathematical Society, Providence (2009)
32. Razakarison, N., Carlsson, M., Beldiceanu, N., Simonis, H.: GAC for a linear inequality and an atleast constraint with an application to learning simple polynomials. In: Helmert, M., Röger, G. (eds.) Proceedings of the Sixth Annual Symposium on Combinatorial Search, SOCS 2013, Leavenworth, Washington, USA, 11–13 July 2013. AAAI Press (2013)

33. Roussel, O., Lecoutre, C.: XML representation of constraint networks format XCSP 2.1. Technical report arXiv:0902.2362v1, Universite Lille-Nord de France, Artois (2009)
34. Schreuder, J.A.M.: Combinatorial aspects of construction of competition Dutch professional football leagues. Discret. Appl. Math. **35**(3), 301–312 (1992)
35. Smith, B.M., Brailsford, S.C., Hubbard, P.M., Williams, H.P.: The progressive party problem: integer linear programming and constraint programming compared. Constraints **1**(1/2), 119–138 (1996)
36. Walser, J.P.: Domain-independent local search for linear integer optimization. PhD thesis, Technical Faculty of the University des Saarlandes, Saarbruecken, Germany, October 1998
37. Watkins, J.J.: Across the Board: The Mathematics of Chessboard Problems. Princeton University Press, Princeton (2004)

Learning Constraint Satisfaction Problems: An ILP Perspective

Luc De Raedt[1], Anton Dries[1(✉)], Tias Guns[1], and Christian Bessiere[2]

[1] DTAI, KU Leuven, Leuven, Belgium
anton.dries@cs.kuleuven.be
[2] CNRS, University of Montpellier, Montpellier, France

Abstract. We investigate the problem of learning constraint satisfaction problems from an inductive logic programming perspective. Constraint satisfaction problems are the underlying basis for constraint programming and there is a long standing interest in techniques for learning these. Constraint satisfaction problems are often described using a relational logic, so inductive logic programming is a natural candidate for learning such problems. So far, there is however only little work on the intersection between learning constraint satisfaction problems and inductive logic programming. In this article, we point out several similarities and differences between the two classes of techniques that may inspire further cross-fertilization between these two fields.

1 Introduction

Constraint programming (CP) is an active research area in the field of artificial intelligence. It is concerned with solving combinatorial problems that are formalised as constraint satisfaction problems (CSPs). CP has been used in numerous applications in domains such as time-tabling, scheduling, packing, bioinformatics, etc.

On the other hand, inductive logic programming (ILP) is a research area that has studied the learning of logic programs and relational descriptions for more than twenty years now. ILP has also been applied in a wide variety of contexts, including bio- and chemo-informatics, natural language processing, engineering, etc.

CP has – like ILP – its origins in the field of logic programming and uses a declarative representation. However, while learning traditional logic programs is popular (thanks to ILP), the learning of constraint programs and CSPs has received much less attention, even though several techniques for learning CSPs have been contributed in the past ten years, cf. [1,5,7,10,19,20]. The motivation for learning is that formulating the CSP for a particular application is a nontrivial task.

Most of the techniques to learn logic programs and to learn constraint satisfaction problems have been developed independently of one another (but see [19]). This is surprising as both problems are – as we will show – essentially

© Springer International Publishing AG 2016
C. Bessiere et al. (Eds.): Data Mining and Constraint Programming, LNAI 10101, pp. 96–112, 2016.
DOI: 10.1007/978-3-319-50137-6_5

logical and relational learning problems. This paper contributes to bridging the gap between CP and ILP by surveying the CP-learning techniques from the perspective of ILP. This will allow us to point out differences and similarities between the two approaches and to also indicate opportunities for further research.

This paper is organized as follows. In Sect. 2, we introduce the relevant context on the modelling of constraint satisfaction problems (CSPs). In Sect. 3, we introduce the task of learning CSPs, and we relate this task to ILP in Sect. 4. In Sect. 5, we give an overview of existing systems for solving this task, and we describe them in terms of ILP concepts. Section 6 provides a summary and discussion of the different systems and Sect. 7 concludes this paper.

2   Constraint Satisfaction Problems

Constraint programming (CP) is concerned with solving constraint satisfaction problems (CSPs). A CSP is a constraint network $p = (\mathcal{V}, D, \mathcal{C})$, defined by

- a finite set of variables $\mathcal{V} = \{v_1, \ldots, v_n\}$;
- a domain D, which maps every variable $v \in \mathcal{V}$ to a set of possible values $D(v)$; and
- a finite set of constraints $\mathcal{C} = \{c_1, \ldots, c_n\}$, where each constraint $c_i \in \mathcal{C}$ is essentially a relation $c_i \subseteq D(v_{i_1}) \times \cdots \times D(v_{i_{m_i}})$, that can be specified extensionally or intensionally.

The key question of constraint satisfaction problems is to find an assignment of values to the variables so that all constraints in the constraint network are satisfied. The constraints hence form one big conjunction. Let us now illustrate CSPs using three well-known examples: n-queens, sudoku and graph coloring.

In n-queens, the goal is to put n queens on an n-by-n board, so that no queen *attacks* another one (queens can attack if they are in the same row, column or diagonal, as per the chess rules), cf. Figs. 1 and 2. The valid solutions of the n-queens problem are completely determined by the value of n.

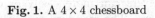

Fig. 1. A 4×4 chessboard   **Fig. 2.** ...and a 4-queens solution.

In Sudoku, one is given a 9×9 grid. The goal of a Sudoku is to enter in each cell a number between 1 and 9, such that no number occurs twice in the same row, column or block. In a Sudoku puzzle, a number of values are already given while guaranteeing that there is a unique solution to the puzzle, cf. Figs. 3 and 4. In a CSP these initial values can be encoded as additional constraints.

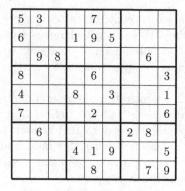

5	3			7				
6			1	9	5			
	9	8					6	
8				6				3
4			8		3			1
7				2				6
	6					2	8	
			4	1	9			5
				8			7	9

5	3	4	6	7	8	9	1	2
6	7	2	1	9	5	3	4	8
1	9	8	3	4	2	5	6	7
8	5	9	7	6	1	4	2	3
4	2	6	8	5	3	7	9	1
7	1	3	9	2	4	8	5	6
9	6	1	5	3	7	2	8	4
2	8	7	4	1	9	6	3	5
3	4	5	2	8	6	1	7	9

Fig. 3. An unsolved 3×3 Sudoku...   **Fig. 4.** ...and its solution.

In graph coloring, one is given a graph and a set of colors. The goal is to assign a color to each node in the graph such that adjacent nodes have a different color. In a CSP the graph structure can be encoded by using auxiliary variables and constraints on them. An example is shown in Figs. 5 and 6.

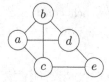

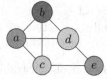

Fig. 5. An uncolored graph   **Fig. 6.** .. and a valid 3-coloring. (Color figure online)

CSPs can be expressed in terms of *local* constraints. These constraints express simple relationships between a bounded number of individual variables, for example, $v_1 = v_2, v_3 \neq v_4, v_5 = v_6 + v_7 + v_8$.

However, the number of such constraints in a CSP can become very large. CSPs are therefore often expressed in terms of *global constraints*. What we call a global constraint is in fact a class of constraints involving any (unbounded) number of variables. The semantics of the global constraint is given by a Boolean function of unbounded arity; thus an instance of the global constraint can be posted on any number of variables and may represent a whole set of local constraints. Global constraints have two main advantages: they simplify the model by reducing the number of constraints, and solvers can more easily exploit the relationships between the constraints in the set. The standard example of a global constraint is the *alldifferent* constraint. For example, the constraint *alldifferent*($[v_1, v_2, v_3]$) is equivalent to the set $v_1 \neq v_2, v_1 \neq v_3, v_2 \neq v_3$. Additionally, higher level languages for expressing CSPs (such as MiniZinc [23] and B-prolog [28]) offer constructs for compactly expressing loops (e.g., *foreach*).

Listing 1.1 shows an example of an n-queens constraint specification in B-prolog. It uses a list of variables Q, with N such variables (line 2), each with a domain of values from 1 to N (line 3). In this representation, the assignment $Q[i] = j$ means that the queen on row i is at column j (using the knowledge that there can be only one queen per row). Lines 5–9 represent the constraints. Line 5 uses the global *alldifferent* constraint and states that no two queens can be on the same column. Line 6 uses a foreach construct to state that queens can not be on a \-diagonal (example: $Q[2] = 2$ and $Q[3] = 3$) while line 8 states the same for a /-diagonal (example: $Q[3] = 3$ and $Q[4] = 2$).

Listing 1.1. "n-queens on rows in B-prolog"

```
1   queens_rows(N, Q) :-
2       length(Q, N),
3       Q :: 1..N,

5       alldifferent(Q),
6       foreach(R1 in 1..N, R2 in (R1+1)..N, (
7           Q[R1] - Q[R2] #\= R1 - R2 )),
8       foreach(R1 in 1..N, R2 in (R1+1)..N, (
9           Q[R1] - Q[R2] #\= R2 - R1 )),
```

This model contains both a choice on how to represent the queens, as well as how to formulate the constraints. Other representations for the queens are also possible, such as one Boolean variable per board position or one integer variable per column. Other ways to formulate the constraints are also possible, such as with a decomposition of the *alldifferent* constraints.

In this paper, we will review different techniques that have been investigated for learning constraints, as well as their relationship to ILP.

3   The Learning Task

In its basic form, the learning task consists in learning the constraints of a CSP from example assignments. Given is

- a finite set of variables $\mathcal{V} = \{v_1, \ldots, v_n\}$;
- a domain D that maps every variable $v \in \mathcal{V}$ to a set of possible values $D(v)$;
- a set of positive and negative examples that take the form of assignments $\theta = \{v_1 = a_1, \ldots, v_n = a_n\}$ to the variables $\mathcal{V}$ that satisfy or violate the CSP;
- a language of possible constraints $\mathcal{L}_C$, such that each $c \in \mathcal{L}_C$ is of the form $r(t_1, \ldots, t_m)$ with r/m a relation of arity m defined in the background, where the terms t_i correspond to a constant, a variable in $\mathcal{V}$ or a list of variables that are a subset of $\mathcal{V}$.[1]

---

[1] Observe that we choose here to represent global constraints as unary predicates, taking a list of variables as its arguments. An alternative would be to introduce one version of the global predicate for any possible arity, e.g. *alldifferent*(X, Y), *alldifferent*(X, Y, Z), ...

The goal then is to find a hypothesis $h \subseteq \mathcal{L}_C$ that is satisfied by all positive examples and no negative examples.

For the n-queens problem in Listing 1.1, the set of variables would have cardinality n, and the domain for each would be the set $\{1, \ldots, n\}$ (note how the variable representation is part of the learning problem). For 4-queens, a positive example would be $\{Q_1 = 3, Q_2 = 1, Q_3 = 4, Q_4 = 2\}$ and a negative example would be $\{Q_1 = 3, Q_2 = 3, Q_3 = 4, Q_4 = 2\}$; example constraints could be binary predicates for equality and inequality; ternary predicates for addition and multiplication and a unary predicate for the `alldifferent` constraint, among others.

Observations. Several observations can be made about this problem statement:

- CSPs are *conjunctive* descriptions and CP is heavily focussed on dealing with conjunctions as these impose strong constraints that – unlike disjunctive descriptions – propagate well in the search;
- The above definition assumes that all variables are explicit, and no new (auxiliary) variables are introduced to formulate certain constraints;
- Unlike in traditional machine learning and ILP, one typically assumes a noise-free setting;
- The *number* of constraints in CSPs can be quite large, especially when considering ground representations in which *foreach* loops are unrolled; it is typically much larger than the typical clauses learned in ILP; for instance, the Sudoku problem involves $927 = 36 \times 27$ constraints.
- Redundancy amongst constraints is a key problem. For instance, the constraints $x = y$ and $y = z$ imply $x = z$. Together with the high number of variables that is available, this causes severe problems for traversing the search space as there are many *syntactic* variants. These are hypotheses that are formulated differently and hence syntactically different (like $x = y \wedge y = z$ and $x = y \wedge y = z \wedge x = z$) but semantically equivalent. Clever ways for dealing with this are necessary.
- Standard ILP systems often start from a large set of positive and negative examples. The number of solutions to a CSP problem is often small and it can already be hard to generate a single positive one. Therefore, several researchers are learning CSPs from queries [7,8] and from small sets of examples [5,19]; these queries, as we shall discuss, do not always ask for the classification of a complete example (consisting of a value assignment to *all* variables in $\mathcal{V}$).

4   Relation to Inductive Logic Programming

Inductive Logic Programming (ILP) is a machine learning methodology that uses first-order logic to represent the data as well as the learned hypotheses. This use of first-order logic sets it apart from other machine learning techniques. It is often used in a concept learning setting, where the goal is to find a hypothesis that covers all of the positive examples and none of the negative ones. See [13] for a gentle introduction to ILP.

In the ILP literature, there is a well-known distinction between learning from entailment and learning from interpretations [12], which is also quite relevant in the present context. When *learning from entailment*, each example is presented as a ground fact and additional knowledge about the examples and the domain is provided as background knowledge. The goal is to find (a set of) clauses that, combined with the background knowledge, logically entail the positive examples, and do not entail the negative examples.

Several state of the art CSP learning approaches (such as Conacq [10] and QuAcq [7]) map directly to this setting. However, in contrast to traditional ILP, they focus on learning a single clause.

The CSP $p = (\mathcal{V}, D, \mathcal{C})$ can be represented by a single conjunctive clause of the following form:

$$p(v_1, \ldots, v_n) :- d(v_1), \ldots, d(v_n),$$
$$c_1(v_{c_{1_1}}, v_{c_{1_2}}, \ldots, v_{c_{1_r}}),$$
$$\ldots,$$
$$c_m(v_{c_{m_1}}, v_{c_{m_2}}, \ldots, v_{c_{m_s}}).$$

where $\mathcal{V} = \{v_1, \ldots, v_n\}$, $d(v_x)$ represents the domain of v_x and there are m constraints c_i, each involving a subset of the variables in $\mathcal{V}$. In this setting, learning a CSP corresponds to learning a single clause for which in addition $vars(head) = vars(body)$ as no existential variables are allowed in the body of the clause. The definition of the c_i is then part of the background knowledge; cf. below. The goal is then to learn the definition of $p(v_1, \ldots, v_n)$ given this background knowledge and positive and negative examples. Observe that this formulation of the constraint learning problem is closely related to the learnability results for single rules by [18] or conjunctive concepts in structural domains by [17], two settings that have been well-studied within the context of ILP.

Example. For the n-queens problem with $n = 4$, we could have the positive example *queens4*$(2, 3, 1, 4)$ and the negative example *queens4*$(2, 1, 3, 4)$. The background knowledge consists of the declaration of the domains, equality operators and simple mathematical functions.

The goal is to find a set of clauses of the form

$$queens4(q_1, q_2, q_3, q_4) :- body(q_1, q_2, q_3, q_4)$$

such that the body of at least one of the clause is satisfied for the substitution $\{q_1/2, q_2/3, q_3/1, q_4/4\}$ and the body of none of the clauses is satisfied for the substitution $\{q_1/2, q_2/1, q_3/3, q_4/4\}$.

The n-queens problem for $n = 4$ can be formulated as

$$queens4(q_1, q_2, q_3, q_4) :- q_1 \in [1, 4], q_2 \in [1, 4], q_3 \in [1, 4], q_4 \in [1, 4],$$
$$q_1 \neq q_2, q_1 \neq q_3, q_1 \neq q_4, q_2 \neq q_3, q_2 \neq q_4, q_3 \neq q_4,$$
$$|q_1 - q_2| \neq 1, |q_1 - q_3| \neq 2, |q_1 - q_4| \neq 3,$$
$$|q_2 - q_3| \neq 1, |q_2 - q_4| \neq 2, |q_3 - q_4| \neq 1.$$

The second line could be replaced with $alldifferent(q_1, q_2, q_3, q_4)$ if that constraint is available.

Alternatively, the learning of CSPs can also be viewed as a *learning from interpretations* task as originally tackled in the Clausal Discovery system Claudien [14]. In this approach an example is a set of ground facts (typically a Herbrand interpretation). This set is a complete description of the knowledge about the example, that is, facts that are not in the example are considered to be false. In this setting, there is no explicit target predicate such as *queens*.

Furthermore, a hypothesis H is said to cover an example e if the example e is a model of H. Hypotheses are represented in clausal logic, that is, Claudien learns a conjunctive set of clauses of the form $h_1 \lor \cdots \lor h_m \leftarrow b_1 \land \cdots \land b_n$, where the h_i and b_j are first-order terms. All variables in such a clause are universally quantified.

Also with this representation, it is possible to represent a CSP $p = (\mathcal{V}, D, \mathcal{C})$. The form that this could take is to add a predicate $v_i/1$ to represent each variable in $\mathcal{V}$ and then employ for each variable $v_i \in \mathcal{V}$ the following clauses:

$$v_i(X) \land v_i(Y) \to X = Y$$

$$v_i(e_{i,1}) \lor \cdots \lor v(e_{i,i_n})$$

with $D(v_i) = \{e_{i,1}, \ldots, e_{i,i_n}\}$ the domain of the variable v_i. These clauses guarantee that any model will have to take exactly one value e for each variable $v \in \mathcal{V}$. Furthermore, for each constraint c_j we add the clause:

$$v_1(X_1) \land v_2(X_2) \land \cdots \land v_n(X_n) \to c_j(X_1, X_2, \ldots, X_n)$$

where X_i are the different variables involved in the constraint c_j. This constraint guarantees that any model of the theory will satisfy all the constraints in the CSP. Notice again that we assumed here that the number of variables is given and fixed. If so, the learning setting is essentially propositional and closely related to that of Valiant's seminal PAC-learning setting [27]. However, it is often possible to generalize the setting towards any number of variables as in the first order extension to j, k-clausal theories [15] of Valiant's setting. It is this setting that formed the basis for learning from interpretations in ILP. We achieve this transformation by replacing each predicate $v_i(X)$ by a predicate $v(i, X)$, that is, by making the variable index a variable itself. We illustrate this on the n-queens example.

Example. We can represent a solution to the 4-queens problem as the following interpretation:

$$\{size(4), q(1, 2), q(2, 4), q(3, 1), q(4, 3)\}$$

where $q(R, C)$ indicates that there is a queen at position (R, C) on the board. The set of clauses that fully determines the n-queens problem for examples represented in this language is

$$q(R_1, C_1) \land q(R_2, C_2) \land R_1 \neq R_2 \rightarrow C_1 \neq C_2$$
$$q(R_1, C_1) \land q(R_1, C_1) \land R_1 \neq R_2 \rightarrow |R_1 - R_2| \neq |Q_1 - Q_2|$$
$$q(R_1, C_1) \land q(R_2, C_2) \land C_1 \neq C_2 \rightarrow R_1 \neq R_2$$
$$size(N) \land q(R, C) \rightarrow between(C, 1, N)$$
$$size(N) \land q(R, C) \rightarrow between(R, 1, N)$$
$$size(N) \land between(R, 1, N) \rightarrow exists_{row}(R)$$

where $exists_{row}$ is defined in the background knowledge as

$$q(R, C) \rightarrow exists_{row}(R).$$

Similar clauses could be added for $exists_{col}$ but they are redundant. In the common CSP formulation of this problem (see Listing 1.1) the last 4 constraints are implicitly encoded in the representation of the variables as a list of row positions of the queens. Note that this definition can be learned from examples of different sizes.

One interesting consequence of these different representations is the following. The single clause representation is essentially a propositional one, while the representation as a conjunctive set of clauses (CNF) also allows for relational descriptions. The propositional techniques will learn constraints for one specific CSP instance (e.g., n-queens for one specific n or graph coloring for one specific graph), while relational approaches have the potential of learning the general CSP (e.g., n-queens for all n at the same time or graph coloring for arbitrary graphs). Indeed, given that in the single clause representation the arity of the target predicate is fixed and $vars(head) = vars(body)$, it is not possible to learn one clause that will work for any number of queens.

As an example, let us examine the graph coloring problem. Using a propositional representation (either the single clause or the propositional CNF one), one will essentially learn the constraints governing a particular graph. This is easy to see when considering the single clause representation. What will be learned will be a set of inequalities. Each such inequality corresponds to one edge in the graph. This is unusual from an ILP perspective, as there it would typically be assumed that the edges are given. If the edges are given, it is possible to learn the overall concept of graph-coloring using a clause such as

$$edge(X, Y), color(X, CX), color(Y, CY) \rightarrow CX \neq CY$$

5   CSP Learning Systems

In the literature there are several examples of learning systems that focus on the problem of learning CSPs. To describe these learning systems, we shall proceed along a number of dimensions, which are often used to characterize ILP systems. It will be convenient to realize this by answering the following questions:

1. What is the representation language for the examples (or instances in the data)?

2. What is the hypotheses space or language ?
3. What type of background knowledge is used ?
4. What search strategy is used ?
5. How are the resulting hypotheses scored or ranked ?

In the remainder of this section we briefly discuss five different constraint learning systems by answering these questions.

5.1   Learning a CNF

Clausal Discovery (Claudien) [14]. Claudien was developed as a general purpose learning system, not focussed in particular on learning CSPs.

1. Examples are represented as Herbrand interpretations. These interpretations can contain additional information about the problem instance, for example the graph structure in the case of graph coloring. Claudien is capable of learning from only positive examples, or both positive and negative examples.
2. Hypotheses are represented as a conjunctive set of clauses.
3. Background knowledge contains global knowledge, for example, definitions of global constraints that are available. This knowledge is typically represented as clauses or predicate definitions.
4. Search in Claudien is performed on a lattice based on θ-subsumption. It is guided by a refinement operator which can be specified in the $\mathcal{DLAB}$ bias specification language, which specifies which literals can be added to a clause during the search.
5. Claudien computes the most specific hypothesis, that is the one that covers the fewest interpretations.[2]

Lallouet et al. [19]. The system proposed by Lallouet et al. essentially solves the same learning task as Claudien. However, instead of learning a set of clauses in universally quantified conjunctive normal form (UCNF) directly, they exploit the duality between UCNF and clauses in existentially quantified disjunctive normal form (EDNF) [11]. This duality can be expressed by the following property:

$$(\exists l_{1,1} \wedge \cdots \wedge l_{1,n_1}) \vee \cdots \vee (\exists l_{k,1} \wedge \cdots \wedge l_{k,n_1 k})$$

is a solution to an EDNF concept learning task with positive examples P and negative examples N if and only if

$$(\forall \neg l_{1,1} \vee \cdots \vee \neg l_{1,n_1}) \wedge \cdots \wedge (\forall \neg l_{k,1} \vee \cdots \vee \neg l_{k,n_1 k})$$

is a solution to an UCNF concept learning task with N as positive examples and P as negative examples. The clausal theory is thus obtained by learning a EDNF on the examples where the class labels are flipped (so positive become negative

---

[2] It is well-known in ILP [12] that when learning from interpretations, a hypothesis G is more general than S if and only if $S \models G$, while when learning from entailment if and only if $G \models S$.

and vice versa) and by then taking the negation of the obtained formula, which is in UCNF. The main motivation for this approach is the availability of EDNF learning algorithm implementations such as Aleph [26].

An important difference between Claudien and this approach is the use of positive versus negative examples. Claudien learns primarily on positive examples with possible additional information from negative examples, while the approach by Lallouet et al. primarily learns from negative examples due to the class flipping step.

1. same as Claudien
2. same as Claudien
3. same as Claudien
4. The authors observe that neither top-down search as bottom-up search provided the necessary scalability. They propose a bi-directional search method that combines top-down and bottom-up search similar to Mitchell's Candidate Elimination [21].
5. Selection is part of the bidirectional search of the DNF learning algorithm.

ModelSeeker [5]. ModelSeeker searches for global constraints starting from an unstructured list of variables. It does not perform search over individual variables but it searches over blocks of variables instead. These blocks are generated by a generator function that extracts certain structures from the example (e.g. rows of a matrix). This approach consists of two steps:

1. Find a generator that can be applied on the given example. The generator will enumerate blocks of variables (e.g. the rows of a matrix).
2. Find a global constraint, defined on all variables in a given block (e.g. row), that holds for all blocks generated by that generator.

ModelSeeker defines a number of generator templates that can be instantiated. For example, $scheme(n, m_1, m_2, size_1, size_2)$ interprets a sample of length n as a matrix of size $m_1 \times m_2$, and extracts non-overlapping blocks of size $size_1 \times size_2$. Valid instances of this template can be found using Prolog as follows

```
scheme( N, M1, M2, S1, S2 ) :-
    factor(N, M1, M2), M1 =< M2,
    factor(M, S1, _),
    factor(M, S2, _).
```

where `factor(N,M1,M2)` computes a pair of integers M_1 and M_2 such that $N = M_1 \times M_2$. For a 9×9 Sudoku with 81 variables, possible generators include $schema(81, 1, 81, 1, 81)$ (a list), $schema(81, 3, 27, 3, 3)$ (a 3-by-27 matrix where blocks of 3-by-3 are extracted), and $schema(81, 9, 9, 1, 9)$ (where the rows of a 9×9 matrix are extracted).

In the second phase, ModelSeeker searches for a constraint that is satisfied by all blocks belonging to the specific generator instance selected in the first phase.

In a Sudoku, for example, the constraint `alldifferent(Vars)` holds for all sets of variables extracted by $schema(81, 9, 9, 1, 9)$ (i.e. the rows of the matrix). The constraints that are considered are a large subset of those available in the global constraint catalog [3].

Table 1 shows the output of ModelSeeker for the Sudoku problem.

Table 1. Model found by ModelSeeker for the standard Sudoku problem.

Generator	Constraint	Comment
scheme(81,9,9,1,9)	permutation*9	rows
scheme(81,9,9,9,1)	permutation*9	columns
scheme(81,9,9,3,3)	permutation*9	3-by-3 blocks

In an ILP formulation, we can write this as

$$scheme(81, 9, 9, 1, 9, Block) \rightarrow permutation(Block)$$
$$scheme(81, 9, 9, 9, 1, Block) \rightarrow permutation(Block)$$
$$scheme(81, 9, 9, 3, 3, Block) \rightarrow permutation(Block)$$

Note that the generator is described using constants. This indicates that ModelSeeker cannot generalize over problems of different sizes. The enumeration of this kind of clauses requires a very specialized language bias. Thanks to its tailored bias, ModelSeeker is capable of learning from a single example.

ModelSeeker can also introduce auxiliary variables as parameters of global constraints. It then assumes that all constraints over a set of variable subsets (e.g. rows of a matrix) share the same parameter. This is for example the case when learning magic squares, where each row (and column) sums to the same number.

1. Examples are unstructured lists of numbers. They are typically the output that one would expect from a CP solver.
2. Hypotheses consist of a generator and a global constraint.
3. Two sets of background knowledge are provided: a set of predefined generator templates and a set of global constraints. Some (handcrafted) meta information is available about subsumption between constraints.
4. The search strategy is a clever generate and test strategy. It consists of finding all combinations of generator and global constraint that are satisfied in the example.
5. ModelSeeker uses a combination of techniques for ranking and selecting hypotheses. The Constraint Seeker returns a ranked list of constraints, where the ranking is based on a number of properties as described in [4]. ModelSeeker essentially selects the most specific hypothesis. However, ModelSeeker considers the global constraints to be black-boxes on which no automated reasoning is possible. Generality tests are therefore purely based on available

hand-crafted meta-data. This meta-data can be used to express, among others, implication (e.g. `permutation` implies `alldifferent`), contractibility and expandability (e.g. `alldifferent(a,b,c)` implies `alldifferent(a,b)`, etc.)

5.2   Learning a Single Clause

Conacq [6,8]. Conacq employs a version-space like approach (Mitchell's FIND-S algorithm [21]). The version space is the space of all possible constraint networks that can be built on a given set of variables with constraints belonging to a given language. Conacq iterates over the examples to reduce the version space.

1. All examples are complete assignments on a set V of variables taking values in a domain D. Each example is labelled positive or negative depending on whether it satisfies or not all the constraints of the problem. For instance for 4-queens: $(Q_1 = 3, Q_2 = 1, Q_3 = 4, Q_4 = 2)$ is a positive example.
2. The hypotheses are subsets of a set B of the basis constraints, that is, all constraints that possibly participate to the definition of the constraint network. For instance, B could contain all binary constraints $Q_i \diamond Q_j$ where $\diamond \in \{=, \neq, <, \leq, \geq, >, \dots\}$.
3. The background knowledge contains the definition of these constraints, and can also include any interdependency between constraints that could hold between subsets of constraints. For instance, $X \leq Y \wedge Y \leq Z \rightarrow X \leq Z$ tells us that each time constraints $Q_i \leq Q_j$ and $Q_j \leq Q_k$ are learned, we can derive $Q_i \leq Q_k$. This is intended to recognize syntactic variants and work more at a semantic level of generalization that is reminiscent of Buntine's generalized subsumption [9] and the notion of semantic closure [16]. Working at the semantic level allows one to significantly decrease the number of candidate hypotheses in the version space.
4. The version space is compactly represented by a SAT formula. Each model is a hypothesis that accepts all positive examples and rejects all negative ones. Strategies for updating/simplifying the SAT formula involve unit propagation or backbone detection (i.e., detecting constraints that belong to all hypotheses of the version space.
5. No ranking/evaluation function was proposed for selecting hypotheses. The by default function is to take the most specific one.

The active version of Conacq (Conacq2) [8] asks membership queries until the version space has converged on a single hypothesis.

1. as in Conacq
2. as in Conacq
3. as in Conacq
4. The strategy usually used for asking membership queries that will produce a fast decrease in the size of the version space is an adaptation of the *near-miss* strategy [25]. For instance, given a negative example e_1 violating a set κ of constraints, we try to ask the user to classify an example e_2 that

violates a single constraint of κ. If the example is classified positive, that constraint is removed from the candidate constraints. If it is negative, it is learned as belonging to the constraint network. This strategy is reminiscent of Mitchell's FIND-S algorithm [22]. Interdependencies between constraints can make impossible the generation of near-miss queries, leading to slower decrease in the size of the version space (Constraints networks have been shown to be non learnable in a polynomial sequence of membership queries).

5. Conacq2 can be stopped at any time, but it has been presented to be run until convergence. In such a case, it is not necessary to rank the hypotheses as there is only remaining one in the version space.

QuAcq [7]. QuAcq is an extension of Conacq2 that is able to ask *partial* queries to reduce the number of queries required for convergence.

Thanks to this feature, for each example classified as negative, QuAcq uses a dichotomic search to elucidate one constraint of the constraint network with a number of queries logarithmic in the size of the negative example. As a result, QuAcq learns the constraint network in a polynomial number of queries (namely, $t \cdot n$, where t is the size of the network and n the number of its variables) and proves convergence in a number of queries linear in the size of the basis $\mathcal{B}$.

1. examples are partial or complete assignments on the set $\mathcal{V}$ of variables. Each example is labelled positive or negative depending on whether it satisfies or not all the constraints of the problem whose variables are involved in the example. For instance for 4-queens: $(Q_1 = 3, Q_2 = 2, Q_3 = 4)$ is a negative example because Q_1 and Q_2 are on the same diagonal;
2. as in Conacq;
3. QuAcq does not use any background knowledge other than the definition of the operators;
4. for each example classified as positive, QuAcq rules out from the candidate constraints all those that are violated by the example. For each example classified as negative, QuAcq uses a dichotomic search to elucidate one constraint of the constraint network with a number of queries logarithmic in the size of the negative example. This step requires the use of partial queries.
5. as Conacq2, QuAcq is supposed to be run until convergence.

Example. Consider the 4-queens problem. Suppose the example $e = (Q_1 = 3, Q_2 = 1, Q_3 = 3, Q_4 = 2)$ has been classified as negative by the user. This means there is at least one constraint of the network to learn that rejects e (actually there are several). To elucidate such a constraint, QuAcq splits e in two parts of equal size (to guarantee logarithmic convergence) and asks the user the query $(Q_1 = 3, Q_2 = 1)$. As the two remaining queens in this example do not attack each other, the user classifies this partial example as positive and QuAcq removes from the set of candidate constraints all those that are violated by $(Q_1 = 3, Q_2 = 1)$ (e.g., $Q_1 = Q_2$). Then QuAcq extends the example to Q_3. The query $(Q_1 = 3, Q_2 = 1, Q_3 = 3)$ is negative. Hence, QuAcq knows that there is a constraint between $\{Q_1, Q_2\}$ and Q_3. QuAcq generates the query $(Q_1 = 3, Q_3 =$

3), which is classified as negative. At this point QuAcq knows there is a constraint on the scope (Q_1, Q_3), and it knows this constraint forbids the tuple $(3, 3)$. What remains to do is to generate queries on (Q_1, Q_3) that will allow QuAcq to find which constraint leads to the rejection of $(Q_1 = 3, Q_3 = 3)$. Suppose that the remaining candidate constraints that could reject $(Q_1 = 3, Q_3 = 3)$ are $\{\neq, <, >\}$. After having asked $(Q_1 = 3, Q_3 = 2)$ and $(Q_1 = 2, Q_3 = 3)$, both classified positive, QuAcq rules out $Q_1 < Q_3$ and $Q_1 > Q_3$ as candidate constraints, and $Q_1 \neq Q_3$ is added to the learned network.

6   Discussion

We will now analyze these different systems based on a number of dimensions. Table 2 gives an overview of this section.

Propositional vs. Relational. For the systems we discussed, Claudien and Lallouet use first-order logic to learn constraint models, while Conacq and QuAcq are based on propositional logic. Many constraint satisfaction problems have some form of global structure that can be captured very well by first-order logic, but would require many constraints in propositional logic.

ModelSeeker is a mix between the two. It allows one to capture global structure using global constraints, but it only learns a restricted form of clauses. Its output can be considered to be a domain-specific language that can be mapped to a clausal theory, but it lacks the expressivity of the latter.

Active vs. Passive Learning. Most of the systems are passive learning systems, that is, they take examples and produce a model without user interaction. However, Conacq2 and QuAcq are based on posing queries to the user, which allows them to quickly converge to the correct solution, even when no positive examples have been seen. In a sense, it allows the system to learn the model and solve it at the same time. From a machine learning point of view, the use of partial queries is new and unexplored in the context of ILP, cf. [2].

Table 2. Categorization of systems. The question answered are (1) Does the system learn a single clause or multiple ones? (2) Is the system propositional or relational? (3) Does it use active learning? (4) What type of examples does it need (positive, negative, partial)? (5) How does it handle redundancy (logic-based, lattice-based, ruleset-based) (6) Can it learn from examples of difference sizes?

	Claudien	Lallouet	ModelSeeker	Conacq	Conacq2	QuAcq
Clauses?	multi	multi	multi	single	single	single
Prop./Rel.?	rel	rel	mixed	prop	prop	prop
Active?	pass	pass	pass	pass	active	active
Examples?	pos(+neg)	neg(+pos)	pos	pos+neg	pos+neg	partial
Redundancy?	logic	logic	ruleset	lattice	lattice	lattice
Different sizes?	yes	yes	no	no	no	no

Requirements on Examples. An important aspect of learning CSPs is the availability of examples. Some CSPs have many solutions, some have only one. The systems discussed in this paper have different requirements for the examples.

Claudien and the approach of Lallouet et al. start from examples as sets of ground facts. This allows them to learn from structured examples (for example, containing a graph). Both approaches typically require a substantial number of examples, depending on the complexity of the input language and the available background information in the form of language bias. The main difference between both approaches is that Claudien learns a theory on positive examples, while Lallouet et al. starts from negative examples. Both approaches can incorporate information from both positive and negative examples.

Conacq, QuAcq and ModelSeeker start from examples in the form of assignments to the variables in the model. This means they fix the number of variables at the start of the learning task.

ModelSeeker assumes that some structure can be imposed on the variables on which global constraints can be found. It often can learn a good model from just a single positive example. ModelSeeker can not incorporate information from a negative example.

QuAcq can start from partial examples, which means that it can be used to learn problems for which no solution has been found yet.

Redundancy. All systems support some form of redundancy elimination. In Claudien, Lallouet et al., Conacq and QuAcq this is based on logical inference. In ModelSeeker, this is based on metadata provided with the constraints.

7   Conclusion and Future Work

We see the learning of CSPs as a modern challenge in which many of the techniques and insights from ILP can play an important role. The connection between constraint programming and inductive logic programming has been made before by Lallouet et al. [19] for learning CSPs and by Abdennadher and Rigotti [1] for learning propagation rules for CSP solvers. In this survey, we have given an overview of techniques for learning CSPs and relate them to concepts from the ILP community, with the intention of inspiring further cross-fertilization between these two fields.

Each of the systems described in this paper contributes its own ideas. The expert driven ModelSeeker [5] introduces the idea of generators and uses a very well developed search strategy which makes it capable of learning relatively complex CSPs from a small number of examples. However, techniques from ILP can still contribute (1) by making it possible to structure the search space better by using more-general-than relations between the available constraints and the generators, (2) by allowing ModelSeeker to incorporate negative examples, and (3) by making it able to learn from examples of different sizes [24].

The propositional systems Conacq [10] and QuAcq [7] are interesting because they start from a sound theoretical basis and can therefore provide guarantees on the complexity of the learning task. The QuAcq system is especially of interest because it can learn from partial queries to the user. This allows it to learn a CSP for which no solutions are known yet. This setting has never been studied in an ILP setting.

In conclusion, we believe that ILP-based techniques can make a valuable contribution for the task of learning CSPs, and that techniques studied for learning CSPs can be used to improve the effectiveness of ILP systems.

Acknowledgements. This work was supported by the European Commission under the project "Inductive Constraint Programming" (FP7- 284715).

References

1. Abdennadher, S., Rigotti, C.: Automatic generation of rule-based solvers for intensionally defined constraints. IJAIT **11**(2), 283–302 (2002)
2. Angluin, D.: Queries and concept learning. Mach. Learn. **2**(4), 319–342 (1988)
3. Beldiceanu, N., Carlsson, M., Rampon, J.-X.: Global constraint catalog. http://www.emn.fr/z-info/sdemasse/gccat/
4. Beldiceanu, N., Simonis, H.: A constraint seeker: finding and ranking global constraints from examples. In: Lee, J. (ed.) CP 2011. LNCS, vol. 6876, pp. 12–26. Springer, Heidelberg (2011). doi:10.1007/978-3-642-23786-7_4
5. Beldiceanu, N., Simonis, H.: A model seeker: extracting global constraint models from positive examples. In: Milano, M. (ed.) CP 2012. LNCS, vol. 7514, pp. 141–157. Springer, Heidelberg (2012)
6. Bessiere, C., Coletta, R., Freuder, E.C., O'Sullivan, B.: Leveraging the learning power of examples in automated constraint acquisition. In: Wallace, M. (ed.) CP 2004. LNCS, vol. 3258, pp. 123–137. Springer, Heidelberg (2004). doi:10.1007/978-3-540-30201-8_12
7. Bessiere, C., Coletta, R., Hebrard, E., Katsirelos, G., Lazaar, N., Narodytska, N., Quimper, C.-G., Walsh, T.: Constraint acquisition via partial queries. In IJCAI, pp. 475–481. AAAI Press (2013)
8. Bessiere, C., Coletta, R., O'Sullivan, B., Paulin, M.: Query-driven constraint acquisition. In: IJCAI, pp. 50–55 (2007)
9. Buntine, W.: Generalized subsumption and its applications to induction and redundancy. Artif. Intell. **36**(2), 149–176 (1988)
10. Coletta, R., Bessiére, C., O'Sullivan, B., Freuder, E.C., O'Connell, S., Quinqueton, J.: Semi-automatic modeling by constraint acquisition. In: Rossi, F. (ed.) CP 2003. LNCS, vol. 2833, pp. 812–816. Springer, Heidelberg (2003). doi:10.1007/978-3-540-45193-8_58
11. De Raedt, L.: Induction in logic. In: Proceedings of the 3rd International Workshop on Multistrategy Learning, pp. 29–38 (1996)
12. De Raedt, L.: Logical and Relational Learning. Springer, Heidelberg (2008)
13. De Raedt, L.: Inductive logic programming. In: Sammut, C., Webb, G.I. (eds.) Encyclopidea of Machine Learning. Springer, New York (2010)
14. De Raedt, L., Dehaspe, L.: Clausal discovery. ML **26**(2–3), 99–146 (1997)
15. De Raedt, L., Džeroski, S.: First-order jk-clausal theories are PAC-learnable. Artif. Intell. **70**(1–2), 375–392 (1994)
16. De Raedt, L., Ramon, J.: Condensed representations for inductive logic programming. KR **4**, 438–446 (2004)

17. Haussler, D.: Learning conjunctive concepts in structural domains. Machine Learning **4**(1), 7–40 (1989)
18. Kietz, J.-U.: Some lower bounds for the computational complexity of inductive logic programming. In: Brazdil, P.B. (ed.) ECML 1993. LNCS, vol. 667, pp. 115–123. Springer, Heidelberg (1993). doi:10.1007/3-540-56602-3_131
19. Lallouet, A., Lopez, M., Martin, L., Vrain, C.: On learning constraint problems. In: ICTAI, pp. 45–52 (2010)
20. Leo, K., Mears, C., Tack, G., Garcia de la Banda, M.: Globalizing constraint models. In: Schulte, C. (ed.) Principles and Practice of Constraint Programming. LNCS, vol. 8124, pp. 432–447. Springer, Berlin Heidelberg (2013)
21. Mitchell, T.M.: Version spaces: a candidate elimination approach to rule learning. In: IJCAI, pp. 305–310. Morgan Kaufmann Publishers Inc (1977)
22. Mitchell, T.M.: Machine Learning. McGraw Hill, New York (1997)
23. Nethercote, N., Stuckey, P.J., Becket, R., Brand, S., Duck, G.J., Tack, G.: MiniZinc: towards a standard CP modelling language. In: Bessière, C. (ed.) CP 2007. LNCS, vol. 4741, pp. 529–543. Springer, Heidelberg (2007). doi:10.1007/978-3-540-74970-7_38
24. Razakarison, N., Carlsson, M., Beldiceanu, N., Simonis, H.: GAC for a linear inequality and an atleast constraint with an application to learning simple polynomials. In: SOCS. AAAI Press (2013)
25. Smith, B.D., Rosenbloom, P.S.: Incremental non-backtracking focusing: a polynomially bounded generalization algorithm for version spaces. In: AAAI, pp. 848–853. Citeseer (1990)
26. Srinivasan, A.: The aleph manual (2001).http://www.cs.ox.ac.uk/activities/machlearn/Aleph/aleph.html
27. Valiant, L.G.: A theory of the learnable. Commun. ACM **27**(11), 1134–1142 (1984)
28. Zhou, N.-F.: The language features and architecture of B-Prolog. Theory Pract. Log. Program. **12**(1–2), 189–218 (2012)

Learning Modulo Theories

Andrea Passerini[✉]

Department of Computer Science and Information Engineering,
University of Trento, Trento, Italy
passerini@disi.unitn.it

Abstract. Many real-world applications require reasoning over hybrid domains involving combinations of continuous and discrete variables and their relationships. Being able to precisely specify all constraints and their respective importance beforehand is often infeasible for the most experienced designer, let alone for a typical decision maker. In this chapter we discuss Learning Modulo Theories (LMT), a learning framework capable of dealing with hybrid domains by combining structured learning with Satisfiability Modulo Theory (SMT) techniques. LMT incorporates SMT solvers and their extensions for optimization as inference engines within learning algorithms. The learning stage automatically identifies the relevant constraints and their respective weights among a set of candidates. The framework can be cast in the structured-output learning paradigm, where the task is learning the structure of the problem from a set of noisy instances, or as a preference elicitation task, where a decision maker is involved in an interactive optimization loop aimed at generating the most preferred solution. We report experimental results highlighting the potential of the method in automated design and recommendation scenarios.

1 Introduction

Many real-world applications require reasoning over hybrid domains involving combinations of continuous and discrete variables and their relationships. A notable example is layout synthesis, where the task is that of finding an optimal layout subject to a set of constraints, with applications to urban pattern layout [70], decorative mosaics [37] and furniture arrangement [72], to mention a few. Fields like design optimization [66] and computational creativity [52] are also typically characterized by hybrid problems with complex constraints (think of harmony rules in music composition).

Reasoning and learning in hybrid domains is a very challenging task. Apart from hybrid Bayesian Networks, for which efficient inference is limited to conditional Gaussian distributions, there is relatively little previous work on *hybrid* methods. The few existing attempts [21,33,36,38,54,69] impose strong limitations on the type of constraints they can handle. Inference is typically run by approximate methods, based on variational approximations or sampling strategies. Exact inference, support for hard numeric (in addition to Boolean)

© Springer International Publishing AG 2016
C. Bessiere et al. (Eds.): Data Mining and Constraint Programming, LNAI 10101, pp. 113–146, 2016.
DOI: 10.1007/978-3-319-50137-6_6

constraints and combination of diverse types of data, like integer and rational numbers, are out of the scope of these approaches.

In order to overcome these limitations, we focused on the most recent advances in automated reasoning over hybrid domains. Researchers in automated reasoning and formal verification have developed logical languages and reasoning tools that allow for *natively* reasoning over mixtures of Boolean *and* numerical variables (or even more complex structures). These languages are grouped under the umbrella term of *Satisfiability Modulo Theories* (SMT) [3]. Each such language corresponds to a decidable fragment of First-Order Logic augmented with an additional background theory $\mathcal{T}$, like linear arithmetic over the rationals or over the integers. SMT is a decision problem, which consists in finding an assignment to both Boolean and theory-specific variables making an SMT formula true. SMT solvers combine satisfiability testing for a Boolean abstraction of the formula, where theory atoms are replaced with Boolean variables, and theory-specific solvers for finding consistent assignments to the theory variables or providing additional constraints to guide the satisfiability solver toward theory-consistent assignments (hence the name Satisfiability *Modulo* Theories). Recently, researchers have leveraged SMT from decision to optimization. *MAX-SAT Modulo Theories* (*MAX-SMT*) [24,53] deal with the problem of finding a theory-consistent truth assignment maximizing the total weight of the satisfied clauses. The most general framework is that of *Optimization Modulo Theories* (*OMT*) [61], which consists in finding a model for a formula which minimizes the value of some (arithmetic) cost function defined over the variables in the formula.

In this chapter we describe *Learning Modulo Theories* (LMT) [19,64], a novel framework for performing learning and inference in hybrid domains. The goal is learning to predict complex structures (e.g. the optimal program schedule for a large conference, the furniture arrangement of an apartment) characterized by hybrid constraints over their components. The quality of a structure is given by an unknown scoring function which is a weighted combination of (soft) constraints, and that has to be learned from experimental data. The framework combines the efficiency of modern OMT solvers to perform inference in hybrid domains with the effectiveness of learning techniques for predicting structured objects [2]. The learning stage automatically identifies the relevant constraints and their respective weights among a set of candidates, while inference returns the minimal cost (or maximal score) configuration according to the learned problem formulation.

The framework can be cast in the structure-output learning paradigm or as a preference elicitation task. In the former case, the task is learning the scoring function from a set of correct input-output pairs, in order to be able to generate novel maximal scoring instances consistent with the problem constraints. We rely on structured-output Support Vector Machines (SVM) [65], a very flexible max-margin structured prediction method, and adapt them to the LMT framework [64]. Preference elicitation, on the other hand, is the task of eliciting the unknown preference of a decision maker in order to generate her most preferred solution. An interactive optimization loop [19] iteratively asks for feedback on candidate solutions based on the currently learned utility model, and refines the model according to the feedback received, until the decision maker is satisfied with the returned solution.

The rest of the chapter is organized as follows. We start by introducing relevant background in Sect. 2. Section 3 describes LMT for structured-output prediction, including experimental results on a representative problem and related work. Section 4 focuses on LMT for preference elicitation, including again an experimental study and some relevant related work. Conclusions are finally drawn in Sect. 5.

2 Background

In this section we will introduce the necessary background, namely Satisfiability Modulo Theory for reasoning over hybrid domains and structured-output learning for predicting structured entities as outputs. Table 1 summarizes the notation we will use throughout the paper.

Table 1. Explanation of the notation used throughout the text.

Symbol	Meaning
above, right, ...	Boolean variables
$x, y, dx, \ldots$	Rational variables
$\varphi_1, \ldots, \varphi_m$	Constraints
(I, O)	Instance; I is the input, O is the output
$\mathbb{1}_k(I, O)$	Indicator for Boolean constraint φ_k over (I, O)
$s_k(I, O)$	Score for arithmetic constraint φ_k over (I, O)
$\psi(I, O)$	Feature representation of the instance
$\psi_k(I, O) := \mathbb{1}_k(I, O)$	Feature associated to Boolean constraint φ_k
$\psi_k(I, O) := s_k(I, O)$	Feature associated to arithmetic constraint φ_k
w	Weights

2.1 Satisfiability Modulo Theory

Given a formula made of Boolean variables and logical connectives, propositional satisfiability (SAT) deals with the problem of deciding whether the formula can be satisfied by a truth value assignment of the Boolean variables. Satisfiability Modulo Theory (SMT) [59] extends SAT to decide about satisfiability of a first-order formula with respect to a *background theory* $\mathcal{T}$, like linear arithmetic over the rationals ($\mathcal{LRA}$) or integers ($\mathcal{LIA}$), or a combination of theories. We will write SMT($\mathcal{T}$) to indicate satisfiability modulo theory $\mathcal{T}$, e.g. SMT($\mathcal{LRA}$) for satisfiability modulo linear arithmetic over the rationals.

Current SMT solvers are based on the so-called *lazy* approach, where an outer SAT-solver interacts with one or more specialized $\mathcal{T}$-solvers (one for each theory) in order to progressively focus the search towards theory-consistent solutions. In the rest of the paper we will assume for ease of exposition to always deal with single theories, but all the machinery can be applied to arbitrary combinations of theories. Let ϕ be an SMT formula. Its *Boolean abstraction* ϕ^- is obtained replacing each theory-specific predicate in ϕ with a Boolean variable. The SAT solver finds a truth value assignment satisfying ϕ^-, and presents it to the $\mathcal{T}$-solver to check for theory consistency. If the $\mathcal{T}$-solver detects an inconsistency, it returns **unsat**, plus a *justification*, i.e. a subset of the assignment which is unsatisfiable according to the theory. This justification is added to the original formula, and the process is repeated until a theory-consistent solution is found, or the refined formula is not satisfiable.

Example 2.1. *Let ϕ be the following SMT($\mathcal{LIA}$) formula:*

$$(x + y + z \leq 5) \wedge (y < 0) \wedge ((x + y > 4) \vee (x + z > 4)) \wedge ((x < 0) \vee (z = 4))$$

Its Boolean abstraction is:

$$\varphi_1 \wedge \varphi_2 \wedge (\varphi_3 \vee \varphi_4) \wedge (\varphi_5 \vee \varphi_6)$$

Suppose the SAT solver finds the following solution:

$$\varphi_1 = \top, \varphi_2 = \top, \varphi_3 = \top, \varphi_4 = \bot, \varphi_5 = \top, \varphi_6 = \bot$$

corresponding to the following SMT($\mathcal{LIA}$) formula:

$$(x + y + z \leq 5) \wedge (y < 0) \wedge (x + y > 4) \wedge (x + z \leq 4) \wedge (x < 0) \wedge (z \neq 4)$$

The formula cannot be satisfied, as e.g. x and y cannot be both negatives if they need to sum to more than 4. When asked to solve it, the $\mathcal{T}$-solver detects unsatisfiability and returns for instance:

$$\neg(\varphi_2 \wedge \varphi_3 \wedge \varphi_5)$$

as a justification. After including it, the SAT solver finds for instance the new solution:

$$\varphi_1 = \top, \varphi_2 = \top, \varphi_3 = \bot, \varphi_4 = \top, \varphi_5 = \top, \varphi_6 = \bot$$

which corresponds to the formula:

$$(x + y + z \leq 5) \wedge (y < 0) \wedge (x + y \leq 4) \wedge (x + z > 4) \wedge (x < 0) \wedge (z \neq 4)$$

The $\mathcal{T}$-solver is now able to detect satisfiability, resulting for instance in the assignment

$$x = -1, y = -1, z = 6$$

Modern lazy SMT solvers introduce a number of refinements to this basic procedure, combining solving techniques from very heterogeneous domains. We refer the reader to [3, 59] for an overview on lazy SMT solving.

MAX-SMT [24, 25, 53] generalizes SMT in the same way as MAX-SAT does with SAT. Rather than finding an assignment satisfying the formula, the task is that of finding an assignment minimizing the number of unsatisfied constraints. In its weighted version, each constraint has a (typically positive) weight, and the task is that of minimizing the cost, that is the weighted sum of the unsatisfied constraints.

Let $\{(\varphi_1, w_1), \ldots, (\varphi_m, w_m)\}$ be a set of constraints with associated (non-negative) weights. The cost of any assignment is clearly smaller than the sum of all weights $W = \sum_{i=1}^{m} w_i$ and larger than or equal to zero. The search for a minimal cost solution follows a branch and bound approach, where the upper and lower cost bounds are progressively tightened and plain SMT is called within these bounds. Consider an upper bound $\hat{W} < W$. A simple approach to enforce the solution to have a cost smaller than $\hat{W}$ is to add a set of m fresh Boolean variables and weights $\{(\bar{\varphi}_1, \bar{w}_1), \ldots, (\bar{\varphi}_m, \bar{w}_m)\}$ combined with the following constraints:

$$\varphi_i \vee \bar{\varphi}_i \quad \forall i \in [1, m]$$
$$\bar{\varphi}_i \rightarrow (\bar{w}_i = w_i) \quad \forall i \in [1, m]$$
$$\neg\bar{\varphi}_i \rightarrow (\bar{w}_i = 0) \quad \forall i \in [1, m]$$
$$\sum_{i=1}^{m} \bar{w}_i \leq \hat{W}$$

which make any assignment with overall weight larger than $\hat{W}$ inconsistent with the theory.

MAX-SMT has been recently generalized to the so-called Optimization Modulo Theories (OMT) [49, 53, 61], where the task is finding a model for a formula minimizing the value of some arithmetic cost function over the variables of the formula. Existing solvers have focused on the $\mathcal{LRA}$ theory and combine lazy SMT-solving with LP minimization techniques.

Example 2.2. *Consider the following OMT($\mathcal{LRA}$) problem:*

$$(cost = x + y + z) \wedge (x + 2y \geq 10) \wedge ((z \geq y) \vee (z \geq x)) \wedge (y > 0) \wedge (x > 0)$$

Depending on the truth value assignment of its Boolean abstraction (omitted here for the sake of conciseness), the sets of constraints to minimize are:

$$(cost = x + y + z) \wedge (x + 2y \geq 10) \wedge (z \geq y) \wedge (z < x) \wedge (y > 0) \wedge (x > 0)$$
$$(cost = x + y + z) \wedge (x + 2y \geq 10) \wedge (z < y) \wedge (z \geq x) \wedge (y > 0) \wedge (x > 0)$$
$$(cost = x + y + z) \wedge (x + 2y \geq 10) \wedge (z \geq y) \wedge (z \geq x) \wedge (y > 0) \wedge (x > 0)$$

having respective solutions:

$$x = 4, y = 3, z = 3, cost = 10$$
$$x = 0, y = 5, z = 0, cost = 5$$
$$x = 0, y = 5, z = 5, cost = 10$$

giving an overall minimal cost of 5.

Note that while OMT focuses on minimizing costs, structured-output learning and preference elicitation typically deal with maximizing scores and utilities. In the rest of the chapter we will thus focus on maximization problems, with the implicit assumption that optimization will actually be addressed by minimizing their negated versions.

2.2   Learning with Structured Outputs

Statistical learning approaches have traditionally focused on learning settings with vectorial representations as inputs and scalar representations as outputs, either classification or regression. However, many real-world scenarios are characterized by more complex types of data, both in the input (e.g. a website, a protein sequence) and in the output (e.g. collective classification for webpages, secondary structure sequential labeling for protein sequences). Dealing with structured inputs is performed by explicit or implicit feature construction approaches, like kernel machines [63] or neural networks [9]. Structured-output prediction [2] is more tricky, as it requires to learn a function producing a structure as its output. A common approach to the problem is that of learning a scoring function over joint input-output pairs:

$$f(\boldsymbol{I}, \boldsymbol{O}) = \boldsymbol{w}^T \boldsymbol{\psi}(\boldsymbol{I}, \boldsymbol{O}) \tag{1}$$

Here $\boldsymbol{I}$ is the *input* (or observed) part, $\boldsymbol{O}$ is the *output* (or query) part[1] and ψ is a function mapping input-output pairs to a *joint* feature space, where linear discrimination is performed. Given an input $\boldsymbol{I}$, the predicted output will be the one maximizing the scoring function:

$$\boldsymbol{O}^* = \underset{\boldsymbol{O}}{\mathrm{argmax}} \, f(\boldsymbol{I}, \boldsymbol{O}) \tag{2}$$

and the problem boils down to finding efficient procedures for computing the maximum. This formulation is also known as energy-based learning [47] and comprises many statistical relational learning [32] approaches to collective prediction. It also corresponds to maximum-a-posteriori inference in probabilistic graphical models [42], where the scoring function is the conditional probability

---

[1] We depart from the conventional x/y notation for indicating input/output pairs to avoid name clashes with the x-y coordinate variables.

of the output given the input. Efficient exact procedures exist for some special cases—like when the space of feasible solutions can be represented in terms of regular or context free grammars—while approximate inference is typically used in the general case. In this paper we are interested in the case where inputs and outputs are combinations of discrete and continuous variables, and we will leverage SMT techniques to perform efficient inference in this setting.

Discriminative learning of the scoring function is based on a generalization of the max-margin algorithm to the structured-output setting. Max-margin learning in binary classification [18] enforces positive examples to be separated from negative ones with a large margin, possibly accounting for margin errors to be penalized in the objective function. In the structured-output setting, this corresponds [65] to enforcing that the correct output structure for a certain input has a score which is higher (by a large margin) than any possible incorrect structure, again admitting margin violations to be penalized in the objective. The resulting optimization problem is as follows:

$$\operatorname*{argmin}_{\boldsymbol{w}, \boldsymbol{\xi}} \frac{1}{2} \|\boldsymbol{w}\|^2 + \frac{C}{n} \sum_{i=1}^{n} \xi_i \tag{3}$$

s.t. $\boldsymbol{w}^\top (\boldsymbol{\psi}(\boldsymbol{I}_i, \boldsymbol{O}_i) - \boldsymbol{\psi}(\boldsymbol{I}_i, \boldsymbol{O}')) \geq \Delta(\boldsymbol{I}_i, \boldsymbol{O}_i, \boldsymbol{O}') - \xi_i \ \forall \, i = 1, \ldots, n; \ \boldsymbol{O}' \neq \boldsymbol{O}_i$

where $\Delta(\boldsymbol{I}, \boldsymbol{O}, \boldsymbol{O}')$ is a non-negative *loss function* that, for any given observation $\boldsymbol{I}$, quantifies the penalty incurred when predicting $\boldsymbol{O}'$ instead of the correct output $\boldsymbol{O}$. Each inequality states that the score of the correct output $\boldsymbol{O}_i$ should be higher than that of an incorrect output $\boldsymbol{O}' \neq \boldsymbol{O}_i$ by at least the value of the loss between the two outputs and, if this is not the case, that a penalty should be paid in the objective function. Note that as ξ_i is shared among all inequalities involving example i, each example contributes to the objective with a cost equal to the maximum among the penalties for all possible wrong outputs. Minimizing the norm of the weights corresponds to maximizing the margin between correct and incorrect outputs. The regularization term C is a hyper-parameter trading off size of the margin and penalties for margin violations.

A problem with this formulation is that the number of candidate output structures is exponential in the size of the output, which makes exhaustive enumeration of the inequalities infeasible. The problem is addressed by the cutting plane algorithm [40], which iteratively adds the inequality corresponding to the most violated condition for each example pair given the current scoring function, and refines it by solving the resulting quadratic problem with a standard SVM solver. The procedure is shown in Algorithm 1. For each training example $(\boldsymbol{I}_i, \boldsymbol{O}_i)$, the algorithm finds the highest scoring incorrect output $\boldsymbol{O}'_i$ by solving the so-called *separation* problem (line 4):

$$\boldsymbol{O}'_i = \operatorname*{argmax}_{\boldsymbol{O}'} \boldsymbol{w}^T \boldsymbol{\psi}(\boldsymbol{I}_i, \boldsymbol{O}') + \Delta(\boldsymbol{I}_i, \boldsymbol{O}_i, \boldsymbol{O}') \tag{4}$$

The penalty currently associated with the training example $(\boldsymbol{I}_i, \boldsymbol{O}_i)$ is stored in the slack variable ξ_i. If the loss-augmented score of $\boldsymbol{O}'_i$ minus the score of

---

Data: Training instances $\{(I_1, O_1), \ldots, (I_n, O_n)\}$, parameters C, ϵ
Result: Learned weights w
1 $\mathcal{W}_i \leftarrow \emptyset$, $\xi_i \leftarrow 0$ for all $i = 1, \ldots, n$
2 **repeat**
3   **for** $i = 1, \ldots, n$ **do**
4          $O'_i \leftarrow \operatorname{argmax}_{O'} w^\top \psi(I_i, O') + \Delta(I_i, O_i, O')$
5   **if** $w^\top \psi(I_i, O'_i) + \Delta(I_i, O_i, O'_i) - w^\top \psi(I_i, O_i) > \xi_i + \epsilon$ **then**
6

$$\mathcal{W}_i \leftarrow \mathcal{W}_i \cup \{O'_i\}$$

$$w, \xi \leftarrow \operatorname*{argmin}_{w, \xi \geq 0} \frac{1}{2}\|w\|^2 + \frac{C}{n} \sum_{i=1}^{n} \xi_i$$

$$\text{s.t. } \forall O'_1 \in \mathcal{W}_1 \ : \ w^\top \left[\psi(I_1, O_1) - \psi(I_1, O'_1) \right] \geq$$
$$\Delta(I_i, O_i, O'_1) - \xi_1$$

$$\vdots$$

$$\forall O'_n \in \mathcal{W}_n \ : \ w^\top \left[\psi(I_n, O_n) - \psi(I_n, O'_n) \right] \geq$$
$$\Delta(I_i, O_i, O'_n) - \xi_n$$

7   **end**
8   **end**
9 **until** *no $\mathcal{W}_i$ has changed during iteration*
10 **return** w

Algorithm 1. Cutting-plane algorithm for training structured-output SVM.

the correct output O_i exceeds ξ_i by more than a tolerance ϵ (line 5), O'_i is added to the set of conditions for the training example and a new set of weights (and slacks) is generated by solving the resulting quadratic problem (line 6). The procedure is repeated until no condition is added for any of the training examples, and it is guaranteed to find an ϵ-approximate solution in a polynomial number of iterations [65].

This learning formulation is generic, meaning that it can be adapted to any structured prediction problem as long as it is provided with: (i) a joint feature space representation $\psi(I, O)$ of input-output pairs (and consequently a scoring function f, see Eq. (1)); (ii) an oracle to perform inference, i.e. to solve Eq. (2); (iii) an oracle to retrieve the most violated condition, i.e. to solve the separation problem (see Eq. (4)). For a more detailed account, and in particular for the derivation of the separation oracle formulation, please refer to [65].

3   LMT for Structured-Output Prediction

Existing structured-output prediction approaches mostly focus on predicting *discrete* structures as outputs, like (labeled) sequences, trees or graphs. In this section we show how to employ OMT technology to adapt the structured-output SVM to deal with the prediction of hybrid output structures.

3.1 An Introductory Example

In order to introduce the LMT framework, we start with a toy learning example. We are given a unit-length bounding box, $[0,1] \times [0,1]$, that contains a given, fixed block (rectangle), as in Fig. 1(a). The block is identified by the four constants (x_1, y_1, dx_1, dy_1), where x_1, y_1 indicate the bottom-left corner of the rectangle, and dx_1, dy_1 its width and height, respectively. Now, suppose that we are assigned the task of fitting another block, identified by the variables (x_2, y_2, dx_2, dy_2), in the same bounding box, so as to maximize the following *scoring* function:

$$score := w_1 \times dx_2 + w_2 \times dy_2 \qquad (5)$$

with the additional requirements that (i) the two blocks "touch" either from above, below, or sideways, and (ii) the two blocks do not overlap.

It is easy to see that the weights w_1 and w_2 control the shape and location of the optimal solution. Assuming positive weights, if $w_1 \gg w_2$, then the optimal block will be placed so as to occupy as much horizontal space as possible, while if $w_1 \ll w_2$ it will prefer to occupy as much vertical space as possible, as in Fig. 1(b, c). If w_1 and w_2 are close, then the optimal solution depends on the relative amount of available vertical and horizontal space in the bounding box.

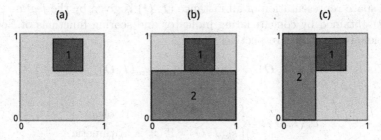

Fig. 1. (a) Initial configuration. (b) Optimal configuration for $w_1 \gg w_2$. (c) Optimal configuration for $w_1 \ll w_2$.

This toy example illustrates two key points. First, the problem involves a mixture of *numerical variables* (coordinates, sizes of block 2) and *Boolean variables* along with *hard rules* that control the feasible space of the optimization procedure (conditions (i) and (ii)), and *soft rules* which control the shape of the optimization landscape. This is the kind of problem that can be solved in terms of Optimization Modulo Linear Arithmetic, OMT($\mathcal{LRA}$). Second, it is possible to estimate the weights w_1, w_2 from data in order to learn what kind of blocks are to be considered optimal. The goal of our learning procedure is precisely to find a good set of weights from examples. In the following we will describe how such a learning task can be framed within the structured-output SVM framework.

3.2 The Method

As previously mentioned, the cutting plane algorithm for structured-output SVM can be adapted to arbitrary structured-output problems by providing a joint input-output feature map and oracles for inference and separation. In the following we will detail how these components are provided within the LMT framework.

Input-Output Feature Map. Recall that in our setting each instance (I, O) is represented as a set of Boolean and rational variables:

$$(I, O) \in \underbrace{(\{\top, \bot\} \times \ldots \times \{\top, \bot\})}_{\text{Boolean part}} \times \underbrace{(\mathbb{Q} \times \ldots \times \mathbb{Q})}_{\text{rational part}}$$

We indicate Boolean variables using predicates[2] such as $\texttt{touching}(i, j)$, and write rational variables as lower-case letters, e.g. *distance*, x, y. Features are represented in terms of *constraints* $\{\varphi_k\}_{k=1}^m$, each constraint φ_k being either a Boolean- or rational-valued function of the instance (I, O). These constraints are constructed using the background knowledge available for the domain. For each Boolean-valued constraint φ_k, we denote its *indicator function* as $\mathbb{1}_k(I, O)$, which evaluates to 1 if the constraint is satisfied and to -1 otherwise (the choice of -1 to represent falsity is customary in the max-margin literature). Similarly, we refer to the *score* of a rational-valued constraint φ_k as $s_k(I, O) \in \mathbb{Q}$. The feature space representation of an instance (I, O) is given by the *feature vector* $\psi(I, O)$ obtained by concatenating indicator and scoring functions of Boolean and rational constraints respectively, i.e.:

$$\psi(I, O) := (\psi_1(I, O), \ldots, \psi_m(I, O))^\top$$

where:

$$\psi_k(I, O) := \begin{cases} \mathbb{1}_k(I, O) & \text{if } \varphi_k \text{ is Boolean} \\ s_k(I, O) & \text{if } \varphi_k \text{ is arithmetic} \end{cases}$$

Note that we are implicitly assuming to have *soft* constraints, whose weights w_k will be learned from data (see Eq. (1)). In most cases, these soft constraints will be complemented by a set of *hard* constraints, which do not contribute to the feature vector but rather define the space of feasible configurations. For a summary of the notation see Table 1.

Inference Oracle. Given the feature vector $\psi(I, O)$, the scoring function $f(I, O)$ is a linear combination of indicator and score functions. Since ψ can be expressed in terms of SMT($\mathcal{LRA}$) formulas, the resulting maximization problem can be readily cast as an OMT($\mathcal{LRA}$) problem and inference is performed by an

---

[2] While we write Boolean variables using a first-order syntax for readability, the OMT solver currently requires the grounding of all Boolean predicates.

OMT-solver (we use the OPTIMATHSAT solver [62]). As previously explained, given that OMT-solvers are conceived to *minimize* cost functions rather than maximize scores, we actually run it on the negated scoring function.

Separation Oracle. The separation problem consists in maximizing the sum of the scoring function and a *loss* function over output pairs (see Eq. (4)). The loss function determines the dissimilarity between output structures, a mixture of Boolean and rational variables in our setting. We observe that by picking a loss function expressible as an OMT($\mathcal{LRA}$) problem, we can readily use the same OMT solver used for inference to also solve the separation oracle. This can be achieved by selecting a loss function such as the following Hamming loss in feature space:

$$\Delta(\boldsymbol{I}, \boldsymbol{O}, \boldsymbol{O}') := \sum_{k \,:\, \varphi_k \text{ is Boolean}} |\mathbb{1}_k(\boldsymbol{I}, \boldsymbol{O}) - \mathbb{1}_k(\boldsymbol{I}, \boldsymbol{O}')| +$$

$$\sum_{k \,:\, \varphi_k \text{ is arithmetic}} |s_k(\boldsymbol{I}, \boldsymbol{O}) - s_k(\boldsymbol{I}, \boldsymbol{O}')|$$

$$= \|\boldsymbol{\psi}(\boldsymbol{I}, \boldsymbol{O}) - \boldsymbol{\psi}(\boldsymbol{I}, \boldsymbol{O}')\|_1$$

This loss function is piecewise-linear, and as such satisfies the desideratum. While this is the loss which was used in all experiments presented in this chapter, LMT can work with any loss function that can be encoded as an SMT formula.

Example 3.1. *Consider the block-world example in Sect. 3.1. Here the input $\boldsymbol{I}$ to the problem is the observed block (x_1, y_1, dx_1, dy_1) while the output $\boldsymbol{O}$ is the generated block (x_2, y_2, dx_2, dy_2). In order to encode the set of constraints $\{\varphi_k\}$, it is convenient to first introduce a background knowledge of predicates expressing facts about the relative positioning of blocks. To this end we add a fresh predicate* left(i, j)*, that encodes the fact that "a block of index i touches a second block j from the left", defined as follows:*

$$\texttt{left}(i, j) := (x_i + dx_i = x_j) \,\wedge$$
$$((y_j \le y_i \le y_j + dy_j) \,\vee\, (y_j \le y_i + dy_i \le y_j + dy_j))$$

Similarly, we add analogous predicates for the other directions: right(i, j), below(i, j), over(i, j). *The hard constraints represent the fact that the output $\boldsymbol{O}$ should be a valid block within the bounding box (all constraints are implicitly conjoined):*

$$0 \le x_2, y_2, dx_2, dy_2 \le 1 \qquad (x_2 + dx_2) \le 1 \,\wedge\, (y_2 + dy_2) \le 1$$

and that the output block $\boldsymbol{O}$ should "touch" the input block $\boldsymbol{I}$:

$$\texttt{left}(1, 2) \vee \texttt{right}(1, 2) \vee \texttt{below}(1, 2) \vee \texttt{over}(1, 2)$$

Note that whenever this rule is satisfied, both conditions (i) and (ii) of the toy example hold, i.e. touching blocks never overlap. The soft constraints here should encode features related to the width and height of the output block, i.e.:

$$\psi(\boldsymbol{I}, \boldsymbol{O}) = (dx_2, dy_2)^{\top}$$

This allows to define a scoring function as a linear combination of features:

$$score := w_1 \times dx_2 + w_2 \times dy_2 = \boldsymbol{w}^{\top}\psi(\boldsymbol{I}, \boldsymbol{O})$$

3.3 Experimental Results

We show the potential of the approach on automatic character drawing, a novel structured-output learning problem that consists in learning to translate any input noisy hand-drawn character into its *symbolic* representation. More specifically, given a black-and-white image of a handwritten letter or digit, the goal is to construct an equivalent symbolic representation of the same character.

In this paper, we assume the character to be representable by a polyline made of a given number m of *directed* segments, i.e. segments identified by a starting point (x^b, y^b) and an ending point (x^e, y^e). The input image $\boldsymbol{I}$ is seen as the set P of coordinates of the pixels belonging to the character, while the output $\boldsymbol{O}$ is a set of m directed segments $\{(x_i^b, y_i^b, x_i^e, y_i^e)\}_{i=1}^m$.

Intuitively, any good output $\boldsymbol{O}$ should satisfy the following requirements: (i) it should be as similar as possible to the noisy input character; and (ii) it should actually "look like" the corresponding symbolic character. Figure 2 shows an example for the "A" character. These requirements will be encoded in two feature vectors: $coverage(\boldsymbol{I}, \boldsymbol{O})$, measuring how many pixels of the input image are covered; $orientation(\boldsymbol{O})$, measuring the resemblance of the output to a symbolic *template* for the corresponding character.

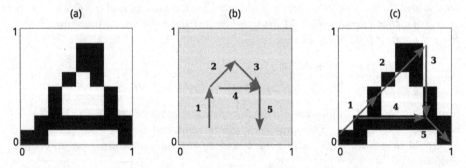

Fig. 2. Left, example bitmap image of an "A". Middle, a set of 5 segments satisfying the "looking like an A" rules in the text. Right, 5 segments satisfying both the rules for character "A" and fitting the underlying image.

Since the output is supposed to be a polyline, we constrain consecutive segments to be connected:

$$\forall i \; \texttt{connected}(i, i+1)$$

We then constrain (without loss of generality) each segment to be oriented from left to right, i.e. $x_i^b \leq x_i^e$, and no larger than the image nor smaller than a pixel: $\forall i \; min\_length \leq length(i) \leq 1$. Finally, we restrict the segments to be either horizontal, vertical or $45°$ diagonal, that is:

$$\forall i \; \texttt{horizontal}(i) \vee \texttt{vertical}(i) \vee \texttt{diagonal}(i)$$

This restriction allows us express all numerical constraints in linear terms. Under these assumptions, we can encode the *coverage* feature as:

$$coverage(\boldsymbol{I}, \boldsymbol{O}) := \frac{1}{|P|} \sum_{p \in P} \mathbb{1}(\texttt{covered}(p))$$

where $\texttt{covered}(p)$ is true if pixel p is covered by at least one of the m segments:

$$\texttt{covered}(p) := \bigvee_{i \in [1,m]} \texttt{covered}(p, i)$$

The fact that a segment $i = (x_i^b, y_i^b, x_i^e, y_i^e)$ covers pixel $p = (x, y)$ depends on the orientation of the segment and is computed using constructs like:

$$\text{If} \quad \texttt{horizontal}(i) \quad \text{then} \quad \texttt{covered}(p, i) := x_i^b \leq x \leq x_i^e \wedge y = y_i^b$$

The coverage formulae for the other segment types can be found in Table 2.

As for the *orientation* vector, it should contain features related to the symbolic representation of characters. These include both the direction of the individual segments and the connections between pairs of segments. Consider the symbolic representation of an "A" in Fig. 2(b). Directions for segments could be encoded as follows:

$$\texttt{increasing}(1) \wedge \texttt{increasing}(2) \wedge \texttt{decreasing}(3) \wedge$$
$$\texttt{horizontal}(4) \wedge \texttt{decreasing}(5)$$

Here $\texttt{increasing}(i)$ and $\texttt{decreasing}(i)$ indicate the direction of segment i, and can be written as:

$$\texttt{increasing}(i) := y_i^e > y_i^b$$
$$\texttt{decreasing}(i) := y_i^e < y_i^b$$

In order to represent connection types, we follow the convention used for Bayesian Networks, where the head of a directed segment is the edge containing the arrow (the ending point (x^e, y^e)) and the tail is the opposite edge (the starting point (x^b, y^b)). For instance, $\texttt{h2t}(i, j)$ indicates that i is head-to-tail with respect to j, $\texttt{h2h}(i, j)$ that they are head-to-head:

$$\texttt{h2t}(i, j) := (x_i^e = x_j^b) \wedge (y_i^e = y_j^b)$$
$$\texttt{h2h}(i, j) := (x_i^e = x_j^e) \wedge (y_i^e = y_j^e)$$

Connections between segments in Fig. 2(b) could then be encoded as follows:

$$\mathtt{h2t}(1,2) \land \mathtt{h2t}(2,3) \land \mathtt{h2h}(3,4) \land \mathtt{h2t}(4,5)$$

For a full list of background knowledge predicates, see Table 2.

Now given the image in Fig. 2(a) and the template in Fig. 2(b), a character drawing algorithm driven by these coverage and orientation criteria could produce an output like the one pictured in Fig. 2(c). However, the formula for the "looking like an A" constraint is not available at test time and should be learned from the data. In order to do so, the *orientation* vector includes possible directions (`increasing`, `decreasing`, `right`) for all m segments and all possible connection types between *all* pairs of segments (`h2t`, `h2h`, `t2t`, `t2h`). Note that we do not include specific segment orientations (i.e., `horizontal`, `vertical`, `diagonal`) in the feature space, to accommodate for alternative symbolic representations of the same letter. For instance, the first segment in an "A" (the lower left one because of the left-to-right rule) is bound to be `increasing`, but may be equally likely `vertical` or `diagonal` (see e.g. Figs. 2(b) and (c)). Summing up, the *orientation* vector can be written as:

$$(\; increasing(1), decreasing(1), right(1),$$
$$\cdots$$
$$increasing(m), decreasing(m), right(m),$$
$$h2t(1,2), t2h(1,2), h2h(1,2), t2t(1,2),$$
$$\cdots$$
$$h2t(1,m), t2h(1,m), h2h(1,m), t2t(1,m),$$
$$\cdots$$
$$h2t(m-1,m), t2h(m-1,m), h2h(m-1,m), t2t(m-1,m) \;)$$

where each feature is the indicator function of the corresponding Boolean variable, e.g. $increasing(1) := \mathbb{1}(\mathtt{increasing}(1))$ (see Table 3).

We evaluated LMT on the character drawing problem by carrying out an extensive experiment using a set of noisy B&W 16×20 character images[3]. Learning to draw characters is a very challenging constructive problem, made even more difficult by the low quality of the noisy images in the dataset (see, e.g. Fig. 4). In this experiment we learn a model for each of the first five letters of the alphabet (A to E), and assess the ability of LMT to generalize over unseen handwritten images of the same character. For each letter, we selected five images at random to be used as training examples. For each of these, we used OPTIMATHSAT to generate a "perfect" symbolic representation according to a human-provided letter template (similar to the "looking like an A" rule above), obtaining a training set of five fully supervised images. The first row of Figs. 3, 4, 5, 6, and 7 report these supervised instances. Note that the resulting supervision is in some cases very noisy, and depends crucially on the quality of

---

[3] Dataset taken from http://cs.nyu.edu/~roweis/data.html.

Table 2. Background knowledge used in the character writing experiment.

Segment types	
Segment i is horizontal	$\mathtt{horizontal}(i) := (x_i^b \neq x_i^e) \wedge (y = y_i^b)$
Segment i is vertical	$\mathtt{vertical}(i) := (x_i^b = x_i^e) \wedge (y_i^e \neq y_i^b)$
Segment i is diagonal	$\mathtt{diagonal}(i) := |x_i^e - x_i^b| = |y_i^e - y_i^b|$
Segment i is increasing	$\mathtt{increasing}(i) := y_i^e > y_i^b$
Segment i is decreasing	$\mathtt{decreasing}(i) := y_i^e < y_i^b$
Segment i is left-to-right	$\mathtt{right}(i) := x_i^e > x_i^b$
Segment i is incr. vert	$\mathtt{incr\_vert}(i) := \mathtt{increasing}(i) \wedge \mathtt{vertical}(i)$
Segment i is decr. vert.	$\mathtt{decr\_vert}(i) := \mathtt{decreasing}(i) \wedge \mathtt{vertical}(i)$
Segment i is incr. diag.	$\mathtt{incr\_diag}(i) := \mathtt{increasing}(i) \wedge \mathtt{diagonal}(i)$
Segment i is decr. diag.	$\mathtt{decr\_diag}(i) := \mathtt{decreasing}(i) \wedge \mathtt{diagonal}(i)$
Segment length	
Length of horiz. segment i	$\mathtt{horizontal}(i) \rightarrow length(i) = |x_i^e - x_i^b|$
Length of vert. segment i	$\mathtt{vertical}(i) \rightarrow length(i) = |y_i^e - y_i^b|$
Lenght of diag. segment i	$\mathtt{diagonal}(i) \rightarrow length(i) = \sqrt{2}\,|y_i^e - y_i^b|$
Connections between segments	
Segments i,j are head-to-tail	$\mathtt{h2t}(i,j) := (x_i^e = x_j^b) \wedge (y_i^e = y_j^b)$
Segments i,j are head-to-head	$\mathtt{h2h}(i,j) := (x_i^e = x_j^e) \wedge (y_i^e = y_j^e)$
Segments i,j are tail-to-tail	$\mathtt{t2t}(i,j) := (x_i^b = x_j^b) \wedge (y_i^b = y_j^b)$
Segments i,j are tail-to-head	$\mathtt{t2h}(i,j) := (x_i^b = x_j^e) \wedge (y_i^b = y_j^e)$
Segments i,j are connected	$\mathtt{connected}(i,j) := \mathtt{h2h}(i,j) \vee \mathtt{h2t}(i,j) \vee$ $\mathtt{t2h}(i,j) \vee \mathtt{t2t}(i,j)$
Whether segment $i = (x^b, y^b, x^e, y^e)$ covers pixel $p = (x,y)$	
Coverage of pixel p	$\mathtt{covered}(p) := \bigvee_i \mathtt{covered}(p,i)$
Coverage of pixel p by seg. i	
$\mathtt{incr\_vert}(i) \rightarrow \mathtt{covered}(p,i) := y_i^b \leq y \leq y_i^e \wedge x = x_i^b$	
$\mathtt{decr\_vert}(i) \rightarrow \mathtt{covered}(p,i) := y_i^e \leq y \leq y_i^b \wedge x = x_i^b$	
$\mathtt{horizontal}(i) \rightarrow \mathtt{covered}(p,i) := x_i^b \leq x \leq x_i^e \wedge y = y_i^b$	
$\mathtt{incr\_diag}(i) \rightarrow \mathtt{covered}(p,i) := y_i^b \leq y \leq y_i^e \wedge x_i^b \leq x \leq x_i^e \wedge x_i^b - y_i^b = x - y$	
$\mathtt{decr\_diag}(i) \rightarrow \mathtt{covered}(p,i) := y_i^e \leq y \leq y_i^b \wedge x_i^b \leq x \leq x_i^e \wedge x_i^b + y_i^b = x + y$	

the character image (see e.g. the "B", the most geometrically complex of the characters).

For each letter, we learned models with a number training examples ranging from 2 to 5 and tested them on 10 randomly chosen test images. We indicate the predictions obtained by models learned with k examples as *pred@k*. The number of segments m was known during both training and inference. The output for all

Table 3. List of all rules used in the character writing problem. Top, hard rules. Middle, soft rules. Bottom, total score of a segment assignment.

(a) Hard constraints	
Left-to-right ordering	$x_i^b \le x_i^e$
Allowed segment types	$\texttt{vertical}(i) \vee \texttt{horizontal}(i) \vee \texttt{diagonal}(i)$
Consecutive segments are connected	$\texttt{connected}(i, i+1)$
Minimum segment size	$min\_length \le length(i) \le 1$
(b) Soft constraints (features)	
Non-zero pixel coverage	$coverage := \frac{1}{|P|} \sum_{p \in P} \mathbb{1}(\texttt{covered}(p))$
Indicator of increasing segment i	$increasing(i) := \mathbb{1}(\texttt{increasing}(i))$
Indicator of decreasing segment i	$decreasing(i) := \mathbb{1}(\texttt{decreasing}(i))$
Indicator of right segment i	$right(i) := \mathbb{1}(\texttt{right}(i))$
Indicator of head-to-tail i, j	$h2t(i,j) := \mathbb{1}(\texttt{h2t}(i,j))$
Indicator of tail-to-head i, j	$t2h(i,j) := \mathbb{1}(\texttt{t2h}(i,j))$
Indicator of head-to-head i, j	$h2h(i,j) := \mathbb{1}(\texttt{h2h}(i,j))$
Indicator of tail-to-tail i, j	$t2t(i,j) := \mathbb{1}(\texttt{t2t}(i,j))$
(c) Score	

$$score := \boldsymbol{w}^\top (\underbrace{increasing(i), decreasing(i), right(i),}_{\text{for all segments } i}$$
$$\underbrace{h2t(i,j), t2h(i,j), h2h(i,j), t2t(i,j),}_{\text{for all segment pairs } (i,j) \text{ with } j>i}$$
$$coverage)$$

letters can be found in Figs. 3, 4, 5, 6, and 7, from the second to the fifth rows of each figure.

In order to speed up inference we extract a set of *hard* constraints from the learned model prior to inference. We add a hard rule for each segment and connection feature with a positive weight. If more than one weight is positive for any given segment/connection, we add the disjunction of the hard rules to the model. Note that this process allows to learn a letter template constraining the search, while the weighted features still allow for some flexibility in the choice of the actual solution.

As a quantitative measure of the quality of the predictions, we also report the distance between the generated symbolic representation O for each letter and a corresponding human-made gold standard O'. Here the error is computed by first aligning the segments using an optimal translation, and then summing the distances between all corresponding segment endpoints. The human generated images respect the same "looking like an X" rule used for generating the training set, i.e. they have the same number of segments, drawn in the same order and within the same set of allowed orientations. The values in Fig. 8 are the average over all instances in the test set, when varying the training set size.

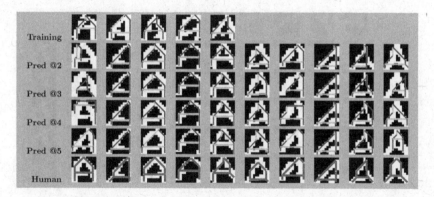

Fig. 3. Results for the "A" character drawing task. The training instances are lined up in the first row. The second to fifth row are the segmentations generated by models learned with the first two training instances, the first three instances, *etc.*, respectively. The last row are the human-made segmentations used in the comparison. The generated symbolic representations are shown overlayed over the corresponding bitmap image. Segments are colored for readability.

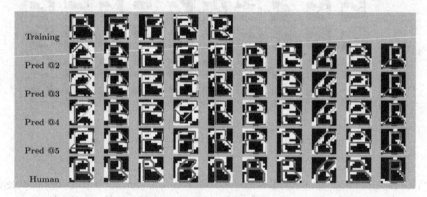

Fig. 4. Results for the "B" character drawing task.

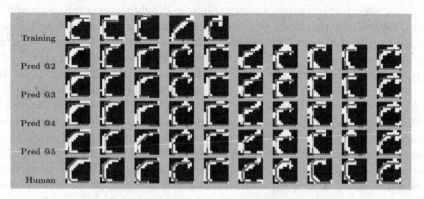

Fig. 5. Results for the "C" character drawing task.

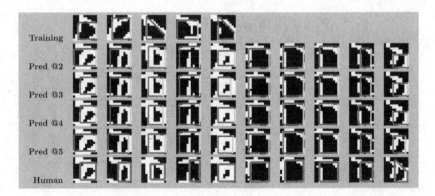

Fig. 6. Results for the "D" character drawing task.

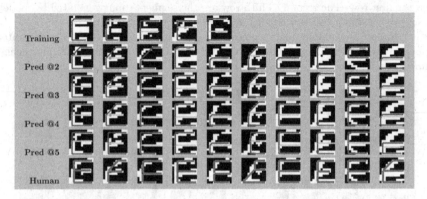

Fig. 7. Results for the "E" character drawing task.

The results show that LMT is able to address the character drawing problem and produce reasonable outputs for all the target letters. It should be stressed here that both the coordinates and the number of character pixels can vary widely between test instances, and our results highlight the generalization ability of our method. Furthermore, the predictions tend to get closer to the human-provided segmentations as the number of training instances increases. For the simplest cases (i.e. "C", "D", and "E", drawn using four to five segments), the outcome is unambiguous. The only real issue is with the "D" which is always represented as a rectangle with no diagonal edge. The main reason is that in none of the test images it is possible to draw a 45° diagonal without sacrificing pixel coverage, and diagonals of different degree are not representable in terms of linear constraints. None of the predictions looks perceptually "wrong". More complex letters like the "A" and "B", with seven and nine segments respectively, also look reasonably similar to the given examples, apart from few predictions for which either coverage (e.g. see the first column in Fig. 3) or "perceptual" representation (e.g. see the first and fourth columns in the fourth row of Fig. 4) are

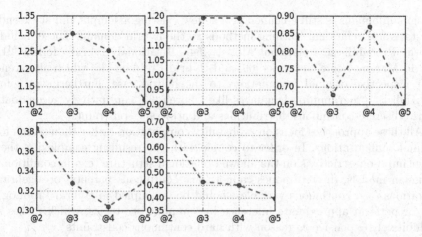

Fig. 8. Average distance between the predicted symbolic images and the human-made ones, while increasing the number of training instances. The **@k** ticks on the x-axis indicate the size of the training set. The y-axis represents the sum of per-segment distances averaged over all images in the test set. From left to right, top to bottom: results for "A", "B", "C", "D", and "E".

sub-optimal. The distance-to-human results in Fig. 8 also show how the algorithm produces more perceptually reasonable predictions as the training set increases, as in almost all cases the distance at *pred@5* is lower than that at *pred@2*. The fluctuations observed in some of the cases are due to the occasional presence of pathologically bad inputs in the training set. The bad quality of the second and third "B" training examples leads to bad performance in the *pred@3* and *pred@4* experiments, while the performance worsening of *pred@4* for "C" is due to the unusual shape of the fourth training example.

Summarizing, excluding cases of pathologically bad inputs, LMT is able to learn an appropriate model for each letter and generalize the learned template over unseen inputs.

3.4   Related Work

There is a body of work concerning integration of relational and numerical data from a feature representation perspective, in order to effectively incorporate numerical features into statistical relational learning models. These include neural networks used as feature generators within Markov Logic Networks [50], T-norms used as continuous relaxations of Boolean constraints in Semantic Based Regularization [26] and Probabilistic Soft Logic [17], and Gaussian Logic [44]. All these approaches aim at incorporating continuous features as *inputs*, while our framework aims at allowing learning and inference over hybrid domains with continuous and discrete variables as *outputs*.

While a number of efficient algorithms have been developed for Relational Continuous Models [1, 22, 23], performing inference over joint continuous-discrete relational domains is still a challenge. The few existing attempts aim at extending statistical relational learning methods to the hybrid domain. These include Hybrid Probabilistic Relational Models [54], Relational Hybrid Models [21], Hybrid Markov Logic Networks [69] and Hybrid ProbLog [36] which all extend their respective original formulations to the hybrid case. Furthermore, some probabilistic programming languages like Church [33] can natively accomodate hybrid discrete-continuous distributions and arbitrary constraints.

All these approaches focus on probability computation rather than search for optimal configurations. In order to couple with the resulting complexity, they either impose restrictions on the allowed relational structure (e.g. in conditional Gaussian models, discrete nodes cannot have continuous parents) or algebraic operations (e.g. continuous variables should be uncoupled in Hybrid ProbLog), and/or perform approximate inference by sampling strategies, which makes it prohibitively expensive to reason with hard continuous constraints[4].

Conversely, LMT can accomodate arbitrary combinations of predicates from the theories for which a solver is available. These currently include linear arithmetic over both rationals and integers as well as a number of other theories like strings, arrays and bit-vectors. Furthermore, the tight integration between theory-specific and SAT solvers [24,49,53,60,61], where the former inform the latter about conflicting constraints and help guiding the search, is widely recognised as a key reason of the success of SMT solvers [3]. This integration is missing in approaches like Hybrid Markov Logic Networks, which use a general-purpose global optimization algorithm (L-BFGS) as a black-box to solve numeric subproblems. Note also that previous attempts to substitute standard SAT solvers with WalkSAT inside an SMT solver have failed, producing dramatic worsening of performance [34].

On the other hand, an advantage of probabilistic inference approaches is that they allow to return marginal probabilities in addition to most probable explanations. This is actually the main focus of these approaches, and the reason why they are less suitable for solving the latter problem when the search space becomes strongly disconnected. As with most structured-output approaches over which it builds, LMT focuses on the task of finding an optimal configuration, which in a probabilistic setting corresponds to generating the most probable explanation. We are planning to extend it to also perform probability computation, as discussed in the conclusions of the paper.

Please refer to the original structured LMT work [64] for a deeper discussion on the above mentioned approaches and their relationship to the LMT framework.

---

[4] Our preliminary experimental studies showed that Church is incapable of solving in reasonable time the simple task of generating a pair of blocks conditioned on the fact that they touch somewhere.

4   LMT for Preference Elicitation

The structured-output formulation discussed so far assumes a training set of input-output pairs is available. This is indeed reasonable in a number of real-world scenarios, where the notion of a correct output is (reasonably) clearly defined and there exist datasets of examples labelled by human experts or experimental procedures (e.g. parse trees for natural language processing, secondary structure for protein structure determination). There is however a relevant class of problems where these data are not available a-priori, because they are very expensive to compute and/or because they depend on some specific condition. A typical example is that of recommendation systems [56], where the scoring function measures the *utility* a user assigns to a certain instance. Existing approaches to user recommendation rely on (combinations of) collaborative filtering on one side, to leverage similarities between users and propagate recommendations, and content-based filtering on the other, to learn a user profile relating item features to user preferences. Content-based filtering usually adopts preference elicitation [15] techniques, to gather feedback from the user in order to refine her utility model. From an optimization viewpoint, this setting can be seen as an instance of *learning while optimizing* [4], where the task is that of learning to optimize a (partially) unknown function. In this section we will discuss how the LMT framework can be adapted to address this type of problems. In the following we will refer to the user as the decision maker (DM).

4.1   An Introductory Example

Consider a customer interested in buying a house. A very clear-headed person could go to a real-estate agency with a (simplified) request like:

> *I would like a house in a safe area, close to my parents and to the kindergarten, with a garden if there are no parks nearby. My maximum budget is 300,000 euro.*

These desiderata can be encoded as an SMT problem as follows:

solve:
$$\varphi_1 \wedge \varphi_2 \wedge \varphi_3 \wedge \varphi_4$$
subject to:
$$\varphi_1 = (\neg O_2 \rightarrow O_1) \qquad \varphi_2 = (O_3 < \theta_1)$$
$$\varphi_3 = (O_4 < \theta_2) \qquad \varphi_4 = (O_5 < \theta_3)$$
$$price(\boldsymbol{O}) \leq 300000$$

where the characteristics of the house are defined in terms of the set of output variables $\boldsymbol{O}$ described in Table 4. Now this problem could have no solution in case none of the houses at the agency disposal satisfies all constraints.

Table 4. Output variables for the housing example.

Name	Description	Type
O_1	Garden	Boolean
O_2	Park nearby	Boolean
O_3	Crime rate	Ordinal
O_4	Distance from parents	Continuous
O_5	Distance from kindergarten	Continuous

A more reasonable alternative is turning the problem into an optimization one, where the task is maximizing the weighted sum of the satisfied constraints (i.e. a MAX-SMT problem). However, assigning weights to requirements is a typically hard task for humans. An exact specification of the set of constraints, like the one above, is also difficult to obtain. The most natural scenario consists of an interactive process, in which the customer is provided with some candidates and the realtor updates her understanding of the customer preferences according to the feedback received. In the following we will present a preference elicitation method automatizing this process (from the recommender side).

4.2 The Method

We will start by introducing the components of the method and then show how these components are combined in the overall algorithm.

Space of Constraints. The first component is the space of candidate constraints, from which the ones which are of interest to the customer will be selected. We will assume there is a *catalogue* of features which can be used to represent entities. These features can be either Boolean (e.g. `there is a garden`), ordinal (e.g. `crime rate`) or continuous (e.g. `distance to kindergarten`) variables (see Table 4 in the previous example for a list). *Atomic* constraints are constructed from these features, by simply taking their values for Boolean variables, and constraining each ordinal and continuous feature to be below a certain (variable-specific) threshold. More complex constraints can be constructed by arbitrary combinations of these building blocks (e.g. `distance to kindergarten` $< \theta \wedge$ `distance to parents` $< \theta$, so that a car is not needed). We distinguish between *hard constraints*, which define the space of feasible configurations and are assumed to be known in advance, and *soft constraints*, unknown desired features which need to be discovered in the elicitation process and can be traded-off in the search for feasible configurations. The space of constructible constraints is clearly exponential in the size of the catalogue. In the following we show how we deal with this complexity.

Utility Function. The utility function takes as input an instance and returns its utility according to the (learned) DM preference:

$$f(\boldsymbol{O}) = \boldsymbol{w}^T \boldsymbol{\psi}^{(d)}(\boldsymbol{O}) \tag{6}$$

As for the structured-output learning scenario, each instance is represented as a vector of values associated with the Boolean and algebraic constraints, and maximization of the utility is cast as an OMT problem and solved by an OMT-solver. There are however some differences in the feature map $\boldsymbol{\psi}^{(d)}$, which we detail below. First, we assume here that all instance features are in the output, i.e. the input $\boldsymbol{I}$ is empty. This is a reasonable assumption in recommendation scenarios, as shown in the previous example. However, our formulation is almost unchanged if a non-empty input is considered. Note also that inputs could be seen as hard constraints on the valid configurations, and incorporated in the set of hard constraints defining the feasible space. Second, the set of features is not specified in advance, but we consider as features all possible conjunctions and/or disjunctions of up to d atomic constraints (the ones in the catalogue). The maximal degree d contributes to limit the size of the feature space, and is grounded on the bounded rationality of humans [51], who can simultaneously handle only a limited number of features. Note that for the same reason, only very few of these candidate features will actually be considered by the DM, so that the utility function will be extremely sparse, with most weights set to zero. This will be accounted for when introducing the learning stage. Third, each feature is computed as the indicator function of the corresponding formula, i.e. $\psi_k(\boldsymbol{O}) := \mathbb{1}_k(\boldsymbol{O})$ regardless of whether the formula containts algebraic atoms or not. The reason for this simplifying choice is that current OMT solvers can only address *linear* cost functions in an efficient way, while combinations of algebraic constraints could generate non-linear functions (e.g. products for conjunctions). The method can be generalized to continuous features by either using mappings like the minimum or the Lukasiewicz t-norms, which current OMT solvers can handle, or leveraging on the research on hybrid non-linear arithmetics [28,41,46], when the solvers will reach the desired level of maturity.

Learning Phase. Learning consists in finding the weights for the utility function matching the unknown DM preferences. Training examples for this phase consist of candidate instances with their evaluation from the DM (see the preference elicitation phase further down). Asking quantitative feedback such as real-valued scores is typically out of reach of human DM. A more realistic scenario consists of asking the DM to rank solutions by preference. We can thus formulate the problem as *learning to rank*, where the task is learning a function returning the same ranking as the one provided by the DM. We focus on the adaptation of SVM for ranking [39], which assumes pairwise ranking preferences, and enforces a (soft) large margin between the two predictions. However, we have an additional requirement, which is the sparsity in the feature weights. Indeed, the feature vector contains all possible constraints (up to a certain complexity), and the learning phase should also perform some form of *constraint learning* by selecting a small set of relevant ones. We favour this behaviour by replacing the 2-norm of SVM with a 1-norm, which is a sparsifying norm encouraging solutions with few non-zero weights [29]. The resulting learning problem is:

$$\min_{w,\xi \geq 0} \quad ||w||_1 + C \sum_{O^i \prec O^j} \xi_{i,j} \tag{7}$$

$$\text{subject to:} \quad w^T(\psi^{(d)}(O^i) - \psi^{(d)}(O^j)) \geq \Delta(O^i, O^j) - \xi_{i,j}$$
$$\forall\, O^i \prec O^j \in \mathcal{D}$$

where $O^i \prec O^j$ indicates that output O^i is ranked before O^j in the DM preference. Constraints enforce pairwise rankings to match DM preferences. A linear penalty is added to the objective function when a less preferred solution gets a score which is not sufficiently smaller than the more preferred one.

Preference Elicitation Phase. The ultimate goal of the algorithm is returning the DM the best possible instance given her utility function. However, given that the utility function is unknown, a preference elicitation phase will be needed in order to gather information on DM preference and use it to refine the current approximation $\hat{f}$ of her utility. In collecting candidates for feedback, one should consider the following principles:

1. the generation of top-quality configurations, consistent with the learnt DM preferences;
2. the generation of diversified configurations, i.e., alternative possibly suboptimal configurations with respect to the learnt utility $\hat{f}$;
3. the search for the DM features which were not recovered by the current approximation $\hat{f}$, i.e., features not appearing in any of the terms in $\hat{f}$.

The rationale for the first principle is focusing on the relevant areas of the utility surface, those of interest to the DM. As a matter of fact, a preference elicitation system that asks to rank low quality configurations will be likely considered useless or annoying by the DM [35]. The second principle favours the exploration of the relationships among the features recovered by the current preference model $\hat{f}$. Finally, as the learnt formulation of $\hat{f}$ may miss some of the user decisional features, their search is promoted by the third principle.

Based on the above principles, our active learning strategy works as follows. First, $\hat{f}$ is maximized (first principle), generating the first candidate configuration O^*. Then, a *hard* constraint is added to the OMT problem as the disjunction of all features not satisfied by O^*, and maximization is run again. This accounts for the second principle, by enforcing a new solution O^{**} which differs from O^* by at least one feature. Finally, each unassigned feature[5] in both O^* and O^{**} is given a random value in its domain, thus incorporating the third principle. Indeed, if these features are truly irrelevant for the DM, setting them at random

---

[5] Unassigned features are catalogue features not appearing in any hard constraint or non-zero weight soft constraint.

Data: Set of catalogue features $\psi_1, \ldots, \psi_m$
Result: Most preferred solution O^*
/* Initialization   */
1 Select three configurations uniformly at random
2 $\mathcal{D} \leftarrow$ ranking of configurations by DM
/* Refinement   */
3 **while** *termination criterion* **do**
        /* learning   */
4

$$\hat{f} \leftarrow \operatorname*{argmin}_{w, \xi \geq 0} \|w\|_1 + C \sum_{O^i \prec O^j} \xi_{i,j}$$
$$\text{s.t.} \ w^T(\psi^{(d)}(O^i) - \psi^{(d)}(O^j)) \geq \Delta(O^i, O^j) - \xi_{i,j}$$
$$\forall \ O^i \prec O^j \in \mathcal{D}$$

        /* preference elicitation   */
5   Collect two configurations optimizing $\hat{f}$
6        $\mathcal{D} \leftarrow \mathcal{D} \cup$ ranking of configurations by DM
7 **end**
   /* final recommendation   */
8 **return** $\operatorname{argmax} \hat{f}$

Algorithm 2. Algorithm for preference elicitation with LMT

should not affect the evaluation of the candidate solutions. If on the other hand some of them are needed to explain the DM preferences, driving their elicitation can allow to identify the deficiencies of the current approximation $\hat{f}$ and recover previously discarded relevant features.

Overall Algorithm. The pseudocode of the full algorithm is shown in Algorithm 2. It takes as input a set of catalogue features (the atomic constraints) and returns the solution which is most preferred to the DM. In the initialization phase, it selects three configurations uniformly at random and asks the DM for feedback on them. The training dataset is initialized with pairwise preferences between these configurations. Then a refinement loop begins, where at each iteration the utility function is first refined using the current feedback, and then used to generate candidates on which to elicit additional feedback. The first step consists in solving the learning to rank problem in Eq. (7), where $\mathcal{D}$ is the dataset of all pairwise preferences collected so far. The regularization parameter C is set to one in the first iteration, and fine-tuned by an internal cross validation on the training set in the following ones. With a slight abuse of notation, we write $\hat{f} \leftarrow \operatorname{argmin}$ to indicate that $\hat{f}$ is the function whose weights w are the result of the minimization. The second step is the preference elicitation phase, where the current utility $\hat{f}$ is used to generate novel candidates according to the up-mentioned rules (see the preference elicitation phase paragraph). The dataset $\mathcal{D}$ is updated according to the feedback the DM gives on these candidates. The process terminates when a stopping condition is met. Being an interactive process involving a human DM, the most obvious termination condition is the DM satisfaction on the current recommendation. Additional conditions could be conceived, for instance, by estimating the improvement one could expect by further refining the utility function.

4.3 Experimental Results

We show the results of the preference elicitation method on a *housing* problem, along the lines of the example in Sect. 4.1. There are different locations available, characterized by different housing values, prices, constraints about the design of the building (e.g., usually in the city center you cannot have a family house with a huge garden and pool), etc. Table 5 reports the full set of catalogue features characterizing locations. As previously specified, atomic constraints are constructed by taking values for Boolean variables and thresholding numeric ones, and the feature vector is constructed with all possible combinations of up to d atomic constraints (we focused on conjunctions of literals and used $d = 3$ in the experiments).

Table 5. Catalogue features for the housing problem.

Num	Feature	Type
1	House type	Ordinal
2	Garden	Boolean
3	Garage	Boolean
4	Commercial facilities in the neighborhood	Boolean
5	Public green areas in the neighborhood	Boolean
6	Cycling and walking facilities in the neighborhood	Boolean
7	Distance from downtown	Continuous
8	Crime rate	Continuous
9	Location-based taxes and fees	Continuous
10	Public transit service quality index	Continuous
11	Distance from high schools	Continuous
12	Distance from nearest free parking	Continuous
13	Distance from working place	Continuous
14	Distance from parents house	Continuous
15	Price	Continuous

Feasible housing locations are defined by a set of hard constraints (Table 6). These hard constraints are stated by the customer (e.g., cost bounds) or by the company (e.g., constraints about the distance of the available locations from user-defined points of interest). Let us note that constraints 5, 6, 7 define a linear bi-objective problem among distances from user-defined points of interest. Prices of potential housing locations are defined as a function of the other features. For example, price increases if a semi-detached house rather than a flat is selected or in the case of green areas in the neighborhood. On the other hand, e.g., when crime index of potential locations increases, price decreases.

Table 6. Hard feasibility constraints for the housing problem. Parameters ρ_i, $i = 1 \ldots 13$, are threshold values specified by the user or by the sales personnel, depending on who states the hard constraint which they refer to.

Num	Hard constraint
1	Price $\leq \rho_1$
2	Location-based taxes and fees $\leq \rho_2$ => *not* public green ares in the neighborhood *and not* public transit service quality index $\leq \rho_3$
3	Commercial facilities in the neighborhood => *not* (garden *and* garage)
4	Crime rate $\leq \rho_4$ => distance from downtown $\geq \rho_5$
5	Distance from working place + distance from parents house $\geq \rho_6$
6	Distance from working place + distance from high schools $\geq \rho_7$
7	Distance from parents house + distance from high schools $\geq \rho_8$
8	Distance from nearest free parking $\leq \rho_9$ => *not* public green areas in the neighborhood
9	Distance from parents house $\leq \rho_{10}$ => distance from downtown $\geq \rho_{11}$ *and* crime rate $\geq \rho_{12}$
10	Garden => house type $\geq \rho_{13}$

We generated a set of 40 literals (i.e. atomic constraints). The target utility function is composed of (soft) constraints combining two or three literals, with at least one combination containing three literals. We assume the maximum number of literals per constraint (three) to be known in advance. Constraint weights are integer values selected uniformly at random in the range $[1, 100]$. Inaccurate feedback from the DM is modelled by a Probit model, a standard noise model in which each evaluation from the DM is assumed to be affected by an additive i.i.d. zero-mean Gaussian noise.

Figure 9 reports the results over a benchmark of 400 randomly generated utility functions for each of the following (*number of literals, number of constraints*) pairs: $\{(5,3),(10,6),(15,9)\}$. Note that this is a very challenging problem due to complex non-linear interactions among the decisional features. When increasing the number of queries asked, the quality of the solution rapidly improves and the algorithm identifies the DM preferred configuration in all the cases. On average, 22 and 69 queries are needed by the algorithm to converge to the DM preferred

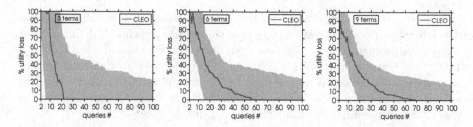

Fig. 9. Performance of LMT for preference elicitation in the housing problem. The y-axis reports the percentage utility loss, while the x-axis contains the number of pairwise-comparison queries asked so far. The curve reports the median values observed over 400 runs of the algorithm, while the shaded area denotes the range between the 25th and the 75th percentiles of the observations.

solution in the case of three and nine[6] constraints respectively. A problem with the current version of the algorithm is the high variance of the performance. This is mainly due to some runs where the elicitation procedure fails to improve the quality of the utility model. As discussed in the conclusions, we plan to adopt smarter query selection strategies to address this issue, possibly also reducing the number of queries to the DM.

4.4   Related Work

The problem of automatically learning utility functions and eliciting preferences from users is widely studied within the Artificial Intelligence community [15,27]. A popular approach consists of modeling the uncertainty about the DM preferences in terms of a set of candidate hypotheses, among which the DM utility is expected to be. An interactive process is conducted in which feedback from the DM is elicited and used to restrict the hypothesis space until the DM is satisfied with the proposed solution. In statistical decision theory, the *minimax regret* criterion [58] is a way to make decisions under uncertainty. Given a certain decision O, the maximum regret is the difference in utility between the DM most preferred solution O^* and O assuming the worst-case scenario, where the DM utility is the one in the feasible set for which this difference is maximal. The minimax regret criterion prescribes to make the decision minimizing this regret. A recent line of research [13,14,16] adapts the minimax regret principle to preference elicitation. Queries to be asked to the DM are selected so as to reduce the minimax regret by restricting the feasible hypothesis set. An advantage of minimax regret approaches with respect to our formulation is that they can provide theoretical guarantees in terms of bounds on the solution quality and convergence to provably-optimal results. On the other hand, these approaches assume perfect feedback from the DM and are not suitable to deal

---

[6] DM utility functions involving nine complex constraints are quite unrealistic and are considered here just for testing the scalability of the algorithm.

with the imprecise and contradictory information which is typical of interactions with human DM.

An alternative solution is that of Bayesian approaches to preference elicitation [11,12,35,67] which define a probability distribution over the hypotheses and query the DM trying to increase the posterior probability of her utility. These approaches can naturally handle the uncertainty in both utility and feedback. Furthermore, the probabilistic framework allows to reason in terms of potential informativeness of queries and select the maximally informative one. The maximum expected loss of taking a decision O is the maximum expected reduction in utility when choosing O instead of the DM most preferred solution O^*, where expectation is taken over the probability distribution of the utility functions. The *expected value of information* criterion prescribes to choose the query producing the largest expected reduction in maximum expected loss. Exact computation of the expected value of information, as well as exact computation of the posterior distribution over utility functions given the feedback, are extremely expensive. Approximate solutions have been proposed in the literature [35]. These approaches are conceived for instances characterized by purely discrete attributes, and cannot be easily extended to hybrid ones. Furthermore, experimental comparisons on purely Boolean datasets showed that even these approximate versions cannot compete with our LMT algorithm on problems with a substantial degree of non-linearity (see [19] for the details). However, adapting Bayesian preference elicitation approaches to our framework is a relevant research direction, as discussed in the conclusions of the chapter.

Finally, closest to our approach from a constraint solving viewpoint is the work on preference elicitation with soft constraints [31]. This work assumes local preferences in terms of weights assigned to each configuration of the variables of a constraint. An algebraic structure called c-semiring allows to combine local preferences into a global one for an entire solution and assign a partial order over candidate solutions. Missing preference values are inferred by an elicitation procedure. In terms of expressiveness, semiring-based soft constraint satisfaction problems can be encoded into weighted MAX-SAT formulations [48]. A weighted MAX-SMT encoding can be used in case of continuous variables. In terms of interaction with the DM, the work assumes perfect knowledge of constraint structure, with missing information limited to part of the local preferences, and quantitative (possibly interval-based [30]) local feedback over assignments for specific soft constraints. Our formulation assumes complete ignorance of constraint structure, limiting initial information to the feature catalogue, and relies on more affordable global pairwise preferences as feedback. Finally, inconsistent preference information is not handled, while our approach trades off preference fitting with complexity of the learned utility model.

5   Conclusions

We described Learning Modulo Theories as an expressive framework for learning in hybrid domains characterized by combinations of Boolean and numerical features and relationships between them. We showed how to cast the problem as a structured-output learning task, in which we learn to generate novel configurations from a training set of correct instances, and a preference elicitation task, where an interactive process with a decision maker suggests candidate configurations and progressively refines the learned utility model according to the feedback received. Both algorithms automatically identify the set of relevant constraints and their weights starting from a catalogue of candidates.

The framework can be extended in a number of directions. First, LMT is currently focused on the task of finding the maximal configuration, and cannot compute marginal probabilities. This gap can be filled thanks to *weighted model integration* [7,8], a recently proposed formulation generalizing weighted model counting [20] to hybrid domains. In terms of expressiveness, OMT is currently limited to quantifier free formulae and linear algebra. Some attempts to extend SMT solvers to quantified formulae [5,43,57] and to non-linear arithmetic [28,41,46] have been presented in the literature. Although the state of the art of these extensions is not yet satisfactory, these techniques are evolving rapidly and we can rather easily extend our framework in these directions as soon as the underlying SMT technology is mature enough. In terms of required knowledge, the current formulation for structured-output prediction assumes knowledge of the desired output for training examples. This requirement can be loosened by introducing latent variables for the unobserved part of the output, to be maximized over during training [71]. In terms of interaction with the user, the preference elicitation strategy we employ is quite simple, and more advanced approaches estimating the value of information of candidate queries could be explored, see e.g. the work by Viappiani and Boutilier on optimal recommendation sets [68]. Adapting these concepts to deal with complex non-linear hybrid formulas is an interesting and challenging direction for future work. Finally, albeit capable of automatically identifying relevant constraints, our approach requires a set of candidates (possibly only atomic ones) to start from and cannot generate completely novel constraints. Adapting constraint acquisition techniques [6,10,45,55] to our framework is a relevant and challenging direction for future research.

Acknowledgments. This chapter builds on a body of work done in collaboration with Paolo Campigotto, Roberto Battiti, Stefano Teso and Roberto Sebastiani.

References

1. Ahmadi, B., Kersting, K., Sanner, S.: Multi-evidence lifted message passing, with application to pagerank and the kalman filter. In: Proceedings of IJCAI 2011, pp. 1152–1158 (2011)
2. Bakir, G.H., Hofmann, T., Schölkopf, B., Smola, A.J., Taskar, B., Vishwanathan, S.V.N.: Predicting Structured Data (Neural Information Processing). The MIT Press, Cambridge (2007)
3. Barrett, C., Sebastiani, R., Seshia, S.A., Tinelli, C.: Satisfiability modulo theories, chap. 26, Frontiers in Artificial Intelligence and Applications, pp. 825–885. IOS Press, February 2009

4. Battiti, R., Brunato, M.: Reactive search optimization: learning while optimizing. In: Gendreau, M., Potvin, J.-Y. (eds.) Handbook of Metaheuristics. International Series in Operations Research & Management Science, vol. 146, pp. 543–571. Springer, New York (2010)
5. Baumgartner, P., Tinelli, C.: Model evolution with equality modulo built-in theories. In: Bjørner, N., Sofronie-Stokkermans, V. (eds.) CADE 2011. LNCS (LNAI), vol. 6803, pp. 85–100. Springer, Heidelberg (2011). doi:10.1007/978-3-642-22438-6_9
6. Beldiceanu, N., Simonis, H.: A model seeker extracting global constraint models from positive examples. In: Milano, M. (ed.) CP 2012. LNCS, vol. 7514, pp. 141–157. Springer, Heidelberg (2012). doi:10.1007/978-3-642-33558-7_13
7. Belle, V., Passerini, A., Van den Broeck, G.: Probabilistic inference in hybrid domains by weighted model integration. In: Proceedings of 24th International Joint Conference on Artificial Intelligence (IJCAI) (2015)
8. Belle, V., Van den Broeck, G., Passerini, A.: Hashing-based approximate probabilistic inference in hybrid domains. In: Proceedings of the 31st Conference on Uncertainty in Artificial Intelligence (UAI) (2015)
9. Bengio, Y.: Learning deep architectures for AI. Found. Trends Mach. Learn. 2(1), 1–127 (2009)
10. Bessiere, C., Coletta, R., Hebrard, E., Katsirelos, G., Lazaar, N., Narodytska, N., Quimper, C.-G., Walsh, T.: Constraint acquisition via partial queries. In: Proceedings of the Twenty-Third International Joint Conference on Artificial Intelligence, IJCAI 2013, pp. 475–481. AAAI Press (2013)
11. Birlutiu, A., Groot, P., Heskes, T.: Efficiently learning the preferences of people. Mach. Learn. 90, 1–28 (2012)
12. Bonilla, E., Guo, S., Sanner, S.: Gaussian process preference elicitation. In: Lafferty, J., Williams, C.K.I., Shawe-Taylor, J., Zemel, R.S., Culotta, A. (eds.) Advances in Neural Information Processing Systems, pp. 262–270 (2010)
13. Boutilier, C., Patrascu, R., Poupart, P., Schuurmans, D.: Constraint-based optimization and utility elicitation using the minimax decision criterion. Artif. Intell. 170(8–9), 686–713 (2006)
14. Boutilier, C., Regan, K., Viappiani, P.: Simultaneous elicitation of preference features and utility. In: Proceedings of the Twenty-Fourth AAAI Conference on Artificial Intelligence (AAAI 2010), Atlanta, GA, USA. AAAI Press, pp. 1160–1167, July 2010
15. Braziunas, D.: Computational approaches to preference elicitation. Technical report, Department of Computer Science, University of Toronto (2006)
16. Braziunas, D., Boutilier, C.: Minimax regret based elicitation of generalized additive utilities. In: Proceedings of the Twenty-Third Conference on Uncertainty in Artificial Intelligence (UAI 2007), Vancouver, pp. 25–32 (2007)
17. Broecheler, M., Mihalkova, L., Getoor, L.: Probabilistic similarity logic. In: Uncertainty in Artificial Intelligence (UAI), pp. 73–82 (2010)
18. Burges. C.: A tutorial on support vector machines for pattern recognition. In: Data Mining and Knowledge Discovery, vol. 2. Kluwer Academic Publishers, Boston (1998)
19. Campigotto, P., Battiti, R., Passerini, A.: Learning modulo theories for preference elicitation in hybrid domains. arXiv (2015)
20. Chavira, M., Darwiche, A.: On probabilistic inference by weighted model counting. Artif. Intell. 172(6–7), 772–799 (2008)

21. Choi, J., Amir, E.: Lifted relational variational inference. In: de Freitas, N., Murphy, K.P. (eds.) UAI 2012: Proceedings of the Twenty-Eighth Conference on Uncertainty in Artificial Intelligence, pp. 196–206. AUAI Press (2012)

22. Choi, J., Guzmn-Rivera, A., Amir, E.: Lifted relational kalman filtering. In: Proceedings of IJCAI 2011, pp. 2092–2099 (2011)

23. Choi, J., Hill, D., Amir, E.: Lifted inference for relational continuous models. In: UAI 2010: Proceedings of the Twenty-Sixth Conference on Uncertainty in Artificial Intelligence (2010)

24. Cimatti, A., Franzén, A., Griggio, A., Sebastiani, R., Stenico, C.: Satisfiability modulo the theory of costs: foundations and applications. In: Esparza, J., Majumdar, R. (eds.) TACAS 2010. LNCS, vol. 6015, pp. 99–113. Springer, Heidelberg (2010). doi:10.1007/978-3-642-12002-2_8

25. Cimatti, A., Griggio, A., Schaafsma, B.J., Sebastiani, R.: A modular approach to MaxSAT modulo theories. In: Järvisalo, M., Van Gelder, A. (eds.) SAT 2013. LNCS, vol. 7962, pp. 150–165. Springer, Heidelberg (2013). doi:10.1007/978-3-642-39071-5_12

26. Diligenti, M., Gori, M., Maggini, M., Rigutini, L.: Bridging logic and kernel machines. Mach. Learn. 86(1), 57–88 (2012)

27. Domshlak, C., Hüllermeier, E., Kaci, S., Prade, H.: Preferences in AI: an overview. Artif. Intell. 175(7–8), 1037–1052 (2011)

28. Fränzle, M., Herde, C., Teige, T., Ratschan, S., Schubert, T.: Efficient solving of large non-linear arithmetic constraint systems with complex boolean structure. JSAT 1(3–4), 209–236 (2007)

29. Friedman, J., Hastie, T., Rosset, S., Tibshirani, R.: Discussion of boosting papers. Ann. Stat. 32, 102–107 (2004)

30. Gelain, M., Pini, M., Rossi, F., Venable, K., Wilson, N.: Interval-valued soft constraint problems. Ann. Math. Artif. Intell. 58, 261–298 (2010)

31. Gelain, M., Pini, M.S., Rossi, F., Venable, K.B., Walsh, T.: Elicitation strategies for soft constraint problems with missing preferences: properties, algorithms and experimental studies. Artif. Intell. J. 174(3–4), 270–294 (2010)

32. Getoor, L., Taskar, B.: Introduction to Statistical Relational Learning (Adaptive Computation and Machine Learning). The MIT Press, Cambridge (2007)

33. Goodman, N.D., Mansinghka, V.K., Roy, D.M., Bonawitz, K., Tenenbaum, J.B.: Church: a language for generative models. In: McAllester, D.A., Myllymäki, P. (eds.) UAI, pp. 220–229. AUAI Press (2008)

34. Griggio, A., Phan, Q.-S., Sebastiani, R., Tomasi, S.: Stochastic local search for SMT: combining theory solvers with WalkSAT. In: Tinelli, C., Sofronie-Stokkermans, V. (eds.) FroCoS 2011. LNCS (LNAI), vol. 6989, pp. 163–178. Springer, Heidelberg (2011). doi:10.1007/978-3-642-24364-6_12

35. Guo, S., Sanner, S.: Real-time multiattribute Bayesian preference elicitation with pairwise comparison queries. J. Mach. Learn. Res. - Proc. Track 9, 289–296 (2010)

36. Gutmann, B., Jaeger, M., Raedt, L.: Extending ProbLog with continuous distributions. In: Frasconi, P., Lisi, F.A. (eds.) ILP 2010. LNCS (LNAI), vol. 6489, pp. 76–91. Springer, Heidelberg (2011). doi:10.1007/978-3-642-21295-6_12

37. Hausner, A.: Simulating decorative mosaics. In: Proceedings of the 28th Annual Conference on Computer Graphics and Interactive Techniques, SIGGRAPH 2001, pp. 573–580. ACM, New York (2001)

38. Islam, M.A., Ramakrishnan, C.R., Ramakrishnan, I.V.: Inference in probabilistic logic programs with continuous random variables. Theory Pract. Log. Program. 12(4–5), 505–523 (2012)

39. Joachims, T.: Optimizing search engines using click through data. In: Proceedings of the Eighth ACM SIGKDD International Conference on Knowledge Discovery and Data Mining, KDD 2002, pp. 133–142. ACM, New York (2002)

40. Joachims, T., Finley, T., Chun-Nam John, Y.: Cutting-plane training of structural SVMs. Mach. Learn. **77**(1), 27–59 (2009)

41. Jovanović, D., Moura, L.: Solving non-linear arithmetic. In: Gramlich, B., Miller, D., Sattler, U. (eds.) IJCAR 2012. LNCS (LNAI), vol. 7364, pp. 339–354. Springer, Heidelberg (2012). doi:10.1007/978-3-642-31365-3_27

42. Koller, D., Friedman, N.: Probabilistic Graphical Models: Principles and Techniques - Adaptive Computation and Machine Learning. The MIT Press, Cambridge (2009)

43. Kruglov, E.: Superposition modulo theory. Ph.D. thesis, Universität des Saarlandes, Postfach 151141, 66041 Saarbrücken (2013)

44. Kuželka, O., Szabóová, A., Holec, M., Železný, F.: Gaussian logic for predictive classification. In: Gunopulos, D., Hofmann, T., Malerba, D., Vazirgiannis, M. (eds.) ECML PKDD 2011. LNCS (LNAI), vol. 6912, pp. 277–292. Springer, Heidelberg (2011). doi:10.1007/978-3-642-23783-6_18

45. Lallouet, A., Lopez, M., Martin, L., Vrain, C.: On learning constraint problems. In: ICTAI (1) 2010, pp. 45–52 (2010)

46. Larraz, D., Oliveras, A., Rodríguez-Carbonell, E., Rubio, A.: Minimal-model-guided approaches to solving polynomial constraints and extensions. In: Sinz, C., Egly, U. (eds.) SAT 2014. LNCS, vol. 8561, pp. 333–350. Springer, Heidelberg (2014). doi:10.1007/978-3-319-09284-3_25

47. LeCun, Y., Chopra, S., Hadsell, R., Ranzato, M., Huang, F.-J.: A tutorial on energy-based learning. In: Bakir, G., Hofman, T., Schölkopf, B., Smola, A., Taskar, B. (eds.) Predicting Structured Data. MIT Press, Cambridge (2006)

48. Leenen, L., Anbulagan, A., Meyer, T., Ghose, A.: Modeling and solving semiring constraint satisfaction problems by transformation to weighted semiring Max-SAT. In: Orgun, M.A., Thornton, J. (eds.) AI 2007. LNCS (LNAI), vol. 4830, pp. 202–212. Springer, Heidelberg (2007). doi:10.1007/978-3-540-76928-6_22

49. Li, Y., Albarghouthi, A., Kincad, Z., Gurfinkel, A., Chechik, M.: Symbolic optimization with SMT solvers. In: Proceedings of the 41st ACM SIGPLAN-SIGACT Symposium on Principles of Programming Languages, POPL 2014, pp. 607–618. ACM, New York (2014)

50. Lippi, M., Frasconi, P.: Prediction of protein-residue contacts by markov logic networks with grounding-specific weights. Bioinformatics **25**(18), 2326–2333 (2009)

51. March, J.G.: Bounded rationality, ambiguity, and the engineering of choice. Bell J. Econ. **9**, 587–608 (1978)

52. McCormack, J., d'Inverno, M. (eds.): Computers and Creativity. Springer, Heidelberg (2012)

53. Nieuwenhuis, R., Oliveras, A.: On SAT modulo theories and optimization problems. In: Biere, A., Gomes, C.P. (eds.) SAT 2006. LNCS, vol. 4121, pp. 156–169. Springer, Heidelberg (2006). doi:10.1007/11814948_18

54. Nrman, P., Buschle, M., Knig, J., Johnson, P.: Hybrid probabilistic relational models for system quality analysis. In: EDOC, pp. 57–66. IEEE Computer Society (2010)

55. Ravkic, I., Ramon, J., Davis, J.: Learning relational dependency networks in hybrid domains. Mach. Learn. **100**(2), 217–254 (2015)

56. Ricci, F., Rokach, L., Shapira, B., Kantor, P.B.: Recommender Systems Handbook, 1st edn. Springer, New York (2010)

57. Rümmer, P.: A constraint sequent calculus for first-order logic with linear integer arithmetic. In: Cervesato, I., Veith, H., Voronkov, A. (eds.) LPAR 2008. LNCS (LNAI), vol. 5330, pp. 274–289. Springer, Heidelberg (2008). doi:10.1007/978-3-540-89439-1_20

58. Savage, L.J.: The theory of statistical decision. J. Am. Stat. Assoc. **46**(253), 55–67 (1951)

59. Sebastiani, R.: Lazy satisfiability modulo theories. J. Satisfiability, Boolean Model. Comput., JSAT **3**(3–4), 141–224 (2007)

60. Sebastiani, R., Tomasi, S.: Optimization in SMT with $\mathcal{LA}(\mathbb{Q})$ cost functions. In: Gramlich, B., Miller, D., Sattler, U. (eds.) IJCAR 2012. LNCS (LNAI), vol. 7364, pp. 484–498. Springer, Heidelberg (2012). doi:10.1007/978-3-642-31365-3_38

61. Sebastiani, R., Tomasi, S.: Optimization modulo theories with linear rational costs. ACM Trans. Comput. Logic **16**(2), 12:1–12:43 (2015)

62. Sebastiani, R., Trentin, P.: OptiMathSAT: a tool for optimization modulo theories. In: Kroening, D., Păsăreanu, C.S. (eds.) CAV 2015. LNCS, vol. 9206, pp. 447–454. Springer, Heidelberg (2015). doi:10.1007/978-3-319-21690-4_27

63. Shawe-Taylor, J., Cristianini, N.: Kernel Methods for Pattern Analysis. Cambridge University Press, New York (2004)

64. Teso, S., Sebastiani, R., Passerini, A.: Structured learning modulo theories. Artif. Intell. (2015)

65. Tsochantaridis, I., Joachims, T., Hofmann, T., Altun, Y.: Large margin methods for structured and interdependent output variables. J. Mach. Learn. Res. **6**, 1453–1484 (2005)

66. Vanderplaats, G.N.: Multidiscipline Design Optimization: Textbook. Vanderplaats Research & Development, Incorporated, Colorado Springs (2007)

67. Viappiani, P.: Monte Carlo methods for preference learning. In: Hamadi, Y., Schoenauer, M. (eds.) LION 6. LNCS, vol. 7219, pp. 503–508. Springer, Heidelberg (2012). doi:10.1007/978-3-642-34413-8_52

68. Viappiani, P., Boutilier, C.: Optimal Bayesian recommendation sets and myopically optimal choice query sets. In: Advances in Neural Information Processing Systems 23 (NIPS), Vancouver, pp. 2352–2360 (2010)

69. Wang, J., Domingos, P.: Hybrid Markov logic networks. In: Proceedings of the 23rd National Conference on Artificial Intelligence, vol. 2, AAAI 2008, pp. 1106–1111. AAAI Press (2008)

70. Yang, Y.-L., Wang, J., Vouga, E., Wonka, P.: Urban pattern: layout design by hierarchical domain splitting. ACM Trans. Graph. **32**(6), 181:1–181:12 (2013)

71. Yu, C.-N.J., Joachims, T.: Learning structural SVMs with latent variables. In: Proceedings of the 26th Annual International Conference on Machine Learning, ICML 2009, pp. 1169–1176. ACM, New York (2009)

72. Yu, L.-F., Yeung, S.-K., Tang, C.-K., Terzopoulos, D., Chan, T.F., Osher, S.J.: Make it home: automatic optimization of furniture arrangement. ACM Trans. Graph. **30**(4), 86:1–86:12 (2011)

Learning to Solve

Algorithm Selection for Combinatorial Search Problems: A Survey

Lars Kotthoff[(✉)]

University of British Columbia, Vancouver, Canada
larsko@cs.ubc.ca

Abstract. The Algorithm Selection Problem is concerned with selecting the best algorithm to solve a given problem on a case-by-case basis. It has become especially relevant in the last decade, as researchers are increasingly investigating how to identify the most suitable existing algorithm for solving a problem instead of developing new algorithms. This survey presents an overview of this work focusing on the contributions made in the area of combinatorial search problems, where Algorithm Selection techniques have achieved significant performance improvements. We unify and organise the vast literature according to criteria that determine Algorithm Selection systems in practice. The comprehensive classification of approaches identifies and analyses the different directions from which Algorithm Selection has been approached. This chapter contrasts and compares different methods for solving the problem as well as ways of using these solutions.

1 Introduction

For many years, Artificial Intelligence research has been focusing on inventing new algorithms and approaches for solving similar kinds of problems. In some scenarios, a new algorithm is clearly superior to previous approaches. In the majority of cases however, a new approach will improve over the current state of the art only for some problems. This may be because it employs a heuristic that fails for problems of a certain type or because it makes other assumptions about the problem or environment that are not satisfied in some cases. Selecting the most suitable algorithm for a particular problem aims at mitigating these problems and has the potential to significantly increase performance in practice. This is known as the Algorithm Selection Problem.

The Algorithm Selection Problem has, in many forms and with different names, cropped up in many areas of Artificial Intelligence in the last few decades. Today there exists a large amount of literature on it. Most publications are concerned with new ways of tackling this problem and solving it efficiently in practice. Especially for combinatorial search problems, the application of Algorithm Selection techniques has resulted in significant performance improvements that leverage the diversity of systems and techniques developed in recent years. This chapter surveys the available literature and describes how research has progressed.

© Springer International Publishing AG 2016
C. Bessiere et al. (Eds.): Data Mining and Constraint Programming, LNAI 10101, pp. 149–190, 2016.
DOI: 10.1007/978-3-319-50137-6_7

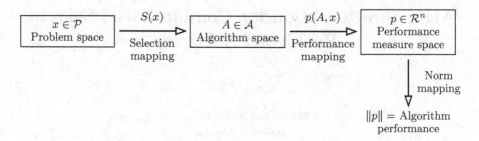

Fig. 1. Basic model for the Algorithm Selection Problem as published in [120].

Researchers have long ago recognised that a single algorithm will not give the best performance across all problems one may want to solve and that selecting the most appropriate method is likely to improve the overall performance. Empirical evaluations have provided compelling evidence for this, e.g. [1,154].

The original description of the Algorithm Selection Problem was published in [120]. The basic model described in the paper is very simple – given a space of problems and a space of algorithms, map each problem-algorithm pair to its performance. This mapping can then be used to select the best algorithm for a given problem. The original figure that illustrates the model is reproduced in Fig. 1. As Rice states,

> "The objective is to determine $S(x)$ [the mapping of problems to algorithms] so as to have high algorithm performance."

He identifies the following four criteria for the selection process.

1. Best selection for all mappings $S(x)$ and problems x. For every problem, an algorithm is chosen to give maximum performance.
2. Best selection for a subclass of problems. A single algorithm is chosen to apply to each of a subclass of problems such that the performance degradation compared to choosing from all algorithms is minimised.
3. Best selection from a subclass of mappings. Choose the selection mapping from a subset of all mappings from problems to algorithms such that the performance degradation is minimised.
4. Best selection from a subclass of mappings and problems. Choose a single algorithm from a subset of all algorithms to apply to each of a subclass of problems such that the performance degradation is minimised.

The first case is clearly the most desirable one. In practice however, the other cases are more common – we might not have enough data about individual problems or algorithms to select the best mapping for everything.

[120] lists five main steps for solving the problem.

Formulation. Determination of the subclasses of problems and mappings to be used.

Existence. Does a best selection mapping exist?

Uniqueness. Is there a unique best selection mapping?

Characterization. What properties characterize the best selection mapping and serve to identify it?

Computation. What methods can be used to actually obtain the best selection mapping?

This framework is taken from the theory of approximation of functions. The questions for existence and uniqueness of a best selection mapping are usually irrelevant in practice. As long as a *good* performance mapping is found and improves upon the current state of the art, the question of whether there is a different mapping with the same performance or an even better mapping is secondary. While it is easy to determine the theoretically best selection mapping on a set of given problems, casting this mapping into a *generalisable* form that will give good performance on new problems or even into a form that can be used in practice is hard. Indeed, [62] shows that the Algorithm Selection Problem in general is undecidable. It may be better to choose a mapping that generalises well rather than the one with the best performance. Other considerations can be involved as well. [28,63] compare different Algorithm selection models and select not the one with the best performance, but one with good performance that is also easy to understand, for example. [146] select their method of choice for the same reason. Similarly, [159] choose a model that is cheap to compute instead of the one with the best performance. They note that,

> "All of these techniques are computationally more expensive than ridge regression, and in our previous experiments we found that they did not improve predictive performance enough to justify this additional cost."

Rice continues by giving practical examples of where his model applies. He refines the original model to include features of problems that can be used to identify the selection mapping. The original figure depicting the refined model is given in Fig. 2. This model, or a variant of it, is what is used in most practical approaches. Including problem features is the crucial difference that often makes an approach feasible.

For each problem in a given set, the features are extracted. The aim is to use these features to produce the mapping that selects the algorithm with the best performance for each problem. The actual performance mapping for each problem-algorithm pair is usually of less interest as long as the individual best algorithm can be identified.

Rice poses additional questions about the determination of features.

- What are the best features for predicting the performance of a specific algorithm?
- What are the best features for predicting the performance of a specific class of algorithms?
- What are the best features for predicting the performance of a subclass of selection mappings?

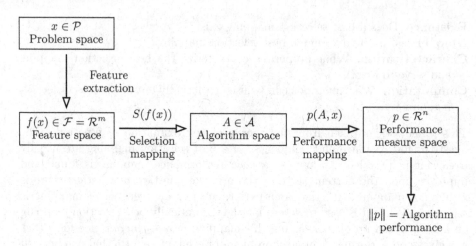

Fig. 2. Refined model for the Algorithm Selection Problem with problem features [120].

He also states that,

> "The determination of the best (or even good) features is one of the most important, yet nebulous, aspects of the algorithm selection problem."

He refers to the difficulty of knowing the problem space. Many problem spaces are not well known and often a sample of problems is drawn from them to evaluate empirically the performance of the given set of algorithms. If the sample is not representative, or the features do not facilitate a good separation of the problem classes in the feature space, there is little hope of finding the best or even a good selection mapping.

[145] note that,

> "While it seems that restricting a heuristic to a special case would likely improve its performance, we feel that the ability to partition the problem space of some $\mathcal{NP}$-hard problems by efficient selectors is mildly surprising."

This sentiment was shared by many researchers and part of the great prominence of Algorithm Selection systems especially for combinatorial search problems can probably be attributed to the surprise that it actually works.

Most approaches employ Machine Learning to learn the performance mapping from problems to algorithms using features extracted from the problems. This often involves a *training phase*, where the candidate algorithms are run on a sample of the problem space to experimentally evaluate their performance. This training data is used to create a *performance model* that can be used to predict the performance on new, unseen problems. The term *model* is used only in the loosest sense here; it can be as simple as a representation of the training data without any further analysis.

1.1   Practical Motivation

[1] notes that in Machine Learning, researchers often perform experiments on a limited number of data sets to demonstrate the performance improvements achieved and implicitly assume that these improvements generalise to other data. He proposes a framework for better experimental evaluation of such claims and deriving rules that determine the properties a data set must have in order for an algorithm to have superior performance. His objective is

"...to derive rules of the form 'this algorithm outperforms these other algorithms on these dependent measures for databases with these characteristics'. Such rules summarize *when* [...] rather than *why* the observed performance difference occurred."

[143] make similar observations and show that there is no algorithm that is universally the best when solving constraint problems. They also demonstrate that the best algorithm-heuristic combination is not what one might expect for some of the surveyed problems. This provides an important motivation for research into performing Algorithm Selection automatically. They close by noting that,

"...research should focus on how to retrieve the most efficient [algorithm-heuristic] combinations for a problem."

The focus of Algorithm Selection is on identifying algorithms with good performance, not on providing explanations for why this is the case. Most publications do not consider the question of "Why?" at all. Rice's framework does not address this question either. The simple reason for this is that explaining the Why? is difficult and for most practical applications not particularly relevant as long as improvements can be achieved. Research into what makes a problem hard, how this affects the behaviour of specific algorithms and how to exploit this knowledge is a fruitful area, but outside the scope of this chapter. However, we present a brief exposition of one of the most important concepts to illustrate its relevance.

The notion of a *phase transition* [26] refers to a sudden change in the hardness of a problem as the value of a single parameter of the problem is changed. Detecting such transitions is an obvious way to facilitate Algorithm Selection. [65] note that,

"In particular, the location of the phase transition point might provide a systematic basis for selecting the type of algorithm to use on a given problem."

While some approaches make use of this knowledge to generate challenging training problems for their systems, it is hardly used at all to facilitate Algorithm Selection. [109] use a set of features that can be used to characterise a phase transition and note that,

"It turns out that [...] this group of features alone suffices to construct reasonably good models."

It remains unclear how relevant phase transitions are to Algorithm Selection in practice. On one hand, their theoretical properties seem to make them highly suitable, but on the other hand almost nobody has explored their use in actual Algorithm Selection systems.

No Free Lunch Theorems. The question arises of whether, in general, the performance of a system can be improved by always picking the best algorithm. The "No Free Lunch" (NFL) theorems [154] state that no algorithm can be the best across all possible problems and that on average, all algorithms perform the same. This seems to provide a strong motivation for Algorithm Selection – if, on average, different algorithms are the best for different parts of the problem space, selecting them based on the problem to solve has the potential to improve performance.

The theorems would apply to Algorithm Selection systems themselves as well though (in particular the version for supervised learning are relevant, see [153]). This means that although performance improvements can be achieved by selecting the right algorithms on one part of the problem space, wrong decisions will be made on other parts, leading to a loss of performance. On average over all problems, the performance achieved by an Algorithm Selection meta-algorithm will be the same as that of all other algorithms.

The NFL theorems are the source of some controversy however. Among the researchers to doubt their applicability is the first proponent of the Algorithm Selection Problem [121]. Several other publications show that the assumptions underlying the NFL may not be satisfied [31,119]. In particular, the distribution of the best algorithms from the portfolio to problems is not random – it is certainly true that certain algorithms are the best on a much larger number of problems than others.

A detailed assessment of the applicability of the NFL theorems to the Algorithm Selection Problem is outside the scope of this chapter. However, a review of the literature suggests that, if the theorems are applicable, the ramifications in practice may not be significant. Most of the many publications surveyed here do achieve performance improvements across a range of different problems using Algorithm Selection techniques. As a research area, it is very active and thriving despite the potentially negative implications of the NFL.

1.2   Scope and Related Work

Algorithm Selection is a very general concept that applies not only in almost all areas of Computer Science, but also other disciplines. However, it is especially relevant in many areas of Artificial Intelligence. This is a large field itself though and surveying all Artificial Intelligence publications that are relevant to Algorithm Selection in a single chapter is infeasible.

In this chapter, we focus on Algorithm Selection for *combinatorial search problems*. This is a large and important subfield of Artificial Intelligence where Algorithm Selection techniques have become particularly prominent in recent years because of the impressive performance improvements that have been achieved by some approaches. Combinatorial search problems include for example satisfiability (SAT), constraint problems, planning, quantified Boolean formulae (QBF), scheduling and combinatorial optimisation.

A combinatorial search problem is one where an initial state is to be transformed into a goal state by application of a series of operators, such as assignment of values to variables. The space of possible states is typically exponential in the size of the input and finding a solution is $\mathcal{NP}$-hard. A common way of solving such problems is to use *heuristics*. A heuristic is a strategy that determines which operators to apply when. Heuristics are not necessarily complete or deterministic, i.e. they are not guaranteed to find a solution if it exists or to always make the same decision under the same circumstances. The nature of heuristics makes them particularly amenable to Algorithm Selection – choosing a heuristic manually is difficult even for experts, but choosing the correct one can improve performance significantly.

There exists a large body of work that is relevant to Algorithm Selection in the Machine Learning literature. [133] presents a survey of many approaches. Repeating this here is unnecessary and outside the scope of this chapter, which focuses on the application of such techniques. The most relevant area of research is that into *ensembles*, where several models are created instead of one. Such ensembles are either implicitly assumed or explicitly engineered so that they complement each other. Errors made by one model are corrected by another. Ensembles can be engineered by techniques such as *bagging* [18] and *boosting* [128]. [9,111] present studies that compare bagging and boosting empirically. [30] provides explanations for why ensembles can perform better than individual algorithms.

There is increasing interest in the integration of Algorithm Selection techniques with programming language paradigms, e.g. [4,68]. While these issues are sufficiently relevant to be mentioned here, exploring them in detail is outside the scope of the chapter. Similarly, technical issues arising from the computation, storage and application of performance models, the integration of Algorithm Selection techniques into complex systems, the execution of choices and the collection of experimental data to facilitate Algorithm Selection are not surveyed here.

1.3 Terminology

Algorithm Selection is a widely applicable concept and as such has cropped up frequently in various lines of research. Often, different terminologies are used.

[15] use the term *algorithm chaining* to mean switching from one algorithm to another while the problem is being solved. [100] call Algorithm Selection *selection by performance prediction*. [145] use the term *hybrid algorithm* for the combination of a set of algorithms and an Algorithm Selection model (which they term *selector*).

In Machine Learning, Algorithm Selection is usually referred to as *meta-learning*. This is because Algorithm Selection models for Machine Learning learn when to use which method of Machine Learning. The earliest approaches also spoke of *hybrid approaches*, e.g. [144]. [1] proposes rules for selecting a Machine Learning algorithm that take the characteristics of a data set into account. He uses the term *meta-learning*. [20] introduces the notion of *selective superiority*. This concept refers to a particular algorithm being best on some, but not all tasks.

In addition to the many terms used for the process of Algorithm Selection, researchers have also used different terminology for the models of what Rice calls *performance measure space*. [2] call them *runtime performance predictors*. [75, 95, 96, 156] coined the term *Empirical Hardness model*. This stresses the reliance on empirical data to create these models and introduces the notion of *hardness* of a problem. The concept of hardness takes into account all performance considerations and does not restrict itself to, for example, runtime performance. In practice however, the described empirical hardness models only take runtime performance into account. In all cases, the predicted measures are used to select an algorithm.

Throughout this chapter, the term *algorithm* is used to refer to what is selected for solving a problem instance. This is for consistency and to make the connection to Rice's framework. An algorithm may be a system, a programme, a heuristic, a classifier or a configuration. This is not made explicit unless it is relevant in the particular context.

1.4   Organisation

An organisation of the Algorithm Selection literature is challenging, as there are many different criteria that can be used to classify it. Each publication can be evaluated from different points of view. The organisation of this chapter follows the main criteria below.

What to select algorithms from
   Section 2 describes how sets of algorithms, or *portfolios*, can be constructed. A portfolio can be *static*, where the designer decides which algorithms to include, or *dynamic*, where the composition or individual algorithms vary or change for different problems.

What to select and when
   Section 3 describes how algorithms from portfolios are selected to solve problems. Apart from the obvious approach of picking a single algorithm, time slots can be allocated to all or part of the algorithms or the execution monitored and earlier decisions revised. We also distinguish between selecting before the solving of the actual problem starts and while the problem is being solved.

How to select
   Section 4 surveys techniques used for making the choices described in Sect. 3. It details how performance models can be built and what kinds of predictions they inform. Example predictions are the best algorithm in the portfolio and the runtime performance of each portfolio algorithm.

How to facilitate the selection

Section 5 gives an overview of the types of analysis different approaches perform and what kind of information is gathered to facilitate Algorithm Selection. This includes the past performance of algorithms and structural features of the problems to be solved.

The order of the material follows a top-down approach. Starting with the high-level idea of Algorithm Selection, as proposed by [120] and described in this introduction, more technical details are gradually explored. Earlier concepts provide motivation and context for later technical details. For example, the choice of whether to select a single algorithm or monitor its execution (Sect. 3) determines the types of predictions required and techniques suitable for making them (Sect. 4) as well as the properties that need to be measured (Sect. 5).

The individual sections are largely self-contained. If the reader is more interested in a bottom-up approach that starts with technical details on what can be observed and measured to facilitate Algorithm Selection, Sects. 2 through 5 may be read in reverse order.

Section 6 again illustrates the importance of the field by surveying the many different application domains of Algorithm Selection techniques with a focus on combinatorial search problems. We close by summarising in Sect. 7.

2   Algorithm Portfolios

For diverse sets of problems, it is unlikely that a single algorithm will be the most suitable one in all cases. A way of mitigating this restriction is to use a *portfolio* of algorithms. This idea is closely related to the notion of Algorithm Selection itself – instead of making an up-front decision on what algorithm to use, it is decided on a case-by-case basis for each problem individually. In the framework presented by [120], portfolios correspond to the algorithm space $\mathcal{A}$.

Portfolios are a well-established technique in Economics. Portfolios of assets, securities or similar products are used to reduce the risk compared to holding only a single product. The idea is simple – if the value of a single security decreases, the total loss is less severe. The problem of allocating funds to the different parts of the portfolio is similar to allocating resources to algorithms in order to solve a computational problem. There are some important differences though. Most significantly, the past performance of an algorithm can be a good indicator of future performance. There are fewer factors that affect the outcome and in most cases, they can be measured directly. In Machine Learning, *ensembles* [30] are instances of algorithm portfolios. In fact, the only difference between algorithm portfolios and Machine Learning ensembles is the way in which its constituents are used.

The idea of algorithm portfolios was first presented by [73]. They describe a formal framework for the construction and application of algorithm portfolios and evaluate their approach on graph colouring problems. Within the Artificial Intelligence community, algorithm portfolios were popularised by [57,58] and a subsequent extended investigation [59]. The technique itself however had been described under different names by other authors at about the same time in different contexts.

[143] experimentally show for a selection of constraint satisfaction algorithms and heuristics that none is the best on all evaluated problems. They do not mention portfolios, but propose that future research should focus on identifying when particular algorithms and heuristics deliver the best performance. This implicitly assumes a portfolio to choose algorithms from. [2] perform a similar investigation and come to similar conclusions. They talk about selecting an appropriate algorithm from an *algorithm family*.

Beyond the simple idea of using a set of algorithms instead of a single one, there is a lot of scope for different approaches. One of the first problems faced by researchers is how to construct the portfolio. There are two main types. *Static portfolios* are constructed offline before any problems are solved. While solving a problem, the composition of the portfolio and the algorithms within it do not change. *Dynamic portfolios* change in composition, configuration of the constituent algorithms or both during solving.

2.1   Static Portfolios

Static portfolios are the most common type. The number of algorithms or systems in the portfolio is fixed, as well as their parameters. In Rice's notation, the algorithm space $\mathcal{A}$ is constant, finite and known. This approach is used for example in SATzilla [109,158,159], AQME [117,118], CPhydra [110], ARGOSMART [108], BUS [72] and Proteus [74].

The vast majority of approaches composes static portfolios from different algorithms or different algorithm configurations. [73] however use a portfolio that contains the same randomised algorithm twice. They run the portfolio in parallel and as such essentially use the technique to parallelise an existing sequential algorithm.

Some approaches use a large number of algorithms in the portfolio, such as ARGOSMART, whose portfolio size is 60. SATzilla uses 19 algorithms, although the authors use portfolios containing only subsets of those for specific applications. BUS uses six algorithms and CPhydra five. [54] select from a portfolio of only two algorithms. AQME has different versions with different portfolio sizes, one with 16 algorithms, one with five and three algorithms of different types and one with two algorithms [118]. The authors compare the different portfolios and conclude that the one with eight algorithms offers the best performance, as it has more variety than the portfolio with two algorithms and it is easier to make a choice for eight than for 16 algorithms. There are also approaches that use portfolios of variable size that is determined by training data [81,157]. [74] combine algorithms and problem encodings in a portfolio – problem instances can be translated into alternative representations, for which other algorithms are available.

As the algorithms in the portfolio do not change, their selection is crucial for its success. Ideally, the algorithms will complement each other such that good performance can be achieved on a wide range of different problems. [66] report that portfolios composed of a random selection from a large pool of diverse algorithms outperform portfolios composed of the algorithms with the best overall performance. They develop a framework with a mathematical model that theoretically justifies this observation. [126] use a portfolio of heuristics for solving quantified Boolean formulae problems that have specifically been crafted to be orthogonal to each other. [157] automatically engineer a portfolio with algorithms of complementary strengths. In [162], the authors analyse the contributions of the portfolio constituents to the overall performance and conclude that not algorithms with the best overall performance, but with techniques that set them apart from the rest contribute most. [81] use a static portfolio of variable size that adapts itself to the training data. They cluster the training problems and choose the best algorithm for each cluster. They do not emphasise diversity, but suitability for distinct parts of the problem space. [157] also construct a portfolio with algorithms that perform well on different parts of the problem space, but do not use clustering.

In financial theory, constructing portfolios can be seen as a quadratic optimisation problem. The aim is to balance expected performance and risk (the expected variation of performance) such that performance is maximised and risk minimised. [37] solve this problem for algorithm portfolios using genetic algorithms.

Most approaches make the composition of the portfolio less explicit. Many systems use portfolios of solvers that have performed well in solver competitions with the implicit assumption that they have complementing strengths and weaknesses and the resulting portfolio will be able to achieve good performance.

2.2 Dynamic Portfolios

Rather than relying on a priori properties of the algorithms in the portfolio, dynamic portfolios adapt the composition of the portfolio or the algorithms depending on the problem to be solved. The algorithm space $\mathcal{A}$ changes with each problem and is a subspace of the potentially infinite super algorithm space $\mathcal{A}'$. This space contains all possible (hypothetical) algorithms that could be used to solve problems from the problem space. In static portfolios, the algorithms in the portfolio are selected from $\mathcal{A}'$ once either manually by the designer of the portfolio or automatically based on empirical results from training data.

One approach is to build a portfolio by combining algorithmic building blocks. An example of this is the Adaptive Constraint Engine (ACE) [35,36]. The building blocks are so-called advisors, which characterise variables of the constraint problem and give recommendations as to which one to process next. ACE combines these advisors into more complex ones. [33,34] use a similar idea to construct search strategies for solving constraint problems. [42,43] proposes CLASS, which combines heuristic building blocks to form composite heuristics for solving SAT problems. In these approaches, there is no strong notion of a portfolio – the algorithm or strategy used to solve a problem is assembled from lower level components.

Closely related is the concept of specialising generic building blocks for the problem to solve. This approach is taken in the SAGE system (Strategy Acquisition Governed by Experimentation) [92,93]. It starts with a set of general operators that can be applied to a search state. These operators are refined by making the preconditions more specific based on their utility for finding a solution. The MULTI-TAC (Multi-tactic Analytic Compiler) system [103–105] specialises a set of generic heuristics for the constraint problem to solve.

There can be complex restrictions on how the building blocks are combined. RT-Syn [131] for example uses a preprocessing step to determine the possible combinations of algorithms and data structures to solve a software specification problem and then selects the most appropriate combination using simulated annealing. [8] model the construction of a constraint solver from components as a constraint problem whose solutions denote valid combinations of components.

Another approach is to modify the parameters of parameterised algorithms in the portfolio. This is usually referred to as automatic tuning and not only applicable in the context of algorithm portfolios, but also for single algorithms. The HAP system [146] automatically tunes the parameters of a planning system depending on the problem to solve. [70] dynamically modify algorithm parameters during search based on statistics collected during the solving process.

Automatic Tuning. The area of automatic parameter tuning has attracted a lot of attention in recent years. This is because algorithms have an increasing number of parameters that are difficult to tune even for experts and because of research into dynamic algorithm portfolios that benefits from automatic tuning. A survey of the literature on automatic tuning is outside the scope of this chapter, but some of the approaches that are particularly relevant to this survey are described below.

Automatic tuning and portfolio selection can be treated separately, as done in the Hydra portfolio builder [157]. Hydra uses ParamILS [78,79] to automatically tune algorithms in a SATzilla [159] portfolio. Autofolio [98] uses ParamILS and SMAC [76] to train a claspfolio [67] portfolio. ISAC [81] uses GGA [5] to automatically tune algorithms for clusters of problem instances.

[105] first enumerates all possible rule applications up to a certain time or size bound. Then, the most promising configuration is selected using beam search, a form of parallel hill climbing, that empirically evaluates the performance of each candidate. [8] use hill climbing to similarly identify the most efficient configuration for a constraint solver on a set of problems. [42,141] use genetic algorithms to evolve promising configurations.

The systems described in the previous paragraph are only of limited suitability for dynamic algorithm portfolios. They either take a long time to find good configurations or are restricted in the number or type of parameters. Interactions between parameters are only taken into account in a limited way. More recent approaches have focused on overcoming these limitations.

The ParamILS system [78,79] uses techniques based on local search to identify parameter configurations with good performance. The authors address over-confidence (overestimating the performance of a parameter configuration on a test set) and over-tuning (determining a parameter configuration that is too specific). SMAC [76] builds a model of the performance response surface in parameter space to predict where the most promising configurations are. [5] use genetic algorithms to discover favourable parameter configurations for the algorithms being tuned. The authors use a racing approach to avoid having to run all generated configurations to completion. They also note that one of the advantages of the genetic algorithm approach is that it is inherently parallel.

Both of these approaches are capable of tuning algorithms with a large number of parameters and possible values as well as taking interactions between parameters into account. They are used in practice in the Algorithm Selection systems Hydra and ISAC, respectively. In both cases, they are only used to construct static portfolios however. More recent approaches focus on exploiting parallelism, e.g. [77, 97].

Dynamic portfolios are in general a more fruitful area for Algorithm Selection research because of the large space of possible decisions. Static portfolios are usually relatively small and the decision space is amenable for human exploration. This is not a feasible approach for dynamic portfolios though. [105] notes that

"MULTI-TAC turned out to have an unexpected advantage in this arena, due to the complexity of the task. Unlike our human subjects, MULTI-TAC experimented with a wide variety of combinations of heuristics. Our human subjects rarely had the inclination or patience to try many alternatives, and on at least one occasion incorrectly evaluated alternatives that they did try."

3   Problem Solving with Portfolios

Once an algorithm portfolio has been constructed, the way in which it is to be used has to be decided. There are different considerations to take into account. The two main issues are as follows.

What to select
   Given the full set of algorithms in the portfolio, a subset has to be chosen for solving the problem. This subset can consist of only a single algorithm that is used to solve the problem to completion, the entire portfolio with the individual algorithms interleaved or running in parallel or anything in between.
When to select
   The selection of the subset of algorithms can be made only once before solving starts or continuously during search. If the latter is the case, selections can be made at well-defined points during search, for example at each node of a search tree, or when the system judges it to be necessary to make a decision.

Rice's model assumes that only a single algorithm $A \in \mathcal{A}$ is selected. It implicitly assumes that this selection occurs only once and before solving the actual problem.

3.1   What to Select

A common and the simplest approach is to select a single algorithm from the portfolio and use it to solve the problem completely. This single algorithm has been determined to be the best for the problem at hand. For example SATzilla [109,158,159], ARGOSMART [108], SALSA [29] and EUREKA [28] do this. The disadvantage of this approach is that there is no way of mitigating a wrong selection. If an algorithm is chosen that exhibits bad performance on the problem, the system is "stuck" with it and no adjustments are made, even if all other portfolio algorithms would perform much better.

An alternative approach is to compute schedules for running (a subset of) the algorithms in the portfolio. In some approaches, the terms portfolio and schedule are used synonymously – all algorithms in the portfolio are selected and run according to a schedule that allocates time slices to each of them. The task of Algorithm Selection becomes determining the schedule rather than to select algorithms.

[122] rank the portfolio algorithms in order of expected performance and allocate time according to this ranking. [72] propose a round-robin schedule that contains all algorithms in the portfolio. The order of the algorithms is determined by the expected run time and probability of success. The first algorithm is allocated a time slice that corresponds to the expected time required to solve the problem. If it is unable to solve the problem during that time, it and the remaining algorithms are allocated additional time slices until the problem is solved or a time limit is reached.

[118] determine a schedule according to three strategies. The first strategy is to run all portfolio algorithms for a short time and if the problem has not been solved after this, run the predicted best algorithm exclusively for the remaining time. The second strategy runs all algorithms for the same amount of time, regardless of what the predicted best algorithm is. The third variation allocates exponentially increasing time slices to each algorithm such that the total time is again distributed equally among them. In addition to the three different scheduling strategies, the authors evaluate four different ways of ordering the portfolio algorithms within a schedule that range from ranking based on past performance to random. They conclude that ordering the algorithms based on their past performance and allocating the same amount of time to all algorithms gives the best overall performance.

[110] optimise the computed schedule with respect to the probability that the problem will be solved. They use the past performance data of the portfolio algorithms for this. However, they note that their approach of using a simple complete search procedure to find this optimal schedule relies on small portfolio sizes and that "for a large number of solvers, a more sophisticated approach would be necessary". Later approaches, e.g. the SUNNY approach [3], improve on this.

[80] formulate the problem of computing a schedule that solves most problems in a training set in the lowest amount of time as a resource constrained set covering integer programme. They pursue similar aims as [110] but note that

their approach is more efficient and able to scale to larger schedules. However, their evaluation concludes that the approach with the best overall performance is to run the predicted best algorithm for 90 % of the total available time and distribute the remaining 10 % across the other algorithms in the portfolio according to a static schedule.

[113] presents a framework for calculating optimal schedules. The approach is limited by a number of assumptions about the algorithms and the execution environment, but is applicable to a wide range of research in the literature. [16,114] compute an optimal static schedule for allocating fixed time slices to each algorithm. [127] propose an algorithm to efficiently compute an optimal schedule for portfolios of fixed size and show that the problem of generating or even approximating an optimal schedule is computationally intractable. [123] explore different strategies for allocating time slices to algorithms. In a serial execution strategy, each algorithm is run once for an amount of time determined by the average time to find a solution on previous problems or the time that was predicted for finding a solution on the current problem. A round-robin strategy allocates increasing time slices to each algorithm. The length of a time slice is based on the proportion of successfully solved training problems within this time. [56] compute round-robin schedules following a similar approach. Not all of their computed schedules contain all portfolio algorithms. [138] compute a schedule with the aim of improving the average-case performance. In later work, they compute theoretical guarantees for the performance of their schedule [140].

[155] approach scheduling the chosen algorithms in a different way and assume a fixed limit on the amount of resources an algorithm can consume while solving a problem. All algorithms are run sequentially for this fixed amount of time. Similar to [56], they simulate the performance of different allocations and select the best one based on the results of these simulations. [41] estimates the performance of candidate allocations through bootstrap sampling. [57,59] also evaluate the performance of different candidate portfolios, but take into account how many algorithms can be run in parallel. They demonstrate that the optimal schedule (in this case the number of algorithms that are being run) changes as the number of available processors increases. [47] investigate how to allocate resources to algorithms in the presence of multiple CPUs that allow to run more than one algorithm in parallel. [165] craft portfolios with the specific aim of running the algorithms in parallel.

[69] consider computing optimal schedules without selection. They note that their approach can be used in a variety of settings, in particular parallel portfolios.

Related research is concerned with the scheduling of restarts of stochastic algorithms – it also investigates the best way of allocating resources. The chapter that introduced algorithm portfolios [73] uses a portfolio of identical stochastic algorithms that are run with different random seeds. There is a large amount of research on how to determine restart schedules for randomised algorithms and a survey of this is outside the scope of this chapter. A few approaches that are particularly relevant to Algorithm Selection and portfolios are mentioned below.

[70] determine the amount of time to allocate to a stochastic algorithm before restarting it. They use dynamic policies that take performance predictions into account, showing that it can outperform an optimal fixed policy.

[27] investigate a restart model that allocates resources to an algorithm proportional to the number of times it has been successful in the past. In particular, they note that the allocated resources should grow doubly exponentially in the number of successes. Allocation of fewer resources results in over-exploration (too many different things are tried and not enough resources given to each) and allocation of more resources in over-exploitation (something is tried for to too long before moving on to something different).

[139] compute restart schedules that take the runtime distribution of the portfolio algorithms into account. They present an approach that does so statically based on the observed performance on a set of training problems as well as an approach that learns the runtime distributions as new problems are solved without a separate training set.

3.2   When to Select

In addition to whether they choose a single algorithm or compute a schedule, existing approaches can also be distinguished by whether they operate before the problem is being solved (offline) or while the problem is being solved (online). The advantage of the latter is that more fine-grained decisions can be made and the effect of a bad choice of algorithm is potentially less severe. The price for this added flexibility is a higher overhead however, as algorithms are selected more frequently.

Examples of approaches that only make offline decisions include [105,110, 131,159]. In addition to having no way of mitigating wrong choices, often these will not even be detected. These approaches do not monitor the execution of the chosen algorithms to confirm that they conform with the expectations that led to them being chosen. Purely offline approaches are inherently vulnerable to bad choices. Their advantage however is that they only need to select an algorithm once and incur no overhead while the problem is being solved.

Moving towards online systems, the next step is to monitor the execution of an algorithm or a schedule to be able to intervene if expectations are not met. [39,40] investigates setting a time bound for the algorithm that has been selected based on the predicted performance. If the time bound is exceeded, the solution attempt is abandoned. More sophisticated systems furthermore adjust their selection if such a bound is exceeded. [15] try to detect behaviour during search that indicates that the algorithm is performing badly, for example visiting nodes in a subtree of the search that clearly do not lead to a solution. If such behaviour is detected, they propose switching the currently running algorithm according to a fixed replacement list.

[125] explore the same basic idea. They switch between two algorithms for solving constraint problems that achieve different levels of consistency. The level of consistency refers to the amount of search space that is ruled out by inference before actually searching it. Their approach achieves the same level of

search space reduction as the more expensive algorithm at a significantly lower cost. This is possible because doing more inference does not necessarily result in a reduction of the search space in all cases. The authors exploit this fact by detecting such cases and doing the cheaper inference. [112,136] also investigate switching propagation methods during solving. [163,164] do not monitor the execution of the selected algorithm, but instead the values of the features used to select it. They re-evaluate the selection function when its inputs change.

Further examples of approaches that monitor the execution of the selected algorithm are [49,118], but also [70] where the offline selection of an algorithm is combined with the online selection of a restart strategy. An interesting feature of [118] is that the authors adapt the model used for the offline algorithm selection if the actual run time is much higher than the predicted runtime. In this way, they are not only able to mitigate bad choices during execution, but also prevent them from happening again.

The approaches that make decisions during search, for example at every node of the search tree, are necessarily online systems. [6] select the best search strategy at checkpoints in the search tree. Similarly, [20] recursively partitions the classification problem to be solved and selects an algorithm for each partition. In this approach, a lower-level decision can lead to changing the decision at the level above. This is usually not possible for combinatorial search problems, as decisions at a higher level cannot be changed easily.

Closely related is the work by [90,91], which partitions the search space into recursive subtrees and selects the best algorithm from the portfolio for every subtree. They specifically consider recursive algorithms. At each recursive call, the Algorithm Selection procedure is invoked. This is a more natural extension of offline systems than monitoring the execution of the selected algorithms, as the same mechanisms can be used. [126] also select algorithms for recursively solving sub-problems.

The PRODIGY system [22] selects the next operator to apply in order to reach the goal state of a planning problem at each node in the search tree. Similarly, [92] learn weights for operators that can be applied at each search state and select from among them accordingly.

Most approaches rely on an offline element that makes a decision before search starts. In the case of recursive calls, this is no different from making a decision during search however. [44,46,49] on the other hand learn the Algorithm Selection model only dynamically while the problem is being solved. Initially, all algorithms in the portfolio are allocated the same (small) time slice. As search progresses, the allocation strategy is updated, giving more resources to algorithms that have exhibited better performance. The expected fastest algorithm receives half of the total time, the next best algorithm half of the remaining time and so on. [7] also rely exclusively on a selection model trained online in a similar fashion. They evaluate different strategies of allocating resources to algorithms according to their progress during search. All of these strategies converge to allocating all resources to the algorithm with the best observed performance.

4   Portfolio Selectors

Research on *how* to select from a portfolio in an Algorithm Selection system has generated the largest number of different approaches within the framework of Algorithm Selection. In Rice's framework, it roughly corresponds to the performance mapping $p(A, x)$, although only few approaches use this exact formulation. Rice assumes that the performance of a particular algorithm on a particular problem is of interest. While this is true in general, many approaches only take this into account implicitly. Selecting the single best algorithm for a problem for example has no explicit mapping into Rice's performance measure space $\mathcal{R}^n$ at all. The selection mapping $S(f(x))$ is also related to the problem of how to select.

There are many different ways a mechanism to select from a portfolio can be implemented. Apart from accuracy, one of the main requirements for such a selector is that it is relatively cheap to run – if selecting an algorithm for solving a problem is more expensive than solving the problem, there is no point in doing so. [145] explicitly define the selector as "an efficient (polynomial time) procedure".

There are several challenges associated with making selectors efficient. Algorithm Selection systems that analyse the problem to be solved, such as SATzilla, need to take steps to ensure that the analysis does not become too expensive. Two such measures are the running of a pre-solver and the prediction of the time required to analyse a problem [159]. The idea behind the pre-solver is to choose an algorithm with reasonable general performance from the portfolio and use it to start solving the problem before starting to analyse it. If the problem happens to be very easy, it will be solved even before the results of the analysis are available. After a fixed time, the pre-solver is terminated and the results of the Algorithm Selection system are used. [118] use a similar approach and run all algorithms for a short time in one of their strategies. Only if the problem has not been solved after that, they move on to the algorithm that was actually selected.

Predicting the time required to analyse a problem is a closely related idea. If the predicted required analysis time is too high, a default algorithm with reasonable performance is chosen and run on the problem. This technique is particularly important in cases where the problem is hard to analyse, but easy to solve. As some systems use information that comes from exploring part of the search space (cf. Sect. 5), this is a very relevant concern in practice. On some problems, even probing just a tiny part of the search space may take a very long time.

[54,55] report that using the misclassification penalty as a weight for the individual problems during training improves the quality of the predictions. The misclassification penalty quantifies the "badness" of a wrong prediction; in this case as the additional time required to solve a problem. If an algorithm was chosen that is only slightly worse than the best one, it has less impact than choosing an algorithm that is orders of magnitude worse. Using the penalty

during training is a way of guiding the learned model towards the problems where the potential performance improvement is large.

There are many different approaches to how portfolio selectors operate. The selector is not necessarily an explicit part of the system. [105] compiles the Algorithm Selection system into a Lisp programme for solving the original constraint problem. The selection rules are part of the programme logic. [43,50] evolve selectors and combinators of heuristic building blocks using genetic algorithms. The selector is implicit in the evolved programme.

4.1   Performance Models

The way the selector operates is closely linked to the way the performance model of the algorithms in the portfolio is built. In early approaches, the performance model was usually not learned but given in the form of human expert knowledge. [15,125] use hand-crafted rules to determine whether to switch the algorithm during solving. [2] also have hand-crafted rules, but estimate the runtime performance of an algorithm. More recent approaches sometimes use only human knowledge as well. [150] select a local search heuristic for solving SAT problems by a hand-crafted rule that considers the distribution of clause weights. [142] model the performance space manually using statistical methods and use this hand-crafted model to select a heuristic for solving constraint problems. [146] learn rules automatically, but then filter them manually.

A more common approach today is to automatically learn performance models using Machine Learning on training data. The portfolio algorithms are run on a set of representative problems and based on these experimental results, performance models are built. This approach is used by [61,81,110,117,159], to name but a few examples. A drawback of this approach is that the training time is usually large. [45] investigate ways of mitigating this problem by using censored sampling, which introduces an upper bound on the runtime of each experiment in the training phase. [85] also investigate censored sampling where not all algorithms are run on all problems in the training phase. Their results show that censored sampling may not have a significant effect on the performance of the learned model.

Models can also be built without a separate training phase, but while the problem is solved. This approach is used by [7,46] for example. While this significantly reduces the time to build a system, it can mean that the result is less effective and efficient. At the beginning, when no performance models have been built, the decisions of the selector might be poor. Furthermore, creating and updating performance models while the problem is being solved incurs an overhead.

The choice of Machine Learning technique is affected by the way the portfolio selector operates. Some techniques are more amenable to offline approaches (e.g. linear regression models used by [159]), while others lend themselves to online methods (e.g. reinforcement learning used by [7]).

Performance models can be categorised by the type of entity whose performance is modelled – the entire portfolio or individual algorithms within it.

There are publications that use both of those categories however, e.g. [134]. In some cases, no performance models as such are used at all. [8,25,105] run the candidates on a set of test problems and select the one with the best performance that way for example. [56,57,155] simulate the performance of different selections on training data.

Per-Portfolio Models. One automated approach is to learn a performance model of the entire portfolio based on training data. Usually, the prediction of such a model is the best algorithm from the portfolio for a particular problem. There is only a weak notion of an individual algorithm's performance. In Rice's notation for the performance mapping $P(A, x)$, A is the (subset of the) portfolio instead of an individual algorithm, i.e. $A \subseteq \mathcal{A}$ instead of Rice's $A \in \mathcal{A}$.

This is used for example by [28,61,108,110,117]. Again there are different ways of doing this. Lazy approaches do not learn an explicit model, but use the set of training examples as a case base. For new problems, the closest problem or the set of n closest problems in the case base is determined and decisions made accordingly. [51,101,108,110,117,151] use nearest-neighbour classifiers to achieve this. Apart from the conceptual simplicity, such an approach is attractive because it does not try to abstract from the examples in the training data. The problems that Algorithm Selection techniques are applied to are usually complex and factors that affect the performance are hard to understand. This makes it hard to assess whether a learned abstract model is appropriate and what its requirements and limitations are.

Explicitly-learned models try to identify the concepts that affect performance for a given problem. This acquired knowledge can be made explicit to improve the understanding of the researchers of the problem domain. There are several Machine Learning techniques that facilitate this, as the learned models are represented in a form that is easy to understand by humans. [20,22,60,146] learn classification rules that guide the selector. [146] note that the decision to use a classification rule leaner was not so much guided by the performance of the approach, but the easy interpretability of the result. [36,92,107] learn weights for decision rules to guide the selector towards the best algorithms. [12,28,54,61,63,122] go one step further and learn decision trees. [63] again note that the reason for choosing decision trees was not primarily the performance, but the understandability of the result. [116] show the set of learned rules in the paper to illustrate its compactness. Similarly, [54] show their final decision tree in the paper.

Some approaches learn probabilistic models that take uncertainty and variability into account. [60] use a probabilistic model to learn control rules. The probabilities for candidate rules being beneficial are evaluated and updated on a training set until a threshold is reached. This methodology is used to avoid having to evaluate candidate rules on larger training sets, which would show their utility more clearly but be more expensive. [29] learn multivariate Bayesian decision rules. [23] learn a Bayesian classifier to predict the best algorithm after a certain amount of time. [137] learn Bayesian models that incorporate collaborative filtering. [32] learn decision rules using naïve Bayes classifiers. [90,113] learn performance models based on Markov Decision Processes. [85] use statistical relational learning to predict the ranking of the algorithms in the portfolio on a particular problem. None of these approaches make explicit use of the uncertainty attached to a decision though.

Other approaches include support vector machines [6,71], reinforcement learning [7], neural networks [44], decision tree ensembles [71], ensembles of general classification algorithms [87], boosting [12], hybrid approaches that combine regression and classification [83], multinomial logistic regression [126], self-organising maps [134] and clustering [81,102,136]. [127,138] compute schedules for running the algorithms in the portfolio based on a statistical model of the problem instance distribution and performance data for the algorithms. This is not an exhaustive list, but focuses on the most prominent approaches and publications. Within a single family of approaches, such as decision trees, there are further distinctions that are outside the scope of this chapter, such as the type of decision tree inducer.

[6] discuss a technical issue related to the construction of per-portfolio performance models. A particular algorithm often exhibits much better performance in general than other algorithms on a particular instance distribution. Therefore, the training data used to learn the performance model will be skewed towards that algorithm. This can be a problem for Machine Learning, as always predicting this best algorithm might have a very high accuracy already, making it very hard to improve on. The authors mention two means of mitigating this problem. The training set can be *under-sampled*, where examples where the best overall algorithm performs best are deliberately omitted. Alternatively, the set can be *over-sampled* by artificially increasing the number of examples where another algorithm is better.

Per-Algorithm Models. A different approach is to learn performance models for the individual algorithms in the portfolio. The predicted performance of an algorithm on a problem can be compared to the predicted performance of the other portfolio algorithms and the selector can proceed based on this. The advantage of this approach is that it is easier to add and remove algorithms from the portfolio – instead of having to retrain the model for the entire portfolio, it suffices to train a model for the new algorithm or remove one of the trained models. Most approaches only rely on the order of predictions being correct. It does not matter if the prediction of the performance itself is wildly inaccurate as long as it is correct relative to the other predictions.

This is the approach that is implicitly assumed in Rice's framework. The prediction is the performance mapping $P(A, x)$ for an algorithm $A \in \mathcal{A}$ on a problem $x \in \mathcal{P}$. Models for each algorithm in the portfolio are used for example by [2,46,72,100,159].

A common way of doing this is to use regression to directly predict the performance of each algorithm. This is used by [64,72,95,123,159]. The performance of the algorithms in the portfolio is evaluated on a set of training problems, and a relationship between the characteristics of a problem and the performance of an algorithm derived. This relationship usually has the form of a simple formula that is cheap to compute at runtime.

[130] on the other hand learn latent class models of unobserved variables to capture relationships between solvers, problems and run durations. Based on the predictions, the expected utility is computed and used to select an algorithm. [129] surveys sampling methods to estimate the cost of solving constraint problems. [148] models the behaviour of local search algorithms with Markov chains.

Another approach is to build statistical models of an algorithm's performance based on past observations. [149] use Bayesian belief propagation to predict the runtime of a particular algorithm on a particular problem. Bayesian inference is used to determine the class of a problem and the closest case in the knowledge base. A performance profile is extracted from that and used to estimate the runtime. The authors also propose an alternative approach that uses neural nets. [39, 40] computes the expected gain for time bounds based on past success times. The computed values are used to choose the algorithm and the time bound for running it. [17] compare algorithm rankings based on different past performance statistics. Similarly, [94] maintain a ranking based on past performance. [27] propose a bandit problem model that governs the allocation of resources to each algorithm in the portfolio. [147] also use a bandit model, but furthermore evaluate a Q-learning approach, where in addition to bandit model rewards, the states of the system are taken into account. [56, 57, 155] use the past performance of algorithms to simulate the performance of different algorithm schedules and use statistical tests to select one of the schedules.

Hierarchical Models. There are some approaches that combine several models into a hierarchical performance model. There are two basic types of hierarchical models. One type predicts additional *properties of the problem* that cannot be measured directly or are not available without solving the problem. The other type makes *intermediate predictions* that do not inform Algorithm Selection directly, but rather the final predictions.

[156] use sparse multinomial logistic regression to predict whether a SAT problem instance is satisfiable and, based on that prediction, use a logistic regression model to predict the runtime of each algorithm in the portfolio. [64] also predict the satisfiability of a SAT instance and then choose an algorithm from a portfolio. Both report that being able to distinguish between satisfiable and unsatisfiable problems enables performance improvements. The satisfiability of a problem is a property that needs to be *predicted* in order to be useful for Algorithm Selection. If the property is *computed* (i.e. the problem is solved), there is no need to perform Algorithm Selection any more.

[55] use classifiers to first decide on the level of consistency a constraint propagator should achieve and then on the actual implementation of the propagator that achieves the selected level of consistency. A different publication that uses the same data set does not make this distinction however [87], suggesting that the performance benefits are not significant in practice.

[74] proposes a hierarchical model that has more than two levels – at the top, the decision is made whether to solve a given constraint problem as a constraint problem or convert it to SAT. At the second level, if the decision to convert to SAT has been made, the encoding for the transformation is chosen. At the third level, the constraint or SAT solver is chosen.

Such hierarchical models are only applicable in a limited number of scenarios, which explains the comparatively small amount of research into them. For many application domains, only a single property needs to be predicted and can be predicted without intermediate steps with sufficient accuracy. [83] proposes a hierarchical approach that is domain-independent. He uses the performance predictions of regression models as input to a classifier that decides which algorithm to choose and demonstrates performance improvements compared to selecting an algorithm directly based on the predicted performance. The idea is very similar to that of *stacking* in Machine Learning [152].

Selection of Model Learner. Apart from the different types of performance models, there are different Machine Learning algorithms that can be used to learn a particular kind of model. While most of the approaches mentioned here rely on a single way of doing this, some of the research compares different methods.

[159] mention that, in addition to the chosen ridge regression for predicting the runtime, they explored using lasso regression, support vector machines and Gaussian processes. They chose ridge regression not because it provided the most accurate predictions, but the best trade-off between accuracy and cost to make the prediction. [149] propose an approach that uses neural networks in addition to the Bayesian belief propagation approach they describe initially. [28] compare different decision tree learners, a Bayesian classifier, a nearest neighbour approach and a neural network. They chose the C4.5 decision tree inducer because even though it may be outperformed by a neural network, the learned trees are easily understandable by humans and may provide insight into the problem domain. [95] compare several versions of linear and non-linear regression. [75] report having explored support vector machine regression, multivariate adaptive regression splines (MARS) and lasso regression before deciding to use the linear regression approach of [95]. They also report experimental results with sequential Bayesian linear regression and Gaussian Process regression. [62,63] explore using decision trees, naïve Bayes rules, Bayesian networks and meta-learning techniques. They also chose the C4.5 decision tree inducer because it is one of the top performers and creates models that are easy to understand and quick to execute. [52] compare nearest neighbour classifiers, decision trees and statistical models. They show that a nearest neighbour classifier outperforms all the other approaches on their data sets.

[71] use decision tree ensembles and support vector machines. [12] investigate alternating decision trees and various forms of boosting, while [117] use decision trees, decision rules, logistic regression and nearest neighbour approaches. They do not explicitly choose one of these methods in the paper, but their Algorithm Selection system AQME uses a nearest neighbour classifier by default. [123] use

32 different Machine Learning algorithms to predict the runtime of algorithms and probability of success. They attempt to provide explanations for the performance of the methods they have chosen in [124]. [130] compare the performance of different latent class models. [55] evaluate the performance of 19 different Machine Learning classifiers on an Algorithm Selection problem in constraint programming. The investigation is extended to include more Machine Learning algorithms as well as different performance models and more problem domains in [85]. They identify several Machine Learning algorithms that show particularly good performance across different problem domains, namely linear regression and alternating decision trees. They do not consider issues such as how easy the models are to understand or how efficient they are to compute.

Only [52,55,63,71,85,117,130] quantify the differences in performance of the methods they used. The other comparisons give only qualitative evidence. Not all comparisons choose one of the approaches over the other or provide sufficient detail to enable the reader to do so. In cases where a particular technique is chosen, performance is often not the only selection criterion. In particular, the ability to understand a learned model plays a significant role.

4.2 Types of Predictions

The way of creating the performance model of a portfolio or its algorithms is not the only choice researchers face. In addition, there are different predictions the performance model can make to inform the decision of the selector of a subset of the portfolio algorithms. The type of decision is closely related to the learned performance model however. The prediction can be a single categorical value – the algorithm to choose. This type of prediction is usually the output of per-portfolio models and used for example in [28,54,61,108,117]. The advantage of this simple prediction is that it determines the choice of algorithm without the need to compare different predictions or derive further quantities. One of its biggest disadvantages however is that there is no flexibility in the way the system runs or even the ability to monitor the execution for unexpected behaviour.

A different approach is to predict the runtime of the individual algorithms in the portfolio. This requires per-algorithm models. For example [70,113,130] do this. [159] do not predict the runtime itself, but the logarithm of the runtime. They note that,

> "In our experience, we have found this log transformation of runtime to be very important due to the large variation in runtimes for hard combinatorial problems."

[85] also compare predicting the runtime itself and the log thereof, but find no significant difference between the two. [83] however also reports better results with the logarithm.

[2] estimate the runtime by proxy by predicting the number of constraint checks. [100] estimate the runtime by predicting the number of search nodes to explore and the time per node. [90] talk of the *cost* of selecting a particular

algorithm, which is equal to the time it takes to solve the problem. [107] uses the *utility* of a choice to make his decision. The utility is an abstract measure of the "goodness" of an algorithm that is adapted dynamically. [142] use the *value of information* of selecting an algorithm, defined as the amount of time saved by making this choice. [160] predict the *penalized average runtime score*, a measure that combines runtime with possible timeouts. This approach aims to provide more realistic performance predictions when runtimes are capped.

More complex predictions can be made, too. In most cases, these are made by combining simple predictions such as the runtime performance. [17,94,135] produce rankings of the portfolio algorithms. [85] use statistical relational learning to directly predict the ranking instead of deriving it from other predictions. [46,49,72,110,122] predict resource allocations for the algorithms in the portfolios. [14,52,99] consider selecting the most appropriate formulation of a constraint problem. [8,19,131,151] select algorithms and data structures to be used in a software system.

Some types of predictions require online approaches that make decisions during search. [7,15,23,125] predict when to switch the algorithm used to solve a problem. [70] predict whether to restart an algorithm. [90,91] predict the cost to solve a sub-problem. However, most online approaches make predictions that can also be used in offline settings, such as the best algorithm to proceed with.

The primary selection criteria and prediction for [135] and [94] is the quality of the solution an algorithm produces rather than the time it takes the algorithm to find that solution. In addition to the primary selection criteria, a number of approaches predict secondary criteria. [40,72,123] predict the probability of success for each algorithm. [149] predict the quality of a solution.

In Rice's model, the prediction of an Algorithm Selection system is the performance $p \in \mathcal{R}^n$ of an algorithm. This abstract notion does not rely on time and is applicable to many approaches. It does not fit techniques that predict the portfolio algorithm to choose or more complex measures such as a schedule however. As Rice developed his approach long before the advent of algorithm portfolios, it should not be surprising that the notion of the performance of individual algorithms as opposed to sets of algorithms dominates. The model is sufficiently general to be able to accommodate algorithm portfolios with only minor modifications to the overall framework however.

5   Features

The different types of performance models described in the previous sections usually use features to inform their predictions. Features are an integral part of systems that do Machine Learning. They characterise the inputs, such as the problem to be solved or the algorithm employed to solve it, and facilitate learning the relationship between the inputs and the outputs, such as the time it will take the algorithm to solve the problem. In Rice's model, features $f(x)$ for a particular problem x are extracted from the feature space $\mathcal{F}$.

The selection of the most suitable features is an important part of the design of Algorithm Selection systems. There are different types of features researchers can use and different ways of computing these. They can be categorised according to two main criteria.

First, they can be categorised according to how much background knowledge a researcher needs to have to be able to use them. Features that require no or very little knowledge of the application domain are usually very general and can be applied to new Algorithm Selection problems with little or no modification. Features that are specific to a domain on the other hand may require the researcher building the Algorithm Selection system to have a thorough understanding of the domain. These features usually cannot be applied to other domains, as they may be non-existent or uninformative in different contexts.

The second way of distinguishing different classes of features is according to when and how they are computed. Features can be computed *statically*, i.e. before the search process starts, or *dynamically*, i.e. during search. These two categories roughly align with the offline and online approaches to portfolio problem solving described in Sect. 3.

[132] present a survey that focuses on what features can be used for Algorithm Selection. This chapter categorises the features used in the literature.

5.1   Low and High-Knowledge Features

In some cases, researchers use a large number of features that are specific to the particular problem domain they are interested in, but there are also publications that only use a single, general feature – the performance of a particular algorithm on past problems. [27,49,113,130,138], to name but a few examples, use this approach to build statistical performance models of the algorithms in their portfolios. The underlying assumption is that all problems are similar with respect to the relative performance of the algorithms in the portfolio – the algorithm that has done best in the past has the highest chance of performing best in the future.

Approaches that build runtime distribution models for the portfolio algorithms usually do not select a single algorithm for solving a problem, but rather use the distributions to compute resource allocations for the individual portfolio algorithms. The time allocated to each algorithm is proportional to its past performance.

Other sources of features that are not specific to a particular problem domain are more fine-grained measures of past performance or measures that characterise the behaviour of an algorithm during search. [93] for example determines whether a search step performed by a particular algorithm is good, i.e. leading towards a solution, or bad, i.e. straying from the path to a solution if the solution is known or revisiting an earlier search state if the solution is not known. [57,59] use the runtime distributions of algorithms over the size of a problem, as measured by the number of backtracks. [40] uses the past success times of an algorithm as candidate time bounds on new problems. [17] do not consider the runtime, but the error rate of algorithms. [56] use both computation time and solution quality.

[11,23,24] evaluate the performance also during search. They explicitly focus on features that do not require a lot of domain knowledge. [11] note that,

"While existing algorithm selection techniques have shown impressive results, their knowledge-intensive nature means that domain and algorithm expertise is necessary to develop the models. The overall requirement for expertise has not been reduced: it has been shifted from algorithm selection to predictive model building."

They do, like several other approaches, assume *anytime* algorithms – after search has started, the algorithm is able to return the best solution found so far at any time. The features are based on how search progresses and how the quality of solutions is improved by algorithms. While this does not require any knowledge about the application domain, it is not applicable in cases when only a single solution is sought.

Most approaches learn models for the performance on particular problems and do not use past performance as a feature, but to inform the prediction to be made. Considering problem features facilitates a much more nuanced approach than a broad-brush general performance model. This is the classic supervised Machine Learning approach – given the correct prediction derived from the behaviour on a set of training problems, learn a model that enables to make this prediction.

The features that are considered to learn the model are specific to the problem domain or even a subset of the problem domain to varying extents. For combinatorial search problems, the most commonly used basic features include,

- the number of variables,
- properties of the variable domains, i.e. the list of possible assignments,
- the number of clauses in SAT, the number of constraints in constraint problems, the number of goals in planning,
- the number of clauses/constraints/goals of a particular type (for example the number of `alldifferent` constraints, [55]),
- ratios of several of the above features and summary statistics.

Such features are used for example in [72,110,117,149,159].

Other sources of features include the generator that produced the problem to be solved [70], the runtime environment [7], structures derived from the problem such as the primal graph of a constraint problem [51,54,61], specific parts of the problem model such as variables [35], the algorithms in the portfolio themselves [71] or the domain of the problem to be solved [22,56] rely on the problem domain as the only problem-specific feature and select based on past performance data for the particular domain. [10] consider not only the values of properties of a problem, but the changes of those values while the problem is being solved. [131] consider features of abstract representations of the algorithms. [163,164] use features that represent technical details of the behaviour of an algorithm on a problem, such as the type of computations done in a loop. [74] consider features not only of the instance being solved, but also of alternative encodings of the same instance.

Most approaches use features that are applicable to all problems of the application domain they are considering. However, [70] use features that are not only specific to their application domain, but also to the specific family of problems they are tackling, such as the variance of properties of variables in different columns of Latin squares. They note that,

"...the inclusion of such domain-specific features was important in learning strongly predictive models."

5.2   Static and Dynamic Features

In most cases, the approaches that use a large number of domain-specific features compute them *offline*, i.e. before the solution process starts (cf. Sect. 3.2). Examples of publications that only use such static features are [61,95,117].

An implication of using static features is that the decisions of the Algorithm Selection system are only informed by the performance of the algorithms on past problems. Only dynamic features allow to take the performance on the current problem into account. This has the advantage that remedial actions can be taken if the problem is unlike anything seen previously or the predictions are wildly inaccurate for another reason.

A more flexible approach than to rely purely on static features is to incorporate features that can be determined statically, but try to estimate the performance on the current problem. Such features are computed by probing the search space. This approach relies on the performance probes being sufficiently representative of the entire problem and sufficiently equal across the different evaluated algorithms. If an algorithm is evaluated on a part of the search space that is much easier or harder than the rest, a misleading impression of its true performance may result.

Examples of systems that combine static features of the problem to be solved with features derived from probing the search space are [54,110,159]. There are also approaches that use only probing features. We term this *semi-static* feature computation because it happens before the actual solving of the problem starts, but parts of the search space are explored during feature extraction. Examples include [2,11,100].

The idea of probing the search space is related to *landmarking* [116], where the performance of a set of initial algorithms (the *landmarkers*) is linked to the performance of the set of algorithms to select from. The main consideration when using this technique is to select landmarkers that are computationally cheap. Therefore, they are usually versions of the portfolio algorithms that have either been simplified or are run only on a subset of the data the selected algorithm will run on.

While the work done during probing explores part of the search space and could be used to speed search up subsequently by avoiding to revisit known areas, almost no research has been done into this. [11] run all algorithms in their (small) portfolio on a problem for a fixed time and select the one that has made the best progress. The chosen algorithm resumes its earlier work, but no attempt is made to avoid duplicating work done by the other algorithms.

To the best of our knowledge, there exist no systems that attempt to avoid redoing work performed by a different algorithm during the probing stage.

For successful systems, the main source of performance improvements is the selection of the right algorithm using the features computed through probing. As the time to compute the features is usually small compared to the runtime improvements achieved by Algorithm Selection, using the results of probing during search to avoid duplicating work does not have the potential to achieve large additional performance improvements.

The third way of computing features is to do so *online*, i.e. while search is taking place. These dynamic features are computed by an execution monitor that adapts or changes the algorithm during search based on its performance. Approaches that rely purely on dynamic features are for example [15,107,136].

There are many different features that can be computed during search. [105] determines how closely a generated heuristic approximates a generic target heuristic by checking the heuristic choices at random points during search. He selects the one with the closest match. Similarly, [107] learn how to select heuristics during the search process based on their performance. [7] use an agent-based model that rewards good actions and punishes bad actions based on computation time. [89] follow a very similar approach that also takes success or failure into account.

[23,24] monitor the solution quality during search. They decide whether to switch the current algorithm based on this by changing the allocation of resources. [150] monitor a feature that is specific to their application domain, the distribution of clause weights in SAT, during search and use it to decide whether to switch a heuristic. [136] monitors propagation events in a constraint solver to a similar aim. [25] evaluate the performance of candidate algorithms in terms of number of calls to a specific high-level procedure. They note that in contrast to using the runtime, their approach is machine-independent.

5.3   Feature Selection

The features used for learning the Algorithm Selection model are crucial to its success. Uninformative features might prevent the model learner from recognising the real relation between problem and performance or the most important feature might be missing. Many researchers have recognised this problem.

[72] manually select the most important features. They furthermore take the unique approach of learning one model per feature for predicting the probability of success and combine the predictions of the models. [95,159] perform automatic feature selection by greedily adding features to an initially empty set. In addition to the basic features, they also use the pairwise products of the features. [117] also perform automatic greedy feature selection, but do not add the pairwise products. [85] automatically select the most important subset of the original set of features, but conclude that in practice the performance improvement compared to using all features is not significant. [151] use genetic algorithms to determine the importance of the individual features. [115] evaluate subsets of the features they use and learn weights for each of them. [124] consider using a

single feature and automatic selection of a subset of all features. [63,88] also use techniques for automatically determining the most predictive subset of features. [83] compares the performance of ten different sets of features.

It is not only important to use informative features, but also features that are cheap to compute. If the cost of computing the features and making the decision is too high, the performance improvement from selecting the best algorithm might be eroded. [160] predict the feature computation time for a given problem and fall back to a default selection if it is too high to avoid this problem. They also limit the computation time for the most expensive features as well as the total time allowed to compute features. [13] consider the computational complexity of calculating problem features when selecting the features to use. They show that while achieving comparable accuracy to the full set of features, the subset of features selected by their method is significantly cheaper to compute. [54] explicitly exclude features that are expensive to compute.

6   Application Domains

The approaches for solving the Algorithm Selection Problem that have been surveyed here are usually not specific to a particular application domain, within combinatorial search problems or otherwise. Nevertheless this survey would not be complete without a brief exposition of the various contexts in which Algorithm Selection techniques have been applied.

Over the years, Algorithm Selection systems have been used in many different application domains. These range from Mathematics, e.g. differential equations [82,149], linear algebra [29] and linear systems [12,89], to the selection of algorithms and data structures in software design [19,21,131,151]. A very common application domain are combinatorial search problems such as SAT [91, 130,159], constraints [36,105,110], Mixed Integer Programming [161], Quantified Boolean Formulae [118,137], planning [22,38,72], scheduling [10,11,27], combinatorial auctions [46,51,95], Answer Set Programming [53,67], the Travelling Salesperson Problem [41,86], graph colouring [106] and general search algorithms [28,93,100].

Other domains include Machine Learning [94,135], the most probable explanation problem [63], parallel reduction algorithms [163,164] and simulation [37,147]. It should be noted that a significant part of Machine Learning research is concerned with developing Algorithm Selection techniques; the publications listed in this paragraph are the most relevant that use the specific techniques and framework surveyed here.

Some publications consider more than one application domain. [137] choose the best algorithm for Quantified Boolean Formulae and combinatorial auctions. [2,88] look at SAT and constraints. [59] consider SAT and Mixed Integer Programming. In addition to these two domains, [81] also investigate set covering problems. [140] apply their approach to SAT, Integer Programming and planning. [48,83,85] compare the performance across Algorithm Selection problems from constraints, Quantified Boolean Formulae and SAT.

In most cases, researchers take some steps to adapt their approaches to the application domain. This is usually done by using domain-specific features, such as the number of constraints and variables in constraint programming. In principle, this is not a limitation of the proposed techniques as those features can be exchanged for ones that are applicable in other application domains. While the overall approach remains valid, the question of whether the performance would be acceptable arises. [85] investigate how specific techniques perform across several domains with the aim of selecting the one with the best overall performance. There are approaches that have been tailored to a specific application domain to such an extent that the technique cannot be used for other applications. This is the case for example in the case of hierarchical models for SAT [64,156].

7 Summary

Over the years, there have been many approaches to solving the Algorithm Selection Problem. Especially in Artificial Intelligence and for combinatorial search problems, researchers have recognised that using Algorithm Selection techniques can provide significant performance improvements with relatively little effort. Most of the time, the approaches involve some kind of Machine Learning that attempts to learn the relation between problems and the performance of algorithms automatically. This is not a surprise, as the relationship between an algorithm and its performance is often complex and hard to describe formally. In many cases, even the designer of an algorithm does not have a general model of its performance.

Despite the theoretical difficulty of Algorithm Selection, dozens of systems have demonstrated that it can be done in practice with great success. In some sense, this mirrors achievements in other areas of Artificial Intelligence. Satisfiability is formally a problem that cannot be solved efficiently, yet researchers have come up with ways of solving very large instances of satisfiability problems with very few resources. Similarly, some Algorithm Selection systems have come very close to always choosing the best algorithm.

This survey presented an overview of the Algorithm Selection research that has been done to date with a focus on combinatorial search problems. A categorisation of the different approaches with respect to fundamental criteria that determine Algorithm Selection systems in practice was introduced. This categorisation abstracts from many of the low level details and additional considerations that are presented in most publications to give a clear view of the underlying principles. We furthermore gave details of the many different ways that can be used to tackle Algorithm Selection and the many techniques that have been used to solve it in practice.

On a high level, the approaches surveyed here can be summarised as follows.

– Algorithms are chosen from portfolios, which can be statically constructed or dynamically augmented with newly constructed algorithms as problems are being solved. Portfolios can be engineered such that the algorithms in it complement each other (i.e. are as diverse as possible), by automatically tuning algorithms on a set of training problems or by using a set of algorithms from the literature or competitions. Dynamic portfolios can be composed of algorithmic building blocks that are combined into complete algorithms by the selection system. Compared to tuning the parameters of algorithms, the added difficulty is that not all combinations of building blocks may be valid.

- A single algorithm can be selected from a portfolio to solve a problem to completion or a set of larger size can be selected that is run in parallel or according to a schedule. Another approach is to select a single algorithm to start with and then decide if and when to switch to another algorithm. Some approaches always select the entire portfolio and vary the resource allocation to the algorithms.
- Algorithm Selection can happen offline, without any interaction with the Algorithm Selection system after solving starts, or online. Some approaches monitor the performance of the selected algorithm and take action if it does not conform to the expectations or some other criteria. Others repeat the selection process at specific points during the search (e.g. every node in the search tree), skew a computed schedule towards the best performers or decide whether to restart stochastic algorithms.
- Performance can be modelled and predicted either for a portfolio as a whole (i.e. the prediction is the best algorithm) or for each algorithm independently (i.e. the prediction is the performance). A few approaches use hierarchical models that make a series of predictions to facilitate selection. Some publications make secondary predictions (e.g. the quality of a solution) that are taken into account when selecting the most suitable algorithm, while others make predictions that the desired output is derived from instead of predicting it directly. The performance models are usually learned automatically using Machine Learning, but a few approaches use hand-crafted models and rules. Models can be learned from separate training data or incrementally while a problem is being solved.
- Learning and using performance models is facilitated by features of the algorithms, problems or runtime environment. Features can be domain-independent or specific to a particular set of problems. Similarly, features can be computed by inspecting the problem before solving or while it is being solved. The use of feature selection techniques that automatically determine the most important and relevant features is quite common.

Given the amount of relevant literature, it is infeasible to discuss every approach in detail. The scope of this survey is necessarily limited to the detailed description of high-level details and a summary overview of low-level traits. Work in related areas that is not immediately relevant to Algorithm Selection for combinatorial search problems has been pointed to, but cannot be explored in more detail.

The proliferation of different approaches, application domains and data sets has stimulated the creation of a common data format and benchmark repository for algorithm selection problems, http://aslib.net. It provides a starting point for researchers wishing to compare their new approach to existing approaches.

A tabular summary of the literature organised according to the criteria introduced here can be found at http://larskotthoff.github.io/assurvey/. This table is updated continuously.

Acknowledgements. Ian Miguel and Ian Gent provided valuable feedback that helped shape this chapter. We also thank the anonymous reviewers of a previous version of this chapter whose detailed comments helped to greatly improve it. This work was supported by an EPSRC doctoral prize and EU FP7 FET project ICON. A shorter version of this chapter has appeared in AI Magazine [84].

References

1. Aha, D.W.: Generalizing from case studies: a case study. In: Proceedings of the 9th International Workshop on Machine Learning, pp. 1–10. Morgan Kaufmann Publishers Inc, San Francisco (1992)
2. Allen, J.A., Minton, S.: Selecting the right heuristic algorithm: runtime performance predictors. In: McCalla, G. (ed.) AI 1996. LNCS, vol. 1081, pp. 41–53. Springer, Heidelberg (1996). doi:10.1007/3-540-61291-2_40
3. Amadini, R., Gabbrielli, M., Mauro, J.: SUNNY: a lazy portfolio approach for constraint solving. TPLP **14**(4–5), 509–524 (2014)
4. Ansel, J., Chan, C., Wong, Y.L., Olszewski, M., Zhao, Q., Edelman, A., Amarasinghe, S.: PetaBricks: a language and compiler for algorithmic choice. SIGPLAN Not. **44**(6), 38–49 (2009)
5. Ansótegui, C., Sellmann, M., Tierney, K.: A gender-based genetic algorithm for the automatic configuration of algorithms. In: Gent, I.P. (ed.) CP 2009. LNCS, vol. 5732, pp. 142–157. Springer, Heidelberg (2009). doi:10.1007/978-3-642-04244-7_14
6. Arbelaez, A., Hamadi, Y., Sebag, M.: Online heuristic selection in constraint programming. In: Symposium on Combinatorial Search (2009)
7. Armstrong, W., Christen, P., McCreath, E., Rendell, A.P.: Dynamic algorithm selection using reinforcement learning. In: International Workshop on Integrating AI and Data Mining, pp. 18–25, December 2006
8. Balasubramaniam, D., Gent, I.P., Jefferson, C., Kotthoff, L., Miguel, I., Nightingale, P.: An automated approach to generating efficient constraint solvers. In: 34th International Conference on Software Engineering, pp. 661–671, June 2012
9. Bauer, E., Kohavi, R.: An empirical comparison of voting classification algorithms: bagging, boosting, and variants. Mach. Learn. **36**(1–2), 105–139 (1999)
10. Beck, J.C., Fox, M.S.: Dynamic problem structure analysis as a basis for constraint-directed scheduling heuristics. Artif. Intell. **117**(1), 31–81 (2000)
11. Beck, J.C., Freuder, E.C.: Simple rules for low-knowledge algorithm selection. In: Régin, J.-C., Rueher, M. (eds.) CPAIOR 2004. LNCS, vol. 3011, pp. 50–64. Springer, Heidelberg (2004). doi:10.1007/978-3-540-24664-0_4
12. Bhowmick, S., Eijkhout, V., Freund, Y., Fuentes, E., Keyes, D.: Application of machine learning in selecting sparse linear solvers. Technical report, Columbia University (2006)
13. Bhowmick, S., Toth, B., Raghavan, P.: Towards low-cost, high-accuracy classifiers for linear solver selection. In: Allen, G., Nabrzyski, J., Seidel, E., Albada, G.D., Dongarra, J., Sloot, P.M.A. (eds.) ICCS 2009. LNCS, vol. 5544, pp. 463–472. Springer, Heidelberg (2009). doi:10.1007/978-3-642-01970-8_45

14. Borrett, J.E., Tsang, E.P.K.: A context for constraint satisfaction problem formulation selection. Constraints **6**(4), 299–327 (2001)
15. Borrett, J.E., Tsang, E.P.K., Walsh, N.R.: Adaptive constraint satisfaction: The quickest first principle. In: ECAI, pp. 160–164 (1996)
16. Bougeret, M., Dutot, P., Goldman, A., Ngoko, Y., Trystram, D.: Combining multiple heuristics on discrete resources. In: IEEE International Symposium on Parallel and Distributed Processing, pp. 1–8. IEEE Computer Society, Washington, DC (2009)
17. Brazdil, P.B., Soares, C.: A comparison of ranking methods for classification algorithm selection. In: López de Mántaras, R., Plaza, E. (eds.) ECML 2000. LNCS (LNAI), vol. 1810, pp. 63–75. Springer, Heidelberg (2000). doi:10.1007/3-540-45164-1_8
18. Breiman, L.: Bagging predictors. Mach. Learn. **24**(2), 123–140 (1996)
19. Brewer, E.A.: High-level optimization via automated statistical modeling. In: Proceedings of the 5th ACM SIGPLAN Symposium on Principles and Practice of Parallel Programming PPOPP 1995, pp. 80–91. ACM, New York (1995)
20. Brodley, C.E.: Addressing the selective superiority problem: automatic algorithm/model class selection. In: ICML, pp. 17–24 (1993)
21. Cahill, E.: Knowledge-based algorithm construction for real-world engineering PDEs. Math. Comput. Simul. **36**(4–6), 389–400 (1994)
22. Carbonell, J., Etzioni, O., Gil, Y., Joseph, R., Knoblock, C., Minton, S., Veloso, M.: PRODIGY: an integrated architecture for planning and learning. SIGART Bull. **2**, 51–55 (1991)
23. Carchrae, T., Beck, J.C.: Low-knowledge algorithm control. In: AAAI, pp. 49–54 (2004)
24. Carchrae, T., Beck, J.C.: Applying machine learning to Low-knowledge control of optimization algorithms. Comput. Intell. **21**(4), 372–387 (2005)
25. Caseau, Y., Laburthe, F., Silverstein, G.: A meta-heuristic factory for vehicle routing problems. In: Jaffar, J. (ed.) CP 1999. LNCS, vol. 1713, pp. 144–158. Springer, Heidelberg (1999). doi:10.1007/978-3-540-48085-3_11
26. Cheeseman, P., Kanefsky, B., Taylor, W.M.: Where the really hard problems are. In: 12th International Joint Conference on Artificial Intelligence, pp. 331–337. Morgan Kaufmann Publishers Inc, San Francisco, CA, USA (1991)
27. Cicirello, V.A., Smith, S.F.: The max k-armed bandit: a new model of exploration applied to search heuristic selection. In: Proceedings of the 20th National Conference on Artificial Intelligence, pp. 1355–1361. AAAI Press (2005)
28. Cook, D.J., Varnell, R.C.: Maximizing the benefits of parallel search using machine learning. In: Proceedings of the 14th National Conference on Artificial Intelligence, pp. 559–564. AAAI Press (1997)
29. Demmel, J., Dongarra, J., Eijkhout, V., Fuentes, E., Petitet, A., Vuduc, R., Whaley, R.C., Yelick, K.: Self-adapting linear algebra algorithms and software. Proc. IEEE **93**(2), 293–312 (2005)
30. Dietterich, T.G.: Ensemble methods in machine learning. In: Kittler, J., Roli, F. (eds.) MCS 2000. LNCS, vol. 1857, pp. 1–15. Springer, Heidelberg (2000). doi:10.1007/3-540-45014-9_1
31. Domingos, P.: How to get a free lunch: a simple cost model for machine learning applications. In: AAAI98/ICML98 Workshop on the Methodology of Applying Machine Learning, pp. 1–7. AAAI Press (1998)
32. Domshlak, C., Karpas, E., Markovitch, S.: To max or not to max: online learning for speeding up optimal planning. In: AAAI (2010)

33. Elsayed, S.A.M., Michel, L.: Synthesis of search algorithms from high-level CP models. In: Proceedings of the 9th International Workshop on Constraint Modelling and Reformulation, September 2010
34. Elsayed, S.A.M., Michel, L.: Synthesis of search algorithms from high-level CP models. In: Lee, J. (ed.) CP 2011. LNCS, vol. 6876, pp. 256–270. Springer, Heidelberg (2011). doi:10.1007/978-3-642-23786-7_21
35. Epstein, S.L., Freuder, E.C.: Collaborative learning for constraint solving. In: Walsh, T. (ed.) CP 2001. LNCS, vol. 2239, pp. 46–60. Springer, Heidelberg (2001). doi:10.1007/3-540-45578-7_4
36. Epstein, S.L., Freuder, E.C., Wallace, R., Morozov, A., Samuels, B.: The adaptive constraint engine. In: Hentenryck, P. (ed.) CP 2002. LNCS, vol. 2470, pp. 525–540. Springer, Heidelberg (2002). doi:10.1007/3-540-46135-3_35
37. Ewald, R., Schulz, R., Uhrmacher, A.M.: Selecting simulation algorithm portfolios by genetic algorithms. In: IEEE Workshop on Principles of Advanced and Distributed Simulation PADS 2010, IEEE Computer Society, Washington, DC (2010)
38. Fawcett, C., Vallati, M., Hutter, F., Hoffmann, J., Hoos, H., Leyton-Brown, K.: Improved features for runtime prediction of domain-independent planners. In: ICAPS (2014)
39. Fink, E.: Statistical selection among problem-solving methods. Technical report CMU-CS-97-101. Carnegie Mellon University (1997)
40. Fink, E.: How to solve it automatically: selection among problem-solving methods. In: Proceedings of the 4th International Conference on Artificial Intelligence Planning Systems, pp. 128–136. AAAI Press (1998)
41. Fukunaga, A.S.: Genetic algorithm portfolios. IEEE Congr. Evol. Comput. 2, 1304–1311 (2000)
42. Fukunaga, A.S.: Automated discovery of composite SAT variable-selection heuristics. In: 18th National Conference on Artificial Intelligence, pp. 641–648. American Association for Artificial Intelligence, Menlo Park (2002)
43. Fukunaga, A.S.: Automated discovery of local search heuristics for satisfiability testing. Evol. Comput. 16, 31–61 (2008)
44. Gagliolo, M., Schmidhuber, J.: A neural network model for inter-problem adaptive online time allocation. In: Duch, W., Kacprzyk, J., Oja, E., Zadrożny, S. (eds.) ICANN 2005. LNCS, vol. 3697, pp. 7–12. Springer, Heidelberg (2005). doi:10.1007/11550907_2
45. Gagliolo, M., Schmidhuber, J.: Impact of censored sampling on the performance of restart strategies. In: Benhamou, F. (ed.) CP 2006. LNCS, vol. 4204, pp. 167–181. Springer, Heidelberg (2006). doi:10.1007/11889205_14
46. Gagliolo, M., Schmidhuber, J.: Learning dynamic algorithm portfolios. Ann. Math. Artif. Intell. 47(3–4), 295–328 (2006)
47. Gagliolo, M., Schmidhuber, J.: Towards distributed algorithm portfolios. In: Corchado, J.M., Rodríguez, S., Llinas, J., Molina, J.M. (eds.) Advances in Soft Computing. AINSC, vol. 50, pp. 634–643. Springer, Heidelberg (2008). doi:10.1007/978-3-540-85863-8_75
48. Gagliolo, M., Schmidhuber, J.: Algorithm portfolio selection as a bandit problem with unbounded losses. Ann. Math. Artif. Intell. 61(2), 49–86 (2011)
49. Gagliolo, M., Zhumatiy, V., Schmidhuber, J.: Adaptive online time allocation to search algorithms. In: Boulicaut, J.-F., Esposito, F., Giannotti, F., Pedreschi, D. (eds.) ECML 2004. LNCS (LNAI), vol. 3201, pp. 134–143. Springer, Heidelberg (2004). doi:10.1007/978-3-540-30115-8_15

50. Garrido, P., Riff, M.: DVRP: a hard dynamic combinatorial optimisation problem tackled by an evolutionary hyper-heuristic. J. Heuristics **16**, 795–834 (2010)
51. Gebruers, C., Guerri, A., Hnich, B., Milano, M.: Making choices using structure at the instance level within a case based reasoning framework. In: CPAIOR, pp. 380–386 (2004)
52. Gebruers, C., Hnich, B., Bridge, D., Freuder, E.: Using CBR to select solution strategies in constraint programming. In: Proceedings of ICCBR 2005, pp. 222–236 (2005)
53. Gebser, M., Kaminski, R., Kaufmann, B., Schaub, T., Schneider, M.T., Ziller, S.: A portfolio solver for answer set programming: preliminary report. In: Delgrande, J.P., Faber, W. (eds.) LPNMR 2011. LNCS (LNAI), vol. 6645, pp. 352–357. Springer, Heidelberg (2011). doi:10.1007/978-3-642-20895-9_40
54. Gent, I., Jefferson, C., Kotthoff, L., Miguel, I., Moore, N., Nightingale, P., Petrie, K.: Learning when to use lazy learning in constraint solving. In: 19th European Conference on Artificial Intelligence, pp. 873–878, August 2010
55. Gent, I., Kotthoff, L., Miguel, I., Nightingale, P.: Machine learning for constraint solver design - a case study for the alldifferent constraint. In: 3rd Workshop on Techniques for implementing Constraint Programming Systems (TRICS), pp. 13–25 (2010)
56. Gerevini, A.E., Saetti, A., Vallati, M.: An automatically configurable portfolio-based planner with macro-actions: PbP. In: Proceedings of the 19th International Conference on Automated Planning and Scheduling, pp. 350–353 (2009)
57. Gomes, C.P., Selman, B.: Algorithm portfolio design: theory vs. practice. In: UAI, pp. 190–197 (1997)
58. Gomes, C.P., Selman, B.: Practical aspects of algorithm portfolio design. In: Proceedings of 3rd ILOG International Users Meeting (1997)
59. Gomes, C.P., Selman, B.: Algorithm portfolios. Artif. Intell. **126**(1–2), 43–62 (2001)
60. Gratch, J., DeJong, G.: COMPOSER: a probabilistic solution to the utility problem in speed-up learning. In: AAAI, pp. 235–240 (1992)
61. Guerri, A., Milano, M.: Learning techniques for automatic algorithm portfolio selection. In: ECAI, pp. 475–479 (2004)
62. Guo, H.: Algorithm selection for sorting and probabilistic inference: a machine learning-based approach. Ph.D. thesis, Kansas State University (2003)
63. Guo, H., Hsu, W.H.: A learning-based algorithm selection meta-reasoner for the real-time MPE problem. In: Webb, G.I., Yu, X. (eds.) AI 2004. LNCS (LNAI), vol. 3339, pp. 307–318. Springer, Heidelberg (2004). doi:10.1007/978-3-540-30549-1_28
64. Haim, S., Walsh, T.: Restart strategy selection using machine learning techniques. In: Kullmann, O. (ed.) SAT 2009. LNCS, vol. 5584, pp. 312–325. Springer, Heidelberg (2009). doi:10.1007/978-3-642-02777-2_30
65. Hogg, T., Huberman, B.A., Williams, C.P.: Phase transitions and the search problem. Artif. Intell. **81**(1–2), 1–15 (1996)
66. Hong, L., Page, S.E.: Groups of diverse problem solvers can outperform groups of high-ability problem solvers. Proc. Natl. Acad. Sci. U.S.A. **101**(46), 16385–16389 (2004)
67. Hoos, H., Lindauer, M., Schaub, T.: claspfolio 2: Advances in algorithm selection for answer set programming. TPLP **14**(4–5), 569–585 (2014)
68. Hoos, H.H.: Programming by optimization. Commun. ACM **55**(2), 70–80 (2012)

69. Hoos, H.H., Kaminski, R., Lindauer, M., Schaub, T.: aspeed: Solver scheduling via answer set programming. Theory Pract. Logic Program. FirstView **15**, 1–26 (2014)

70. Horvitz, E., Ruan, Y., Gomes, C.P., Kautz, H.A., Selman, B., Chickering, D.M.: A Bayesian approach to tackling hard computational problems. In: Proceedings of the 17th Conference in Uncertainty in Artificial Intelligence, pp. 235–244. Morgan Kaufmann Publishers Inc., San Francisco (2001)

71. Hough, P.D., Williams, P.J.: Modern machine learning for automatic optimization algorithm selection. In: Proceedings of the INFORMS Artificial Intelligence and Data Mining Workshop, November 2006

72. Howe, A.E., Dahlman, E., Hansen, C., Scheetz, M., Mayrhauser, A.: Exploiting competitive planner performance. In: Biundo, S., Fox, M. (eds.) ECP 1999. LNCS (LNAI), vol. 1809, pp. 62–72. Springer, Heidelberg (2000). doi:10.1007/10720246_5

73. Huberman, B.A., Lukose, R.M., Hogg, T.: An economics approach to hard computational problems. Science **275**(5296), 51–54 (1997)

74. Hurley, B., Kotthoff, L., Malitsky, Y., O'Sullivan, B.: Proteus: a hierarchical portfolio of solvers and transformations. In: CPAIOR, May 2014

75. Hutter, F., Hamadi, Y., Hoos, H.H., Leyton-Brown, K.: Performance prediction and automated tuning of randomized and parametric algorithms. In: Benhamou, F. (ed.) CP 2006. LNCS, vol. 4204, pp. 213–228. Springer, Heidelberg (2006). doi:10.1007/11889205_17

76. Hutter, F., Hoos, H.H., Leyton-Brown, K.: Sequential model-based optimization for general algorithm configuration. In: Coello, C.A.C. (ed.) LION 2011. LNCS, vol. 6683, pp. 507–523. Springer, Heidelberg (2011). doi:10.1007/978-3-642-25566-3_40

77. Hutter, F., Hoos, H.H., Leyton-Brown, K.: Parallel algorithm configuration. In: Hamadi, Y., Schoenauer, M. (eds.) LION. LNCS, vol. 7219, pp. 55–70. Springer, Heidelberg (2012). doi:10.1007/978-3-642-34413-8_5

78. Hutter, F., Hoos, H.H., Leyton-Brown, K., Stützle, T.: ParamILS: an automatic algorithm configuration framework. J. Artif. Int. Res. **36**(1), 267–306 (2009)

79. Hutter, F., Hoos, H.H., Stützle, T.: Automatic algorithm configuration based on local search. In: Proceedings of the 22nd National Conference on Artificial Intelligence, pp. 1152–1157. AAAI Press (2007)

80. Kadioglu, S., Malitsky, Y., Sabharwal, A., Samulowitz, H., Sellmann, M.: Algorithm selection and scheduling. In: 17th International Conference on Principles and Practice of Constraint Programming, pp. 454–469 (2011)

81. Kadioglu, S., Malitsky, Y., Sellmann, M., Tierney, K.: ISAC instance-specific algorithm configuration. In: 19th European Conference on Artificial Intelligence, pp. 751–756. IOS Press (2010)

82. Kamel, M.S., Enright, W.H., Ma, K.S.: ODEXPERT: an expert system to select numerical solvers for initial value ODE systems. ACM Trans. Math. Softw. **19**(1), 44–62 (1993)

83. Kotthoff, L.: Hybrid regression-classification models for algorithm selection. In: 20th European Conference on Artificial Intelligence, pp. 480–485, August 2012

84. Kotthoff, L.: Algorithm selection for combinatorial search problems: a survey. AI Mag. **35**(3), 48–60 (2014)

85. Kotthoff, L., Gent, I.P., Miguel, I.: An evaluation of machine learning in algorithm selection for search problems. AI Commun. **25**(3), 257–270 (2012)

86. Kotthoff, L., Kerschke, P., Hoos, H., Trautmann, H.: Improving the state of the art in inexact TSP solving using per-instance algorithm selection. In: Dhaenens, C., Jourdan, L., Marmion, M.-E. (eds.) LION 2015. LNCS, vol. 8994, pp. 202–217. Springer, Heidelberg (2015). doi:10.1007/978-3-319-19084-6_18

87. Kotthoff, L., Miguel, I., Nightingale, P.: Ensemble classification for constraint solver configuration. In: 16th International Conference on Principles and Practices of Constraint Programming, pp. 321–329, September 2010

88. Kroer, C., Malitsky, Y.: Feature filtering for Instance-Specific algorithm configuration. In: Proceedings of the 23rd International Conference on Tools with Artificial Intelligence (2011)

89. Kuefler, E., Chen, T.-Y.: On using reinforcement learning to solve sparse linear systems. In: Bubak, M., Albada, G.D., Dongarra, J., Sloot, P.M.A. (eds.) ICCS 2008. LNCS, vol. 5101, pp. 955–964. Springer, Heidelberg (2008). doi:10.1007/978-3-540-69384-0_100

90. Lagoudakis, M.G., Littman, M.L.: Algorithm selection using reinforcement learning. In: Proceedings of the 17th International Conference on Machine Learning, pp. 511–518. Morgan Kaufmann Publishers Inc., San Francisco (2000)

91. Lagoudakis, M.G., Littman, M.L.: Learning to select branching rules in the DPLL procedure for satisfiability. In: LICS/SAT, pp. 344–359 (2001)

92. Langley, P.: Learning effective search heuristics. In: IJCAI, pp. 419–421 (1983)

93. Langley, P.: Learning search strategies through discrimination. Int. J. Man-Mach. Stud. 18, 513–541 (1983)

94. Leite, R., Brazdil, P., Vanschoren, J., Queiros, F.: Using active testing and meta-level information for selection of classification algorithms. In: 3rd PlanLearn Workshop, August 2010

95. Leyton-Brown, K., Nudelman, E., Shoham, Y.: Learning the empirical hardness of optimization problems: the case of combinatorial auctions. In: Hentenryck, P. (ed.) CP 2002. LNCS, vol. 2470, pp. 556–572. Springer, Heidelberg (2002). doi:10.1007/3-540-46135-3_37

96. Leyton-Brown, K., Nudelman, E., Shoham, Y.: Empirical hardness models: methodology and a case study on combinatorial auctions. J. ACM 56, 1–52 (2009)

97. Lindauer, M., Hoos, H., Hutter, F.: From sequential algorithm selection to parallel portfolio selection. In: Dhaenens, C., Jourdan, L., Marmion, M.-E. (eds.) LION 2015. LNCS, vol. 8994, pp. 1–16. Springer, Heidelberg (2015). doi:10.1007/978-3-319-19084-6_1

98. Lindauer, M., Hoos, H.H., Hutter, F., Schaub, T.: AutoFolio: algorithm configuration for algorithm selection. In: Twenty-Ninth AAAI Workshops on Artificial Intelligence, January 2015

99. Little, J., Gebruers, C., Bridge, D., Freuder, E.: Capturing constraint programming experience: a case-based approach. In: Modref (2002)

100. Lobjois, L., Lemaître, M.: Branch and bound algorithm selection by performance prediction. In: Proceedings of the 15th National/10th Conference on Artificial Intelligence/Innovative Applications of Artificial Intelligence, pp. 353–358. American Association for Artificial Intelligence, Menlo Park (1998)

101. Malitsky, Y., Sabharwal, A., Samulowitz, H., Sellmann, M.: Non-model-based algorithm portfolios for SAT. In: Sakallah, K.A., Simon, L. (eds.) SAT 2011. LNCS, vol. 6695, pp. 369–370. Springer, Heidelberg (2011). doi:10.1007/978-3-642-21581-0_33

102. Malitsky, Y., Sabharwal, A., Samulowitz, H., Sellmann, M.: Algorithm portfolios based on cost-sensitive hierarchical clustering. In: IJCAI, August 2013

103. Minton, S.: An analytic learning system for specializing heuristics. In: Proceedings of the 13th International Joint Conference on Artifical Intelligence IJCAI 1993, pp. 922–928. Morgan Kaufmann Publishers Inc., San Francisco (1993)

104. Minton, S.: Integrating heuristics for constraint satisfaction problems: a case study. In: Proceedings of the 11th National Conference on Artificial Intelligence, pp. 120–126. AAAI (1993)

105. Minton, S.: Automatically configuring constraint satisfaction programs: a case study. Constraints 1, 7–43 (1996)

106. Musliu, N., Schwengerer, M.: Algorithm selection for the graph coloring problem. In: Nicosia, G., Pardalos, P. (eds.) LION 2013. LNCS, vol. 7997, pp. 389–403. Springer, Heidelberg (2013). doi:10.1007/978-3-642-44973-4_42

107. Nareyek, A.: Choosing search heuristics by non-stationary reinforcement learning. In: Nareyek, A. (ed.) Metaheuristics: Computer Decision-Making. Applied Optimization, vol. 86, pp. 523–544. Kluwer Academic Publishers, New York (2001)

108. Nikolić, M., Marić, F., Janičić, P.: Instance-based selection of policies for SAT solvers. In: Kullmann, O. (ed.) SAT 2009. LNCS, vol. 5584, pp. 326–340. Springer, Heidelberg (2009). doi:10.1007/978-3-642-02777-2_31

109. Nudelman, E., Leyton-Brown, K., Hoos, H.H., Devkar, A., Shoham, Y.: Understanding random SAT: beyond the clauses-to-variables ratio. In: Wallace, M. (ed.) CP 2004. LNCS, vol. 3258, pp. 438–452. Springer, Heidelberg (2004). doi:10.1007/978-3-540-30201-8_33

110. O'Mahony, E., Hebrard, E., Holland, A., Nugent, C., O'Sullivan, B.: Using case-based reasoning in an algorithm portfolio for constraint solving. In: Proceedings of the 19th Irish Conference on Artificial Intelligence and Cognitive Science (2008)

111. Opitz, D., Maclin, R.: Popular ensemble methods: an empirical study. J. Artif. Intell. Res. 11, 169–198 (1999)

112. Paparrizou, A., Stergiou, K.: Evaluating simple fully automated heuristics for adaptive constraint propagation. In: ICTAI (2012)

113. Petrik, M.: Statistically optimal combination of algorithms. In: Local Proceedings of SOFSEM 2005 (2005)

114. Petrik, M., Zilberstein, S.: Learning parallel portfolios of algorithms. Ann. Math. Artif. Intell. 48(1–2), 85–106 (2006)

115. Petrovic, S., Qu, R.: Case-based reasoning as a heuristic selector in hyper-heuristic for course timetabling problems. In: KES, pp. 336–340 (2002)

116. Pfahringer, B., Bensusan, H., Giraud-Carrier, C.G.: Meta-Learning by landmarking various learning algorithms. In: 17th International Conference on Machine Learning ICML 2000, pp. 743–750, Morgan Kaufmann Publishers Inc., San Francisco (2000)

117. Pulina, L., Tacchella, A.: A multi-engine solver for quantified Boolean formulas. In: Bessière, C. (ed.) CP 2007. LNCS, vol. 4741, pp. 574–589. Springer, Heidelberg (2007). doi:10.1007/978-3-540-74970-7_41

118. Pulina, L., Tacchella, A.: A self-adaptive multi-engine solver for quantified boolean formulas. Constraints 14(1), 80–116 (2009)

119. Rao, R.B., Gordon, D., Spears, W.: For every generalization action, is there really an equal and opposite reaction? Analysis of the conservation law for generalization performance. In: Proceedings of the 12th International Conference on Machine Learning, pp. 471–479. Morgan Kaufmann (1995)

120. Rice, J.R.: The algorithm selection problem. Adv. Comput. 15, 65–118 (1976)

121. Rice, J.R., Ramakrishnan, N.: How to get a free lunch (at no cost). Techical report 99–014, Purdue University, April 1999

122. Roberts, M., Howe, A.E.: Directing a portfolio with learning. In: AAAI 2006 Workshop on Learning for Search (2006)

123. Roberts, M., Howe, A.E.: Learned models of performance for many planners. In: ICAPS 2007 Workshop AI Planning and Learning (2007)

124. Roberts, M., Howe, A.E., Wilson, B., des Jardins, M.: What makes planners predictable? In: ICAPS, pp. 288–295 (2008)

125. Sakkout, H., Wallace, M.G., Richards, E.B.: An instance of adaptive constraint propagation. In: Freuder, E.C. (ed.) CP 1996. LNCS, vol. 1118, pp. 164–178. Springer, Heidelberg (1996). doi:10.1007/3-540-61551-2_73

126. Samulowitz, H., Memisevic, R.: Learning to solve QBF. In: Proceedings of the 22nd National Conference on Artificial Intelligence, pp. 255–260. AAAI Press (2007)

127. Sayag, T., Fine, S., Mansour, Y.: Combining multiple heuristics. In: Durand, B., Thomas, W. (eds.) STACS 2006. LNCS, vol. 3884, pp. 242–253. Springer, Heidelberg (2006). doi:10.1007/11672142_19

128. Schapire, R.E.: The strength of weak learnability. Mach. Learn. 5(2), 197–227 (1990)

129. Sillito, J.: Improvements to and estimating the cost of solving constraint satisfaction problems. Master's thesis, University of Alberta (2000)

130. Silverthorn, B., Miikkulainen, R.: Latent class models for algorithm portfolio methods. In: Proceedings of the 24th AAAI Conference on Artificial Intelligence (2010)

131. Smith, T.E., Setliff, D.E.: Knowledge-based constraint-driven software synthesis. In: Knowledge-Based Software Engineering Conference, pp. 18–27, September 1992

132. Smith-Miles, K., Lopes, L.: Measuring instance difficulty for combinatorial optimization problems. Comput. Oper. Res. 39(5), 875–889 (2012)

133. Smith-Miles, K.A.: Cross-disciplinary perspectives on meta-learning for algorithm selection. ACM Comput. Surv. 41, 6: 1–6: 25 (2008)

134. Smith-Miles, K.A.: Towards insightful algorithm selection for optimisation using meta-learning concepts. In: IEEE International Joint Conference on Neural Networks, pp. 4118–4124, June 2008

135. Soares, C., Brazdil, P.B., Kuba, P.: A meta-learning method to select the kernel width in support vector regression. Mach. Learn. 54(3), 195–209 (2004)

136. Stergiou, K.: Heuristics for dynamically adapting propagation in constraint satisfaction problems. AI Commun. 22(3), 125–141 (2009)

137. Stern, D.H., Samulowitz, H., Herbrich, R., Graepel, T., Pulina, L., Tacchella, A.: Collaborative expert portfolio management. In: AAAI, pp. 179–184 (2010)

138. Streeter, M.J., Golovin, D., Smith, S.F.: Combining multiple heuristics online. In: Proceedings of the 22nd National Conference on Artificial Intelligence, pp. 1197–1203. AAAI Press (2007)

139. Streeter, M.J., Golovin, D., Smith, S.F.: Restart schedules for ensembles of problem instances. In: Proceedings of the 22nd National Conference on Artificial Intelligence, pp. 1204–1210. AAAI Press (2007)

140. Streeter, M.J., Smith, S.F.: New techniques for algorithm portfolio design. In: UAI, pp. 519–527 (2008)

141. Terashima-Marín, H., Ross, P., Valenzuela-Rendón, M.: Evolution of constraint satisfaction strategies in examination timetabling. In: Proceedings of the Genetic and Evolutionary Computation Conference, pp. 635–642. Morgan Kaufmann (1999)

142. Tolpin, D., Shimony, S.E.: Rational deployment of CSP heuristics. In: IJCAI, pp. 680–686 (2011)

143. Tsang, E.P.K., Borrett, J.E., Kwan, A.C.M.: An attempt to map the performance of a range of algorithm and heuristic combinations. In: Proceedings of AISB 1995, pp. 203–216. IOS Press (1995)

144. Utgoff, P.E.: Perceptron trees: a case study in hybrid concept representations. In: National Conference on Artificial Intelligence, pp. 601–606 (1988)

145. Vassilevska, V., Williams, R., Woo, S.L.M.: Confronting hardness using a hybrid approach. In: Proceedings of the 17th Annual ACM-SIAM Symposium on Discrete Algorithms SODA 2006, pp. 1–10. ACM, New York (2006)

146. Vrakas, D., Tsoumakas, G., Bassiliades, N., Vlahavas, I.: Learning rules for adaptive planning. In: Proceedings of the 13th International Conference on Automated Planning and Scheduling, pp. 82–91 (2003)

147. Wang, J., Tropper, C.: Optimizing time warp simulation with reinforcement learning techniques. In: Proceedings of the 39th Conference on Winter simulation WSC 2007, pp. 577–584. IEEE Press, Piscataway (2007)

148. Watson, J.: Empirical modeling and analysis of local search algorithms for the job-shop scheduling problem. Ph.D. thesis, Colorado State University, Fort Collins, CO, USA (2003)

149. Weerawarana, S., Houstis, E.N., Rice, J.R., Joshi, A., Houstis, C.E.: PYTHIA: a knowledge-based system to select scientific algorithms. ACM Trans. Math. Softw. $22(4)$, 447–468 (1996)

150. Wei, W., Li, C.M., Zhang, H.: Switching among non-weighting, clause weighting, and variable weighting in local search for SAT. In: Stuckey, P.J. (ed.) CP 2008. LNCS, vol. 5202, pp. 313–326. Springer, Heidelberg (2008). doi:10.1007/978-3-540-85958-1_21

151. Wilson, D., Leake, D., Bramley, R.: Case-based recommender components for scientific problem-solving environments. In: Proceedings of the 16th International Association for Mathematics and Computers in Simulation World Congress (2000)

152. Wolpert, D.H.: Stacked generalization. Neural Netw. **5**, 241–259 (1992)

153. Wolpert, D.H.: The supervised learning no-free-lunch theorems. In: Proceedings of the 6th Online World Conference on Soft Computing in Industrial Applications, pp. 25–42 (2001)

154. Wolpert, D.H., Macready, W.G.: No free lunch theorems for optimization. IEEE Trans. Evol. Comput. **1**(1), 67–82 (1997)

155. Wu, H., van Beek, P.: On portfolios for backtracking search in the presence of deadlines. In: Proceedings of the 19th IEEE International Conference on Tools with Artificial Intelligence, pp. 231–238. IEEE Computer Society, Washington, DC (2007)

156. Xu, L., Hoos, H.H., Leyton-Brown, K.: Hierarchical hardness models for SAT. In: CP, pp. 696–711 (2007)

157. Xu, L., Hoos, H.H., Leyton-Brown, K.: Hydra: automatically configuring algorithms for portfolio-based selection. In: 24th Conference of the Association for the Advancement of Artificial Intelligence (AAAI 2010), pp. 210–216 (2010)

158. Xu, L., Hutter, F., Hoos, H.H., Leyton-Brown, K.: SATzilla-07: the design and analysis of an algorithm portfolio for SAT. In: CP, pp. 712–727 (2007)

159. Xu, L., Hutter, F., Hoos, H.H., Leyton-Brown, K.: SATzilla: portfolio-based algorithm selection for SAT. J. Artif. Intell. Res. (JAIR) **32**, 565–606 (2008)

160. Xu, L., Hutter, F., Hoos, H.H., Leyton-Brown, K.: SATzilla2009: an automatic algorithm portfolio for SAT. In: 2009 SAT Competition (2009)

161. Xu, L., Hutter, F., Hoos, H.H., Leyton-Brown, K.: Hydra-MIP: automated algorithm configuration and selection for mixed integer programming. In: RCRA Workshop on Experimental Evaluation of Algorithms for Solving Problems with Combinatorial Explosion at the International Joint Conference on Artificial Intelligence (IJCAI) (2011)
162. Xu, L., Hutter, F., Hoos, H., Leyton-Brown, K.: Evaluating component solver contributions to portfolio-based algorithm selectors. In: Cimatti, A., Sebastiani, R. (eds.) SAT 2012. LNCS, vol. 7317, pp. 228–241. Springer, Heidelberg (2012). doi:10.1007/978-3-642-31612-8_18
163. Yu, H., Rauchwerger, L.: An adaptive algorithm selection framework for reduction parallelization. IEEE Trans. Parallel Distrib. Syst. **17**(10), 1084–1096 (2006)
164. Yu, H., Zhang, D., Rauchwerger, L.: An adaptive algorithm selection framework. In: Proceedings of the 13th International Conference on Parallel Architectures and Compilation Techniques, pp. 278–289. IEEE Computer Society, Washington, DC (2004)
165. Yun, X., Epstein, S.L.: Learning algorithm portfolios for parallel execution. In: Hamadi, Y., Schoenauer, M. (eds.) Proceedings of the 6th International Conference Learning and Intelligent Optimisation LION. LNCS, vol. 7219, pp. 323–338. Springer, Heidelberg (2012)

Advanced Portfolio Techniques

Barry Hurley[1], Lars Kotthoff[2]([⊠]), Yuri Malitsky[1], Deepak Mehta[1],
and Barry O'Sullivan[1]

[1] Insight Centre for Data Analytics, University College Cork, Cork, Ireland
[2] University of British Columbia, Vancouver, Canada
larsko@cs.ubc.ca

Abstract. There exists a proliferation of different approaches to using
portfolios and algorithm selection to make solving combinatorial search
and optimisation problems more efficient, as surveyed in the previous
chapter. In this chapter, we take a look at a detailed case study that leverages transformations between problem representations to make portfolios
more effective, followed by extensions to the state of the art that make
algorithm selection more robust in practice.

1 Outline

In the previous chapter, a number of different portfolio and algorithm selection
techniques was introduced. At the start of this chapter, we take a detailed look at
a specific example system, Proteus, that leverages not only algorithm selection
techniques, but also the fact that there exist polynomial time transformations
between different representations of NP-complete problems. Using this novel
methodology, we combine the best of the worlds of SAT and CSP solving to create
Proteus, a hierarchical portfolio approach that achieves significant performance
improvements over approaches that do not use transformations.

After that, we present a series of techniques that can be used to improve the
performance of portfolios and algorithm selection. Specifically, we discuss two
novel algorithm portfolio techniques, ISAC+ and EISAC, showing how to first
iteratively train solvers for a better portfolio and then how to dynamically adapt
to changes in the observed problem instances. Finally, the chapter will show how
the performance of portfolios can be improved through the introduction of better
features, and will present a technique for automatically identifying the properties
of these desired features.

2 Proteus: A Hierarchical Portfolio of Solvers and Transformations

The pace of development in both CSP and SAT solver technology has been
rapid. Combined with portfolio and algorithm selection technology, impressive performance improvements over systems that have been developed only a

© Springer International Publishing AG 2016
C. Bessiere et al. (Eds.): Data Mining and Constraint Programming, LNAI 10101, pp. 191–225, 2016.
DOI: 10.1007/978-3-319-50137-6_8

few years previously have been demonstrated. Constraint satisfaction problems and satisfiability problems are both NP-complete and, therefore, there exist polynomial-time transformations between them. We can leverage this fact to convert CSPs into SAT problems and solve them using SAT solvers.

In this chapter we exploit the fact that different SAT solvers have different performances on different encodings of the same CSP. In fact, the particular choice of encoding that will give good performance with a particular SAT solver is dependent on the problem instance to be solved. We show that, in addition to using dedicated CSP solvers, to achieve the best performance for solving a CSP the best course of action might be to translate it to SAT and solve it using a SAT solver. We name our approach Proteus, after the Greek god Proteus, the shape-shifting water deity that can foretell the future.

Our approach offers a novel perspective on using SAT solvers for constraint solving. The idea of solving CSPs as SAT instances however is not new; the solvers Sugar, Azucar, and CSP2SAT4J are three examples of SAT-based CSP solving. Sugar [42] has been very competitive in recent CSP solver competitions. It converts the CSP to SAT using a specific encoding, known as the order encoding, which will be discussed in more detail later in this chapter. Azucar [43] is a related SAT-based CSP solver that uses the compact order encoding. However, both Sugar and Azucar use a single predefined solver to solve the encoded CSP instances. Our work does not assume that conversion using a specific encoding to SAT is the best way of solving a problem, but considers multiple candidate encodings and solvers. CSP2SAT4J [23] uses the SAT4J library as its SAT back-end and a set of static rules to choose either the direct or the support encoding for each constraint. For intensional and extensional binary constraints that specify the supports, it uses the support encoding. For all other constraints, it uses the direct encoding. Our approach does not have predefined rules but instead chooses the encoding and solver based on features of the problem instance to solve.

Our approach employs algorithm selection techniques to dynamically choose whether to translate to SAT, and if so, which SAT encoding and solver to use, otherwise it selects which CSP solver to use. However, the Proteus approach is not a straightforward application of portfolio techniques. In particular, there is a series of decisions to make that affect not only the solvers that will be available, but also the information that can be used to make the decision. Because of this, the different choices of conversions, encodings and solvers cannot simply be seen as different algorithms or different configurations of the same algorithm.

2.1   Multiple Encodings and Solvers

To motivate our work, we performed a detailed investigation for two solvers to assess the relationship between solver and problem encoding with features of the problem to be solved. For this experiment we considered uniform random binary (URB) CSPs with a fixed number of variables, domain size and number of constraints, and varied the constraint tightness. The constraint tightness t is a measure of the proportion of forbidden to allowed possible assignments to the variables in the scope of the constraint. We vary it from 0 to 1, where 0 means that all assignments are allowed and 1 that no assignments are part of a solution, in increments of 0.005. At each tightness the mean run-time of the solver on 100 random CSP instances is reported. Each instance contains 30 variables with domain size 20 and 300 constraints. This allowed us to study the performance of SAT encodings and solvers across the phase transition.

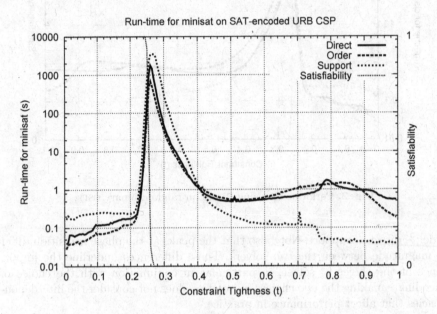

Fig. 1. Performance using `MiniSat` on random binary CSPs.

Figures 1 and 2 plot the run-time for `MiniSat` and `Clasp` on uniformly random binary CSPs that have been translated to SAT using three different encodings. Observe that in Fig. 1 there is a distinct difference in the performance of `MiniSat` on each of the encodings, sometimes an order of magnitude. Before the phase transition, we see that the order encoding achieves the best performance and maintains this until the phase transition. Beginning at constraint tightness 0.41, the order encoding gradually starts achieving poorer performance and the support encoding now achieves the best performance.

Notably, if we rank the encodings based on their performance, the ranking changes after the phase transition. This illustrates that there is not just a single encoding that will perform best overall and that the choice of encoding matters, but also that this choice is dependent on problem characteristics such as constraint tightness.

Around the phase transition, we observe contrasting performance for `Clasp`, as illustrated in Fig. 2. Using `Clasp`, the ranking of encodings around the phase transition is direct $\succ$ support $\succ$ order; whereas for `MiniSat` the ranking is

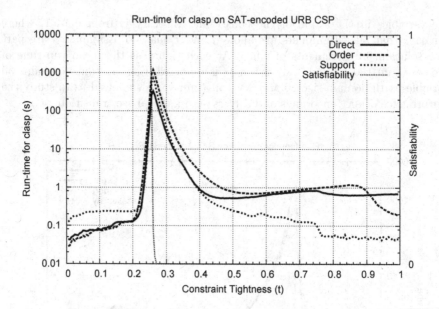

Fig. 2. Performance using Clasp on random binary CSPs.

order ≻ direct ≻ support. Note also that the peaks at the phase transition differ in magnitude between the two solvers. These differences underline the importance of the choice of solver, in particular in conjunction with the choice of encoding – making the two choices in isolation does not consider the interdependencies that affect performance in practice.

In addition to the random CSP instances, our analysis also comprises 1493 challenging benchmark problem instances from the CSP solver competitions that involve global and intensional constraints. Figure 3 illustrates the respective performance of the best CSP-based and SAT-based methods on these instances. Unsurprisingly the dedicated CSP methods often achieve the best performance. There are, however, numerous cases where considering SAT-based methods has the potential to yield significant performance improvements. In particular, there are a number of instances that are unsolved by any CSP solver but can be solved quickly using SAT-based methods. The Proteus approach aims to unify the best of both worlds and take advantage of the potential performance gains.

Setup. The hierarchical model we present in this chapter consists of a number of layers to determine how the instance should be solved. At the top level, we decide whether to solve the instance using a CSP or using a SAT-based method. If we choose to leave the problem as a CSP, then one of the dedicated CSP solvers must be chosen. Otherwise, we must choose the SAT encoding to apply, followed by the choice of SAT solver to run on the SAT-encoded instance.

Each decision of the hierarchical approach aims to choose the direction which has the potential to achieve the best performance in that sub-tree. For example,

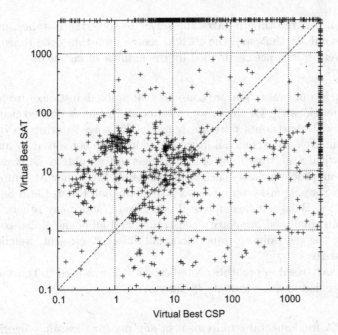

Fig. 3. Performance of the virtual best CSP portfolio and the virtual best SAT-based portfolio. Each point represents the time in seconds of the two approaches. A point below the dashed line indicates that the virtual best SAT portfolio was quicker, whereas a point above means the virtual best CSP portfolio was quicker. Clearly the two approaches are complementary: there are numerous instances for which a SAT-based approach does not perform well or fails to solve the instance but a CSP solver does extremely well, and vice-versa.

for the decision to choose whether to solve the instance using a SAT-based method or not, we choose the SAT-based direction if there is a SAT solver and encoding that will perform faster than any CSP solver would. Whether this particular encoding-solver combination will be selected subsequently depends on the performance of the algorithm selection models used in that sub-tree of our decision mechanism. For regression models, the training data is the best performance of any solver under that branch of the tree. For classification models, it is the label of the sub-branch with the virtual best performance.

This hierarchical approach presents the opportunity to employ different decision mechanisms at each level. We consider 6 regression, 19 classification, and 3 clustering algorithms, which are listed below. For each of these algorithms, we evaluate the performance using 10-fold cross-validation. The dataset is split into 10 partitions with approximately the same size and the same distribution of the best solvers. One partition is used for testing and the remaining 9 partitions as the training data for the model. This process is repeated with a different partition considered for testing each time until every partition has been used for testing. We measure the performance in terms of PAR10 score. The PAR10 score for an instance is the time it takes the solver to solve the instance, unless the solver times out. In this case, the PAR10 score is ten times the timeout value. The sum over all instances is divided by the number of instances.

Instances. In our evaluation, we consider CSP problem instances from the CSP solver competitions [1]. Of these, we consider all instances defined using global and intensional constraints that are not trivially solved during 2 s of feature computation. We also exclude all instances which were not solved by any CSP or SAT solver within the time limit of 1 h. Altogether, we obtain 1,493 non-trivial instances from problem classes such as Timetabling, Frequency Assignment, Job-Shop, Open-Shop, Quasi-group, Costas Array, Golomb Ruler, Latin Square, All Interval Series, Balanced Incomplete Block Design, and many others. This set includes both small and large arity constraints and all of the global constraints used during the CSP solver competitions: all-different, element, weighted sum, and cumulative.

For the SAT-based approaches, Numberjack [15] was used to translate a CSP instance specified in XCSP format [38] into SAT (CNF).

Features. A fundamental requirement of any machine learning algorithm is a set of representative features. We explore a number of different feature sets to train our models: *(i)* features of the original CSP instance, *(ii)* features of the direct-encoded SAT instance, *(iii)* features of the support-encoded SAT instance, *(iv)* features of the direct-order-encoded SAT instance and *(v)* a combination of all four feature sets. These features are described in further detail below.

We computed the 36 features used in CPHYDRA for each CSP instance using `Mistral`; for reasons of space we will not enumerate them all here. The set includes static features like statistics about the types of constraints used, average and maximum domain size; and dynamic statistics recorded by running `Mistral` for 2 s: average and standard deviation of variable weights, number of nodes, number of propagations and a few others. Instances which are solved by `Mistral` during feature computation are filtered out from the dataset.

In addition to the CSP features, we computed the 54 SAT features used by SATZILLA [53] for each of the encoded instances and different encodings. The features encode a wide range of different information on the problem such as problem size, features of the graph-based representation, balance features, the proximity to a Horn formula, DPLL probing features and local search probing features.

Solvers. Our CSP models are able to choose from 4 complete CSP solvers:

- `Abscon` [24],   - `Gecode` [11], and
- `Choco` [44],   - `Mistral` [14].

We additionally considered the following 6 complete SAT solvers:

– clasp [10],
– cryptominisat [40],
– glucose [4],

– lingeling [5],
– riss [30], and
– MiniSat [9].

The performance of each of the 6 SAT solvers was evaluated on the three SAT encodings of 1,493 CSP competition benchmarks with a time-out of 1 h and limited to 2 GB of RAM. The 4 CSP solvers were evaluated on the original CSPs. Our results report the PAR10 score and number of instances solved for each of the algorithms we evaluate. The PAR10 is the sum of the runtimes over all instances, counting 10 times the timeout if that was reached. Data was collected on a cluster of Intel Xeon E5430 Processors (2.66 Ghz) running CentOS 6.4. This data is available as ASlib[1] scenario PROTEUS-2014[2].

Learning Algorithms. We evaluate a number of regression, classification, and clustering algorithms using WEKA [13]. All algorithms, unless otherwise stated use the default parameters. The regression algorithms we used were Linear-Regression, PaceRegression, REPTree, M5Rules, M5P, and SMOreg. The classification algorithms were BayesNet, BFTree, ConjunctiveRule, DecisionTable, FT, HyperPipes, IBk (nearest neighbour) with 1, 3, 5 and 10 neighbours, J48, J48graft, JRip, LADTree, MultilayerPerceptron, OneR, PART, RandomForest, RandomForest with 99 random trees, RandomTree, REPTree, and SimpleLogistic. For clustering, we considered EM, FarthestFirst, and SimplekMeans. The FarthestFirst and SimplekMeans algorithms require the number of clusters to be given as input. We evaluated with multiples of 1 through 5 of the number of solvers in the respective data set given as the number of clusters. The number of clusters is represented by $1n$, $2n$ and so on in the name of the algorithm, where n stands for the number of solvers.

We use the LLAMA toolkit [22] to train and test the algorithm selection models.

Portfolio Results. The performance of a number of hierarchical approaches is given in Table 1. The hierarchy of algorithms which produced the best overall results for our dataset involves M5P regression with CSP features at the root node to choose SAT or CSP, M5P regression with CSP features to select the CSP solver, LinearRegression with CSP features to select the SAT encoding, LinearRegression with CSP features to select the SAT solver for the direct encoded instance, LinearRegression with CSP features to select the SAT solver for the direct-order encoded instance, and LinearRegression with the direct-order features to select the SAT solver for the support encoded instance. The hierarchical tree of specific machine learning approaches we found to deliver the best overall performance on our data set is labelled Proteus and is depicted in Fig. 4.

We would like to point out that in many solver competitions the difference between the top few solvers is fewer than 10 additional instances solved. In the

---

[1] http://aslib.net.
[2] https://github.com/coseal/aslib_data/tree/master/PROTEUS-2014.

Table 1. Performance of the learning algorithms for the hierarchical approach. The 'Category Bests' consists of the hierarchy of algorithms where at each node of the tree of decisions we take the algorithm that achieves the best PAR10 score for that particular decision.

Classifier	Mean PAR10	Number solved
VBS	97	1493
Proteus	1774	1424
M5P with csp features	2874	1413
Category Bests	2996	1411
M5Rules with csp features	3225	1398
M5P with all features	3405	1397
LinearRegression with all features	3553	1391
LinearRegression with csp features	3588	1383
MultilayerPerceptron with csp features	3594	1382
lm with csp features	3654	1380
RandomForest99 with csp features	3664	1379
IBk10 with csp features	3720	1377
RandomForest99 with all features	3735	1383

2012 SAT Challenge for example, the difference between the first and second place single solver was only 3 instances and the difference among the top 4 solvers was only 8 instances. The results we present in Table 1 are therefore very significant in terms of the gains we are able to achieve.

Our results demonstrate the power of Proteus. The performance it delivers is very close to the virtual best (VBS), that is the best performance possible if an oracle could identify the best choice of representation, encoding, and solver, on an instance by instance basis. The improvements we achieve over other approaches are similarly impressive. The results conclusively demonstrate that having the option to convert a CSP to SAT does not only have the potential to achieve significant performance improvements, but also does so in practice.

An interesting observation is that the CSP features are consistently used in each of the top performing approaches. One reason for this is that it is quicker to compute only the CSP features instead of the CSP features, then converting to SAT and computing the SAT features in addition. The additional overhead of computing SAT features is worthwhile in some cases though, for example for LinearRegression, which is at its best performance using all the different feature sets. Note that for the best tree of models (cf. Fig. 4), it is better to use the features of the direct-order encoding for the decision of which solver to choose for a support-encoded SAT instance despite the additional overhead.

We also compare the hierarchical approach to that of a flattened setting with a single portfolio of all solvers and encoding solver combinations. The flattened portfolio includes all possible combinations of the 3 encodings and the 6 SAT solvers and the 4 CSP solvers for a total of 22 solvers. Table 2 shows these

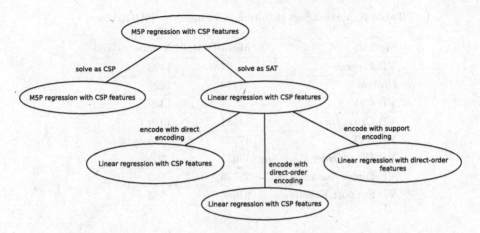

Fig. 4. Overview of the machine learning models used in the hierarchical approach.

Table 2. Ranking of each classification, regression, and clustering algorithm to choose the solving mechanism in a flattened setting. The portfolio consists of all possible combination of the 3 encodings and the 6 SAT solvers and the 4 CSP solvers for a total of 22 solvers.

Classifier	Mean PAR10	Number solved
VBS	97	1493
Proteus	1774	1424
LinearRegression with all features	2144	1416
M5P with csp features	2315	1401
LinearRegression with csp features	2334	1401
lm with all features	2362	1407
lm with csp features	2401	1398
M5P with all features	2425	1404
RandomForest99 with all features	2504	1401
SMOreg with all features	2749	1391
RandomForest with all features	2859	1386
IBk3 with csp features	2877	1378

results. The regression algorithm LinearRegression with all features gives the best performance using this approach. However, it is significantly worse than the performance achieved by the hierarchical approach of Proteus.

Greater Than the Sum of Its Parts. Given the performance of Proteus, the question remains as to whether a different portfolio approach that considers just CSP or just SAT solvers could do better. Table 3 summarizes the virtual best performance that such portfolios could achieve. We use all the CSP and SAT solvers

Table 3. Virtual best performances ranked by PAR10 score.

Method	Mean PAR10	Number solved
VB Proteus	97	1493
Proteus	1774	1424
VB CSP	3577	1349
VB CPHydra	4581	1310
VB SAT	17373	775
VB DirectOrder Encoding	17637	764
VB Direct Encoding	21736	593
VB Support Encoding	21986	583

for the respective portfolios to give us VB CSP and VB SAT, respectively. The former is the approach that always chooses the best CSP solver for the current instance, while the latter chooses the best SAT encoding/solver combination. VB Proteus is the portfolio that chooses the best overall approach/encoding. We show the actual performance of Proteus for comparison. Proteus is better than the virtual bests for all portfolios that consider only one encoding. This result makes a very strong point for the need to consider encoding and solver in combination.

Proteus outperforms four other VB portfolios. Specifically, the VB CPHYDRA is the best possible performance that could be obtained from that portfolio if a perfect choice of solver was made. Neither SATzILLA nor ISAC-based portfolios consider different SAT encodings. Therefore, the best possible performance either of them could achieve for a specific encoding is represented in the last three lines of Table 3.

These results do not only demonstrate the benefit of considering the different ways of solving CSPs, but also eliminate the need to compare with existing portfolio systems since we are computing the best possible performance that any of those systems could theoretically achieve. Proteus impressively demonstrates its strengths by significantly outperforming oracle approaches that use only a single encoding.

3   Advanced Portfolio Techniques

3.1   Automated Portfolio Generation

Algorithm selection is a great tool in practice, but in order for it to work, a diverse set of solvers is essential. This is not usually a problem for well established benchmark problem sets where solvers compete on a regular basis, but for the more esoteric domains, it is possible for only a handful, if not a single solver to exist. The question then becomes, if the sole existing solver is parameterized, is it possible to effectively generate multiple configurations of this solver to comprise a portfolio.

Note that it is certainly possible to create a portfolio that simply incorporates all possible configurations, thus choosing to rely on the algorithm selector to make the correct decisions. This approach of course also quickly degrades when the parameter space grows, not to mention that each of the possible parameterizations will need to be evaluated on all the instances. Even choosing a random subset of parameterizations will lead to problems if those parameterizations are not diverse or simply not particularly good. This section therefore shows how the existing ISAC portfolio methodology can be adjusted to this problem.

Note how ISAC solves a core problem of instance-specific algorithm tuning, namely the selection of a parameterization out of a very large and possibly even infinite pool of possible parameter settings. In algorithm portfolios we are dealing with a small set of solvers, and all methods devised for algorithm selection make heavy use of that fact. Clearly, this approach will not work when the number of solvers explodes. ISAC overcomes this problem by *clustering* the training instances. This is a key step in the ISAC methodology as described in [21]: Training instances are first clustered into groups and then a high-performance parameterization is computed for each of the clusters. That is, in ISAC clustering is used *both* for the *generation* of high-quality solver parameterizations, and then for the subsequent *selection* of the parameterization for a given test instance.

Beyond Cluster-Based Algorithm Selection. While [28] showed that cluster-based solver selection outperforms SATzilla-2009, this alone does not fully explain why ISAC often outperforms other instance-specific algorithm configurators like Hydra. Clustering instances upfront appears to give us an advantage when tuning individual parameterizations. Not only do we save a lot of tuning time with this methodology, since the training set for the instance-oblivious tuner is much smaller than the whole set. We also bundle instances together, hoping that they are somewhat similar and thus amenable for being solved efficiently with just one parameterization.

Consequently, we want to keep clustering in ISAC. However, and this is the core observation in this section, *once the parameterizations for each cluster have been computed, there is no reason why we would need to stick to these clusters for selecting the best parameterization for a given test instance.* Consequently, we propose to use an alternate state-of-the-art algorithm selector to choose the best parameterization for the instance we are to solve.

To this end, after ISAC finishes clustering and tuning the parameters of existing solvers on each cluster, we can then use any algorithm selector to choose one of the parameterizations, *independent of the cluster* an instance belongs to. For this final stage, we can use any efficient algorithm selector rather than continuing to rely on the original clusters. We call this overall process of configuring multiple solvers through clustering and effectively utilizing those solvers with state-of-the-art algorithm selection, as ISAC+.

Comparison of ISAC+ with ISAC and Hydra. Let us first compare the ISAC+ methodology with the popular alternatives, ISAC and Hydra. Note that

Table 4. SAT experiments

	Average	PAR1	PAR10	Solved	% solved
BS	28.71	289.3	2753	93	54.39
Hydra	19.80	260.7	2503	100	58.48
ISAC-GGA	18.79	297.5	2887	89	52.05
ISAC-MSC	18.24	273.4	2642	96	56.14
ISAC+	22.09	251.9	2395	103	60.23
VBS	16.40	228.0	2186	109	64.33

(a) HAND

	Average	PAR1	PAR10	Solved	% solved
BS	27.37	121.0	1004	486	83.64
Hydra	20.88	75.7	586.9	526	90.53
ISAC-GGA	22.11	154.4	1390	448	77.11
ISAC-MSC	27.47	79.7	572.3	528	90.88
ISAC+	24.77	71.1	506.3	534	91.91
VBS	15.96	61.2	479.5	536	92.25

(b) RAND

Hydra takes an iterative approach to portfolio generation. It first configures a solver over all of the instances, and subsequently trains new solvers with the objective of best complementing the existing set. Once the set of solvers is comprised, a SATzilla based portfolio algorithm is run to determine when each of them should be used. Note that unlike the ISAC based approaches, Hydra can potentially suffer since each configuration run is done in sequence, which can take a very long time.

For our comparison we use the benchmark set from [51] where Hydra was first introduced. In particular, there are two non-trivial sets of instances: Random (RAND) and Crafted (HAND).

Following the previously established methodology, we start our portfolio construction with 11 local search solvers: paws [45], rsaps [19], saps [47], agwsat0 [49], agwsat+ [50], agwsatp [48], gnovelty+ [34], g2wsat [27], ranov [33], vw [36], and anov09 [17]. We augment these solvers by adding six fixed parameterizations of SATenstein to this set, giving us a total of 17 constituent solvers.

We cluster the training instances of each dataset and add GGA [2] trained versions of SATenstein for each cluster, resulting in 11 new solvers for Random and 8 for Crafted. We use a timeout of 50 s when training these solvers, but employ a 600 s timeout to evaluate the solvers on each respective dataset. The times were measured on dual Intel Xeon 5540 (2.53 GHz) quad-core Nehalem processors and 24 GB of DDR-3 memory (1333 GHz).

Table 4a shows the test performance of various solvers on the HAND benchmark set (342 train and 171 test instances). We conduct 5 runs on each instance for each solver. When referring to a value as 'Average', given is the mean time it takes to solve only those instances that do not timeout. Like always, the value 'PAR1' includes the timeout instances when computing the average. 'PAR10', then gives a penalized average, where every instance that times out is treated as having taken 10 times the timeout to complete. Finally presented is the number of instances solved and the corresponding percentage of solved instances in the test set.

The best single solver (BS) is one of the SATenstein parameterizations tuned by GGA and is able to solve about 54% of all instances. Hydra solves 58% while ISAC-GGA (using only SATenstein) solves only 52%. Using the whole set of solvers for tuning (not just SATenstein), ISAC-MSC solves about 56% of all instances, which is worse than always selecting the best base solver. Of course, we only know a posteriori that this parameterization of SATenstein is the best solver for this test set. However, ISAC's performance is still not convincing. By augmenting the approach using a final portfolio selection stage, we can boost performance. ISAC+ solves ~60% of all test instances, outperforming all other approaches and closing almost 30% of the gap between Hydra and the Virtual Best Solver (VBS), an imaginary perfect oracle that always correctly picks the best solver and parameterization for each instance which marks an upper bound on the performance we may realistically hope for.

The second benchmark we present here is RAND. There are 581 test and 1141 train instances in this benchmark. In Table 4b we see that the best single solver (BS – gnovelty+) solves ~84% of the 581 instances in this test set. Hydra improves this to ~91%, roughly equal in performance to ISAC-MSC. ISAC+ improves performance again and leads to almost 92% of all instances solved within the time limit. The improved approach outperforms all other methods, and ISAC+ closes over 37% of the gap between the original ISAC and the VBS.

Note that using portfolios of the untuned SAT solvers only is in general not competitive as shown in [21,51]. To verify this finding we also ran a comparison using untuned base solvers only. On the SAT RAND data set, for example, we find that the portfolio algorithm using only 17 base solvers can only solve 520 instances, which is not competitive.

Extended Applicability. In the preceding section we demonstrated the potential effectiveness of the ISAC+ approach on SAT problems. We now apply this methodology to a larger problem than what the original Hydra and ISAC were tested on: the MaxSAT problem. Formally, the MaxSAT problem is the optimization version of the regular SAT problem. A *weighted clause* is a pair (C, w), where C is a clause and w is a natural number or infinity, indicating the penalty for falsifying the clause C. A *Weighted Partial MaxSAT formula* (WPMS) is a multiset of weighted clauses $\varphi = \{(C_1, w_1), \ldots, (C_m, w_m), (C_{m+1}, \infty), \ldots, (C_{m+m'}, \infty)\}$ where the first m clauses are *soft* and the last m' clauses are *hard*. Here, a *hard* clause is one that must be satisfied, while also satisfying the maximum combined weight of *soft* clauses. A *Partial MaxSAT formula* (PMS) is a WPMS formula where the weights of soft clauses are equal. The set of variables occurring in a formula φ is noted as var(φ).

A *(total) truth assignment* for a formula φ is a function $I : \text{var}(\varphi) \rightarrow \{0, 1\}$, that can be extended to literals, clauses, SAT formulas. For MaxSAT formulas is defined as $I(\{(C_1, w_1), \ldots, (C_m, w_m)\}) = \sum_{i=1}^{m} w_i (1 - I(C_i))$. The *optimal cost* of a formula is $\text{cost}(\varphi) = \min\{I(\varphi) \mid I : \text{var}(\varphi) \rightarrow \{0, 1\}\}$ and an *optimal assignment* is an assignment I such that $I(\varphi) = \text{cost}(\varphi)$.

The *Weighted Partial MaxSAT problem* for a Weighted Partial MaxSAT formula φ is the problem of finding an optimal assignment.

Given this setup, we conduct a 10-fold cross validation on the four categories of the 2012 MaxSAT Evaluation [3]. These are plain MaxSAT instances, weighted MaxSAT, partial MaxSAT, and weighted partial MaxSAT. The results of the cross validation are presented in Tables 5a–d. Specifically, each data set is broken uniformly at random into non overlapping subsets. Each of these subsets is then used as the test set (one at a time) while the instances from all other folds are used as training data. The tables present the average performance over 10-folds. Furthermore, all experiments were run with a 2,100 s timeout, on the same machines we used in the previous section. We use the following solvers: akmaxsat_ls, akmaxsat, bincd2, WPM1-2012, pwbo2.1, wbo1.6-cnf, QMaxSat-g2, ShinMaxSat, WMaxSatz09, and WMaxSatz+. We also employ the highly parameterized solver QMaxSat-g2 for the configuration aspect of the approach.

The MS data set has 600 instances, split among random, crafted and industrial. Each fold has 60 test instances. Results in Table 5a confirm the findings observed in previous experiments. In this case, ISAC-MSC struggles to improve over the best single solver. At the same time ISAC+ nearly completely closes the gap between BS and VBS.

The partial MaxSAT dataset is similar to the one used in the previous section, but in this case we also augment it with randomly generated instances bringing the count up to 1,086 instances. The Weighted MaxSAT problems consist of only crafted and random instances creating a dataset of size 277. Finally, the weighted partial MaxSAT problems number 718.

All in all, we observe that ISAC+ always outperforms the original ISAC methodology significantly, closing the gap between ISAC-MSC and the VBS by 90%, 74%, 100%, and 52%. The tables give the average performance of the single best solver for each fold (which may of course differ from fold to fold) in the row indexed BS. Note this value is better than what the previous best single MaxSAT solver had to offer. Still, on plain MaxSAT, ISAC+ solves 8% more instances, 58% more on partial MaxSAT, 6% more on weighted MaxSAT, and 29% more instances on weighted partial MaxSAT instances within the timeout. This was a significant improvement in the communities ability to solve MaxSAT instances in practice.

These results were subsequently independently confirmed at the 2013 MaxSAT Evaluation where our portfolios, built based on the methodology described in this paper, won six out of eleven categories and came in second in another three.

Table 5. MaxSAT cross-validation

	Average	PAR1	PAR10	Solved	% solved
BS	117.0	600.5	5199	45.4	75.7
ISAC-MSC	146.3	603.3	4887	47.2	78.7
ISAC+	134.5	487.7	3952	49.0	81.7
VBS	115.9	473.8	3876	49.2	82.0

(a) MS MIX has 60 test instances per fold

	Average	PAR1	PAR10	Solved	% solved
BS	68.0	822.3	7834	68.0	63.0
ISAC-MSC	100.1	328.3	2398	96.1	89.0
ISAC+	98.4	232.7	1713	99.6	92.2
VBS	69.9	206.2	1476	100.8	93.3

(b) PMS MIX has 108 test instances per fold

	Average	PAR1	PAR10	Solved	% solved
BS	50.2	302.7	2633	23.7	87.9
SAC-MSC	65.6	323.5	2653	23.7	87.9
ISAC+	58.8	184.3	1349	25.3	93.8
VBS	58.6	184.3	1349	25.3	93.8

(c) WMS MIX has 27 test instances per fold

	Average	PAR1	PAR10	Solved	% solved
BS	56.3	632.1	5949	51.1	72.0
ISAC-MSC	47.1	229.0	1914	64.7	91.1
ISAC+	54.6	168.6	1511	66.0	92.9
VBS	15.5	131.8	1185	67.1	94.5

(d) WPMS MIX has 71 test instances per fold

3.2 Dynamically Adapting Portfolios

As has been made clear in the previous chapter, there are already many successful portfolio based approaches that vary the employed solver depending on the problems. Yet the scenarios we have explored so far can be categorized as one-shot learning approaches, as the construction of a portfolio optimizing a given objective function is done only once on a chosen set of training instances and then used without change.

In a real world setting it is possible that a distribution of instances chosen for the training phase might not reflect the instances that are being solved in the online phase. Consequently it might be possible to improve the initially constructed portfolio. Therefore, one might desire to have a portfolio approach that evolves based on the set of instances that are solved in the online-phase. One possible direct solution to this is simply periodically relaunching the training approach. However, this can be very computationally expensive. This section

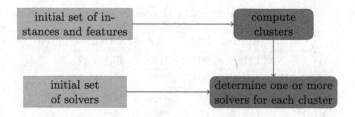

Fig. 5. Initial offline phase of EISAC

therefore shows how to augment an existing portfolio approach, ISAC, to one that evaluates the current portfolio and provides ways to continuously improve it by taking advantage of a growing set of available instances, enriched feature set and more up to-date set of available solvers. This extended approach will be referred to as EISAC as it evolves over time.

For the purposes of this section, let $\mathcal{I} = \{I_1, \ldots, I_n\}$ be the set of problem instances chosen as a training set. Let $\mathcal{F} = \{f_1, \ldots, f_m\}$ be the set of features that are associated with each problem instance during the initial training phase. Let $\mathcal{S} = \{s_1, \ldots, s_k\}$ be the set of solvers chosen for the initial training phase. As was shown in the previous chapter, the ISAC based portfolio approach uses the feature set $\mathcal{F}$ to partition the set $\mathcal{I}$. Let $\mathcal{C} = \{C_1, \ldots, C_s\}$ be such a partition of $\mathcal{I}$. It then determines a best solver from the set $\mathcal{S}$ for each cluster C_i. When an instance is given to be solved in the online phase, its closest cluster is determined and the corresponding solver is used to solve it.

Let $\mathcal{I}_t$, $\mathcal{F}_t$, and $\mathcal{S}_t$ denote the set of instances, the feature set and the set of solvers known at time t and let $\mathcal{C}_t$ denotes the clustering of instances at time t. We assume that each time period is associated with one or more changes in these sets. Following this notation $\mathcal{I}_0$, $\mathcal{F}_0$, and $\mathcal{S}_0$ denotes the initial set of training instances, features, and solvers. The offline phase of EISAC is divided in to 2 phases: initial phase and evolving phase. Given a initial set of instances, $\mathcal{I}_0$, and a initial set of features $\mathcal{F}_0$, EISAC finds a initial set of clusters $\mathcal{C}_0$ and determines one or more best solvers from set $\mathcal{S}_0$ for each cluster in the initial offline phase as shown in Fig. 5. In the evolving phase EISAC updates the current clusters if the current set of features or instances changes, and it updates the set of solvers for each cluster if the current set of solvers or clusters changes as shown in Fig. 6. In the online phase, given a new instance, EISAC determines the best cluster and run one or more solvers associated with the selected cluster for solving the instance as shown in Fig. 7. The following section provides the details of how and when EISAC updates the clusters and the set of solvers for each cluster.

Updating Clusters. EISAC updates the current clustering of the instances in the following scenarios:

– When a new instance is made available in the online phase the instance is added to the current set of instances.

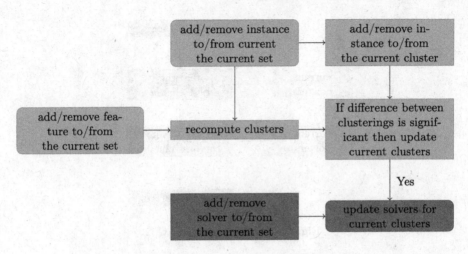

Fig. 6. Evolving offline phase of EISAC

- If at most m instances are maintained at any point, then one or more instances are removed when the number of instances exceeds the value of m
- A new feature is added to the current set of features or if an existing feature is removed for some reason.

In each of these cases one would like to determine whether the existing clustering is still appropriate or should it be modified and so EISAC recomputes the clusters for the entire set of instances. In most cases, the two clusterings one obtained by re-clustering the entire set and another by modifying the current clusters will be similar, so nothing needs to be done. But as the number of modification increases the similarity between the two clusterings might decrease.

Let δ be the time difference between the last time when EISAC was activated and the current time. Given a value δ, EISAC recomputes the partition of the instances and compares it with the current partition. In the following we describe one way of comparing the similarity between two partitions.

The Rand index [18, 37] or Rand measure (named after William M. Rand) is a measure of the similarity between two data clusterings. Given a set of instances $\mathcal{I}_t$ and two partitions of $\mathcal{I}_t$ to compare, $X = \{X_1, \ldots, X_k\}$ a partition of $\mathcal{I}_t$ into k subsets, and, $Y = \{Y_1, \ldots, Y_s\}$ a partition of $\mathcal{I}_k$ into s subsets, the Rand index is defined as follows:

- Let N_{11} denotes the number of pairs of instances in $\mathcal{I}_t$ that are in the same set in X and in the same set in Y.
- Let N_{00} denotes the number of pairs of instances in $\mathcal{I}_t$ that are in different sets in X and in different sets in Y.
- Let N_{10} denotes the number of pairs of instances in $\mathcal{I}_t$ that are in the same set in X and in different sets in Y.
- Let N_{01} denotes the number of pairs of instances in $\mathcal{I}_t$ that are in different sets in X and in the same set in Y.

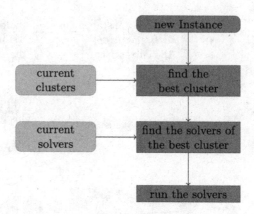

Fig. 7. Online phase of EISAC

The Rand index is defined as follows:

$$R \doteq \frac{N_{11} + N_{00}}{N_{11} + N_{00} + N_{10} + N_{01}} = \frac{2(N_{11} + N_{00})}{n * (n - 1)}$$

Intuitively, $N_{11} + N_{00}$ can be considered as the number of agreements between X and Y and $N_{10} + N_{01}$ as the number of disagreements between X and Y.

If X is the current partition and Y is the new partition and if the Rand index is less than some chosen threshold λ then we replace the current partition of the instances with the new partition. If the current partition is replaced by the new partition then we may need to update the solvers for one or more clusters of the new partition. In the next section we describe how the solvers are updated for the clusters.

Updating Solver Selection for a Cluster. Once the clustering of instances is known, ISAC determines the best solver for each cluster. As the current trend is to create computers with 2, 4 or even 8 cores, it is unusual to find single core machines still in use, and the number of cores will probably continue to double in the coming years. It is for this reason that for EISAC we also consider scenarios where κ cores are available. Therefore instead of just finding one best solver we solve the optimization problem as described below to find κ solvers for each cluster.

Let x_{ij} be a Boolean variable that determines if solver $j \in S_t$ is chosen for instance i of some cluster C. Let T_{ij} denotes the time required for solving instance i using solver $j \in S_i$. Let y_j be a Boolean that denotes whether solver j is selected for the cluster C. For each instance $i \in C$ exactly one solver is selected:

$$\forall_{i \in C} : \sum_{j \in S_k} x_{ij} = 1 \tag{1}$$

The solver j is selected if it is chosen for any instance i:

$$\forall_{i \in C} \forall_{j \in S_t} : x_{ij} \Rightarrow y_j \tag{2}$$

The number of selected solvers should be equal to κ:

$$\sum_{j \in S_t} y_j = \kappa \tag{3}$$

The objective is to minimize the time required to solve all the instances of the cluster, i.e.,

$$\min \sum_{i \in C} \sum_{j \in S_t} T_{ij} \cdot x_{ij} \tag{4}$$

Given a cluster C, a constant κ, the set of solvers, S_t at time t, and the function T, computeBestSolvers(C, κ, S_t, T) denotes the κ best solvers obtained by solving the MIP problem composed of constraints (1)–(3) and the objective function (4). In the following we consider four cases when EISAC might update the set of κ best solvers for a cluster.

Removing Solvers. It may happen that for some reason a previously available solver is no longer available now. This could happen when a solver is no longer supported by the developers, or when a new release is made then one would like to discard the previous version and update it with the new version. If the removed solver is used by any cluster then one can re-solve the above optimization problem for finding the current κ solvers.

Adding Solvers. When a new solver s is added to the set of solvers S_t at time t it can have an impact on the current portfolio. One way to determine this is to run the solver s for all the instances $\mathcal{I}_t$ and then reconstruct the portfolio for each cluster based on the above described optimization problem. Another way could be to select a sample of an appropriate size from each cluster and run the solver s for only those samples. If adding the new solver to S_t improves the total execution time of solving the sample instances then we run the solver s for all the instances of the corresponding cluster, otherwise we avoid running the solver s for remaining instances of each cluster. In EISAC the sample size for a cluster C is set to $(|\mathcal{I}_0|/|\mathcal{I}_t|) * |C|$. The idea behind this is to always maintain full matrix of all the runtimes for at least $|\mathcal{I}_0|$ number of instances.

Removing Instances. If the clustering changes because of removing instances or because of change in the feature set, we just need to re-solve the above optimization problem for each cluster to update the current set of solvers associated with each cluster.

Algorithm 1. updateBestSolvers(C, κ)

1: **loop**
2:    $C_u \leftarrow \{i | i \in C \wedge |\mathcal{S}_{ti}| < |\mathcal{S}_t|\}$
3:    **for all** $i \in C_u$ **do**
4:       $p \leftarrow |\mathcal{S}_t| - |\mathcal{S}_{ti}| + 1$
5:       $b \leftarrow \arg\min_{j \in \mathcal{S}_{ti}}(T_{ij})$
6:       $\forall j \in \mathcal{S}_t - \mathcal{S}_{ti} : \quad T_{ij} \leftarrow e_p \times t_{ib}$
7:    $N_C \leftarrow \texttt{computeBestSolvers}(C, \kappa, \mathcal{S}_t)$
8:    **if** $N_C \neq B_C$ **then**
9:       **for all** $i \in C_u \wedge j \in N_C - \mathcal{S}_{ti}$ **do**
10:          $T_{ij} \leftarrow \texttt{computeRuntime}(i, j)$
11:          $\mathcal{S}_{ti} \leftarrow \mathcal{S}_{ti} \cup \{j\}$
12:       $B_C \leftarrow N_C$
13:    **else**
14:       **return** B_C

Adding Instances. If the clustering changes because of adding new instances to the current set of instances then the current κ best solvers for one or more clusters might change. An obvious way is to determine the run-times of the solvers for each new instance of the cluster and then recompute the κ best solver for each cluster using the above discussed optimization problem. However, it would be more desirable if the κ best solvers for a cluster can be computed without solving all the instances of a cluster using all the solvers. In order to do so we propose an approach that under-estimates the run-times of the solvers for computing κ best solvers, and runs the solvers only if required as described in Algorithm 1.

Let $\mathcal{S}_{ti}$ be the set of solvers at time t for which we know the run-times for solving an instance i. When a new instance i is added to the current set of instances we assume that $\mathcal{S}_{ti}$ is initialised to κ solvers which are used in the on-line phase to solve i. Let B_C be the currently known κ best solvers for the cluster C. Let $C_u \subseteq C$ be the set of instances for which the run-times of one or more solvers in the set $\mathcal{S}_t$ are unknown. The general idea is to under-estimate the run-time for solving each instance in C_u using $\mathcal{S}_t - \mathcal{S}_{ti}$ solvers, and use it to compute κ best solvers for a cluster C, denoted by N_C, until N_C is same as B_C. If N_C is same as B_C then it means that the currently known κ solvers are best even when the runtimes are under-estimated for each solver in $\mathcal{S}_t - \mathcal{S}_{ti}$. If N_C is different to B_C then each instance $i \in C_u$ is solved with each solver $j \in N_C - \mathcal{S}_{ti}$ for computing the actual runtime, denoted by $\texttt{computeRuntime}(i, j)$, and B_C is set to N_C. Notice that the actual run-time of each solver in B_C for each instance in C is always known.

The estimated run-time of a solver $j \in \mathcal{S}_t - \mathcal{S}_{ti}$ for an instance $i \in C_u$ is computed as described below. Let $\mathcal{I}_{ta}$ be the set of instances at time t for which we know the run-times of all the solvers. Let r_{ip} denotes the runtime of pth best solver for instance i. Let e_p denotes the average ratio between the runtimes of the best solver and the pth best solver for instance i, which is computed as follows:

$$e_p = \frac{1}{|\mathcal{I}_{ta}|} \sum_{i \in \mathcal{I}_{ta}} \frac{r_{i1}}{r_{ip}}$$

For each instance $i \in C_u$, if we assume that $\mathcal{S}_{ti}$ is the set of the $|\mathcal{S}_{ti}|$ worst solvers for i then the runtime of the best solver, b, of $\mathcal{S}_{it}$ would have the pth best runtime over all the solvers, where $p = |\mathcal{S}_t| - |\mathcal{S}_{it}| + 1$. The expected best runtime of a solver in $j \in \mathcal{S}_t - \mathcal{S}_{it}$ would be then $t_{ib} \cdot e_p$. Different values of p would result in different performance of EISAC. If $p = 1$ then it means we are optimistic and the current best solvers will never change, and if $p = |\mathcal{S}_t| - |\mathcal{S}_{ti}| + 1$ then we are pessimistic and assuming that the known runtimes are the $|\mathcal{S}_{ti}|$ worst runtimes.

Numerical Results. As for the sections in the last chapter, for numerical results, we use the SAT portfolio data made available by the SATzilla team after the 2011 SAT Competition [8]. This dataset provides the runtimes of 31 top-tier SAT solvers with a 1,200 s timeout on over 3,000 instances spread across the Random, Crafted and Industrial categories. After filtering out the instances where every solver times-out, we are left with 2,524 instances. For each of these instances the dataset also provides all of the known SAT features, but we restrict our study to the 52 standard features [32] that do not rely on local search probing.

We use this dataset to simulate the scenario where instances are made available one at a time. Specifically, we start with a set of $\mathcal{I}_0$ instances for which we know the performance of every solver. Based on this initial set, we generate our initial clusters and select the solver that minimizes the PAR10 score of each cluster. We then add δ instances to our dataset, evaluate them with the current portfolio, and then evaluate whether we should retrain. We use two thresholds for the adjusted rand index, 0.5 and 0.95. Simulating the scenario where we can only keep a certain number of instances for the retraining, once we add the δ new instances, we also remove the oldest δ instances.

Lets first consider the scenario where all the instances are shuffled and come randomly. We then also consider an ordering on the data, where first we iterate through the industrial instances, followed by the crafted, and finally the instances that were randomly generated. This last experiment is meant to simulate the case where instances change over time. This is also the case where traditional portfolio approaches would fail because eventually they are tasked to solve instances they have never observed during training.

Table 6 presents our first test case where the instances come from a shuffled dataset. This is the typical case observed in competitions, where a representative set of the test data is available for training. The table presents the performance of a portfolio which have been given 200 or 500 training instances. The single best solver (BS) is chosen as a single solver during training and then always using it during the test phase. Alternatively, the virtual best solver (VBS) is an oracle portfolio that for every instance always runs the best solver. The VBS represents the limit of what can be achieved by a portfolio. We also evaluate ISAC-c50 and ISAC-c100, trained with a minimum of 50 (respectively 100) instances in each cluster. Note that in this setting ISAC is performing better than BS. It is also

Table 6. Comparison of performance of ISAC and EISAC on shuffled and ordered datasets using 200 or 500 training instances. We set the minimum cluster size to be either 50 or 100 and the adjusted rand index to either 0.5 or 0.95.

Shuffled		BS	ISAC	EISAC	EISAC	ISAC	EISAC	EISAC	VBS
			c50	c50-λ0.5	c50-λ0.95	c100	c100-λ0.5	c100-λ0.95	
200	Solved	1760	1776	1753	1759	1776	1752	1752	2324
	% solved	75.7	76.0	75.4	75.7	76.0	75.4	75.4	100
	PAR10	3001	2923	3037	3006	2923	3038	3038	75.2
	# Train	1	1	275	329	1	166	166	-
500	Solved	1532	1548	1548	1539	1548	1548	1544	2024
	% solved	75.7	76.4	76.4	76.0	76.4	76.4	76.3	100
	PAR10	3004	2912	2912	2962	2912	2912	2935	74.82
	# Train	1	1	1	674	1	1	104	-
Ordered		BS	ISAC	EISAC	EISAC	ISAC	EISAC	EISAC	VBS
			c50	c50-λ0.5	c50-λ0.95	c100	c100-λ0.5	c100-λ0.95	
200	Solved	1078	1078	1725	1793	1078	1741	1741	2324
	% solved	46.3	46.3	74.2	77.2	46.3	74.9	74.9	100
	PAR10	6484	6484	3160	2821	6484	3084	3084	70.42
	# Train	1	1	49	160	1	9	9	-
500	Solved	791	795	1261	1606	817	817	1373	2024
	% solved	39.1	39.3	62.3	79.3	40.4	40.4	67.8	100
	PAR10	7357	7334	4578	2556	7205	7205	3910	70.79
	# Train	1	1	4	611	1	1	97	-

important to note here that in the 2012 SAT Competition, the difference between the winning and second placed single engine solver was 3 instances and only 8 instances between the top 4 solvers. Therefore the improvement of 16 instances when training on 500 instances is significant. When compared to ISAC on this shuffled data, we see that EISAC is performing comparably to ISAC, although requiring significantly more training sessions. For each version of EISAC in the table we present the minimum cluster size and the adjusted rand index threshold. So EISAC-c100-λ0.95 has clusters with at least 100 instances and retrains as soon as the adjusted rand index drops below 0.95.

This comparable performance on shuffled data is to be expected. As the data is coming randomly, the initial training data was representative enough to capture the diversity. And even if the clusters change a little overtime, the basic assignment of solvers to instances doesn't really change. Note that the slight degradation between for the higher threshold in EISAC-c100 for 500 training instance, can likely be attributed to over-tuning (or overfitting) in the numerous re-training steps. Also note that the lower performance for 500 training instances is misleading, since by adding 300 instances to our training set, we are removing 300 instances from the test set.

Table 7. Comparison of performance on ordered dataset using an approximation learning technique.

		BS	ISAC	EISAC	EISAC+	VBS
			c50	c50-λ0.5	c50-λ0.5	
200	Solved	1078	1078	1725	1671	2324
	PAR10	6484	6484	3160	3440	70.42
	# train	1	1	49	44	-
	% eval	100	100	100	59.5	-
500	Solved	791	795	1261	1264	2024
	PAR10	7357	7334	4578	4561	70.79
	# train	1	1	4	3	-
	% eval	100	100	100	83.8	-

The story becomes significantly different if the data is not shuffled as is the case at the bottom of Table 6. Here we see that the clusters and solvers chosen by ISAC initially are ill equipped to solve the future instances. EISAC on the other hand, is able to adapt to the changes and outperform ISAC by almost a factor of 2 in terms of the instances solved. What is also interesting is that for the case of 500 training instances and small clusters, this performance is achieved with only four re-training steps.

Table 7 examines the effectiveness of our proposed training technique. Instead of computing the time of every solver on every training instance during re-tuning, we lazily fill in this data until we converge on the expected best solver for a cluster. We call this approach EISAC+. Due to space limitations, we only present a comparison on the ordered dataset and for algorithms tuned with a minimum cluster size of 50. What we observe is that while performance is maintained with this lazy selection, we cut down the number of evaluations we need to 80% and occasionally to as low as 50%. This means that we can potentially speed up each training stage by a factor of 2 while still maintaining nearly identical performance.

3.3   Feature Generation

Up to this point we have now seen a myriad of ways in which algorithm portfolios can be created, expanded, and utilized. In all cases portfolios were shown to significantly outperform any single solver. Yet while there is now a plethora of competing approaches, all of them are dependent on the quality of a set of structural features they use to distinguish amongst the instances. Over the years, each domain has defined and refined its own set of features, yet at their core they are mostly a collection of everything that was considered useful in the past. As an alternative to this shotgun generation of features, this section will instead show a more systematic approach. Specifically, this section will show how latent features gathered from matrix decomposition are enough for a linear model to achieve a level of performance comparable to a perfect Oracle portfolio.

The reason we emphasize the fact that a linear model can achieve great performance is because while the performance of algorithm selection techniques is continually improving, it does so at the cost of transparency of the employed models. The version of SATzilla that won in 2012, trains a tree to predict the winner between every pair of solvers [52]. CSHC, the 2013 winner, takes an alternate approach of introducing a new splitting criterion for trees that makes sure that each subtree is more consistent on the preferred solver than its root [29]. But in order to make the approach competitive, many of these trees need to be grouped into a forest. Yet other approaches create schedules of solvers to improve the chances of solving each instance [16,20]. Unfortunately, even though all these approaches are highly effective at solving instances, once they are trained they are nearly impossible to use to get a better understanding of the fundamental issues of a particular problem domain. In short, we can now answer *what* we should do when we see a new instance, the new question should therefore be *why* a particular solver is chosen and we should use this information to spur the development of a new wave of solvers. The focus of this section is therefore to present a new portfolio approach that can achieve similar performance to leading portfolios while also presenting a human interpretable model.

In this section, as our running example, we consider the three standard datasets we have seen before of SAT, MaxSAT, and CSP instances. Specifically, here the SAT dataset is comprised of 1,098 industrial instances gathered from SAT Competitions dating back to 2006 and considers 28 solvers from 2012 each run with a 5,000 s timeout. The MaxSAT dataset is comprised of 1,077 instances gathered from the 2013 MaxSAT Evaluation, evaluated by the top 15 solvers from 2013 with a 1,800 s timeout. Finally the CSP dataset considers the 2,207 instances used to train the Proteus portfolio. For all experiments the presented numbers are based on 10-fold cross validation.

Latent Features. A latent variable is by definition something that is not directly observable but rather inferred from observations. This is a concept that is highly related to that of hidden variables, and is employed in a number of disciplines including economics [39], medicine [54], and machine learning [12]. This section introduces the idea of collecting latent variables that best describe the changes in the actual performance of solvers on instances. Thereby instead of composing a large set of structural features that might possibly correlate with the performance, here we present a top down approach.

Singular Value Decomposition. The ideas behind Singular Value Decomposition herald back to the late 1800's when they were independently discovered by five mathematicians: Eugenio Beltrami, Camille Jordan, James Joseph Sylvester, Erhard Schmidt, and Hermann Weyl. In practice, the technique is now currently embraced for tasks like image compression [35] and data mining [31] by reducing massive systems to manageable problems by eliminating redundant information and retaining data critical to the system.

At its essence, Singular Value Decomposition is a method for identifying and ordering the dimensions along which data points exhibit the most variation, which is mathematically represented by the following equation:

$$M = U\Sigma V^T,$$

where M is the $m \times n$ matrix representing the original data. Here, there are m instances each described by n values. The columns of U are the orthonormal eigenvectors of MM^T, the columns of V are orthonormal eigenvectors of $M^T M$, and Σ is a diagonal matrix containing the square roots of eigenvalues from U or V in descending order.

Note that if $m > n$ then, being a diagonal matrix, most of the rows in Σ will be zeros. This means that after multiplication, only the first n columns of U are needed. So for all intents and purposes, for $m > n$, U is an $m \times n$ matrix, while both Σ and V^T are $n \times n$.

Because U and V are orthonormal, intuitively one can interpret the columns of these matrices as a linear vector in the problem space that captures most of the variance in the original matrix M. The values in Σ then specify how much of the variance each column captures. The lower the value in Σ, the less important a particular column is. This is where the concept of compression comes into play, when the amount of columns in U and V can be reduced while still capturing most of the variability in the original matrix.

From the perspective of data mining, the columns of the U matrix and the rows of the V matrix have an additional interpretation. Let us assume that our matrix M records the performance of n solvers over m instances. In such a case it is usually safe to assume that $m > n$. So each row of the U matrix still describes each of the original instances in M. But now each column can be interpreted as a latent topic or feature that describes that instance. Meanwhile, each column of the V^T matrix refers to each solver, while each row presents how active, or important a particular topic is for that solver.

These latent features in U give us exactly the information necessary to determine the runtime of each solver. This is because once the three matrices are multiplied out we are able to reconstruct the original performance matrix M. So if we are given a new instance i, if we are able to identify its latent features, we could multiply by the existing Σ and V^T matrices to get back the performance of each solver.

Therefore, if we had the latent features for an instance as computed after the Singular Value Decomposition, it would be possible to train a linear model to accurately predict the performance of every solver. A linear model where we can see exactly which features influence the performance of each solver. Table 8 demonstrates this idea. For simplicity, here we consider very basic regression based portfolios that predict the runtime of each solver, choosing the one expected to take the least amount of time. Naturally, if we are using the standard features available for each dataset, a random forest with 500 trees is very good at picking the best solver. Now for each training set we also compute matrices U, V and Σ and train each a number of models to use the latent features in U

Table 8. Performance of algorithm selection techniques using the latent features computed after singular value decomposition on SAT, MaxSAT and CSP datasets. The algorithms are compared using the average runtime (AVG), the timeout penalized runtime (PAR10), and the number of instances not solved (NS). We therefore observe that a linear model using the SVD features could potentially perform as well as an oracle.

		SAT			MaxSAT			CSP		
		AVG	PAR10	NS	AVG	PAR10	NS	AVG	PAR10	NS
Standard portfolio	BSS	672	3,620	69	739	6,919	412	1,156	9,851	362
	Forest-500	381	1,382	23	47.1	227	12	225	994	32
	VBS	245	245	0	26.6	26.6	0	131	131	0
SVD-based portfolio	Tree	508	1,635	26	98.0	563	31	167	287	5
	Linear	245	245	0	26.6	26.6	0	131	131	0
	SVM (radial)	286	373	2	38.8	114	5	134	134	0
	K-NN	331	589	6	34.1	109	5	135	159	1
	Forest-500	300	386	2	32.0	77.0	3	135	231	4

to predict the solver performances. For the test instances, we use the runtimes, P, to compute what the values of U should be by computing $PV\Sigma^{-1}$. These latent features are then used by the trained models to predict the best solver.

Unfortunately, these latent features are only available by decomposing the original performance matrix. This is information that we only have available *after* all the solvers have been run on an instance. Information that once computed means we already know which solver should have been run.

Yet, note that the performance of the models is much better than it was using the original features, especially for the linear model. This is again a completely unfair comparison, but it is not as obvious as it first appears. What we can gather from these results is that the matrix V and Σ are still relevant even when applied to previously unseen instances. This means that the differences between solvers can in fact be differentiated by a linear model, provided it has the correct structural information about the instance. This also means that if we are able to replicate the latent features of a new instance, the supporting matrices computed by the decomposition will be able to establish the performances.

Recall also that the values in Σ are based on the eigenvalues of M. This means that the columns associated with the lower valued entries in Σ encapsulate less of the variance in the data than the higher valued entries. Figure 8 therefore shows the performance of the linear model as more of the latent features are removed under the MaxSAT WPMS dataset. We just use the MaxSAT dataset for the example because the CSP dataset only has 4 solvers and the results for the SAT dataset are similar to those presented. Note that while all of the latent features are necessary to recreate the performance of the VBS, it is possible to remove over half the latent features and still be able to solve all but 4 instances.

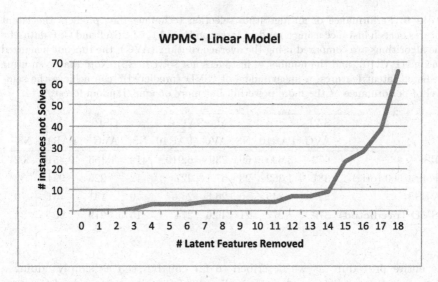

Fig. 8. Number of unsolved instances remaining after using a linear model trained on the latent features after singular value decomposition. The features were removed with those with lowest eigenvalues first. The data is collected on the WPMS dataset. This means that even removing over half the latent features, a portfolio can be trained that solves all but 4 instances.

Estimating Latent Features. Although we do not have direct access to the latent features for a previously unseen instance, we can still estimate them using the original set of features.

Note that while it is possible to use the result of $\Sigma V'$ as a means of computing the final times, training a linear model on top of the latent features is the better option. True, both approaches would be performing a linear transformation of the features, but the linear model will also be able to automatically take into account any small errors in the predictions of the latent features. Therefore, the method proposed in this section would use a variety of models to predict each latent feature using the original features. The resulting predicted features will then be used to train a set of linear models to predict the runtime of each solver. The solver with the best predicted runtime will be evaluated.

To predict each of our latent features it is of course possible to use any regression based approach available in machine learning. From running just the five approaches that we have utilized in the previous sections, unsurprisingly we observe that a random forest provides the highest quality prediction. The results are presented in Table 9. Here SVD_predicted uses a random forest to predict the values of each latent feature and then trains a secondary linear model over the latent features to predict the runtime of each solver.

From the numbers we observe in Table 9, we see that a linear model portfolio using latent features behaves similarly to the original Random Forest approach that simply predicts the runtime of each solver. This is to be expected since

Table 9. Performance of an algorithm selection technique that predicts the latent features of each instance using a random forest on the SAT, MaxSAT and CSP datasets. The algorithms are compared using the average runtime (AVG), the timeout penalized runtime (PAR10), and the number of instances not solved (NS). Note that even using predicted latent features, a linear model of "SVD (predicted)" can achieve the same level of performance as the more powerful, but more opaque, random forest.

	SAT			MaxSAT			CSP		
	AVG	PAR10	NS	AVG	PAR10	NS	AVG	PAR10	NS
BSS	672	3,620	69	739	6,919	412	1,156	9,851	362
Forest-500 (orig)	381	1,382	23	47.1	227	12	225	994	32
VBS	245	245	0	26.6	26.6	0	131	131	0
SVD (predicted)	379	1277	21	49.6	274	15	219	964	31

the entire procedure as we described so far can be seen as simply adding a single meta-layer to the model. After all, one of the nice properties of forests is that they are able to capture highly nonlinear relations between the features and target value. All we are doing here is adding several forests that are then linearly combined into a single value. But this procedure does provide one crucial piece of new information.

Whereas before there was little feedback as to which features were causing the issues, we now know that if we have a perfect prediction of the latent features we can dramatically improve the performance of the resulting portfolio. Furthermore, we know that we don't even need to focus on all of the latent features equally, since Fig. 8 revealed that we can survive with less than half of them.

Therefore, using singular value decomposition we can now identify the latent features that are hard to predict, the ones resulting in the highest error. We can then subsequently use this information to claim that the reason we are unable to predict this value is because the regular features we have available are not properly capturing all of the structural nuances that are needed to distinguish instances. This observation can subsequently be used to split the instances into two groups, one where the random forest over predicts and one where it under predicts. This is information that can then help guide researchers to identify new features that do capture the needed value to differentiate the two groups. This therefore introduces a more systematic approach to generating new features.

Just from the results in Table 9 we know that our current feature vectors are not enough when compared to what is achievable in Table 8. We also see that for the well studied SAT and CSP instances, the performance is better than for MaxSAT where the feature vectors have only recently been introduced.

We can then just aim to observe the next latent feature to focus on. This can be simply done by iterating over each latent feature and artificially assigning it the "correct" value while maintaining all the other predictions. Whichever feature thus corrected results in the most noticeable gains is the one that should be focused on next. Whenever two latent features tie in the possible gains, we should also focus on matching the one with the lower index, since mathematically, that is the feature that captures more of the variance.

If we go by instance names as a descriptive marker, surprisingly following our approach results in a separation where both subsets have instances with the same names. So following the latent feature suggested for the MaxSAT dataset we observe that there is a difference between "ped2.G.recomb10-0.10-8.wcnf" and "ped2.G.recomb1-0.20-14.wcnf". For CSP, we are told that "fapp26-2300-8.xml" and "fapp26-2300-3.xml" should be different. This means that the performance of a solver on an instance goes beyond just the way that instance was generated. There are still some fundamental structural differences between instances that our current features are not able to identify. This only highlights the need for a systematic way in which to continue to expand our feature sets.

Practical Application. One of the underlying messages of this section has been that one of the things that makes algorithm portfolios so successful in practice is the presence of highly descriptive features. Yet as we have shown so far, the features we typically use could be refined using a systematic approach that helps us identify when certain information is missing and helps us define the properties of those missing features. In this part we will therefore go through the full process on a domain where portfolios are only beginning to be introduced and no high quality feature set exists. Specifically we will use the container pre-marshaling process as an example of how features should be developed.

The container pre-marshalling problem (CPMP) is a well-known NP-hard problem in the container terminals and stacking literature [7, 26, 41], first introduced in [25]. The CPMP deals with the sorting of containers in a set of stacks (called a *bay*) of intermodal containers based on their exit times from the stacks, such that containers that must leave the stacks first are placed on top of containers that must leave later. This prevents *mis-overlaid* containers from blocking the timely exit of other containers. The goal of the CPMP is to find the minimal number of container movements necessary to ensure that all of the stacks are sorted by the exit time of each container without exceeding the maximum height of each stack. Solving the CPMP assists container terminals in reducing delays and increasing the efficiency of their operations.

Given an initial layout of a bay with a fixed number of stacks and tiers (stack height), the goal of the CPMP is to find the minimal number of container movements (or *rehandles*) necessary to eliminate all mis-overlays in the bay. Every container is assigned a group that indicates when it must leave the bay. A mis-overlaid container is defined as a container with a group that is higher than the group of any container underneath it, or a container above a mis-overlaid container.

Consider the simple example of Fig. 9, which shows a bay composed of three stacks of containers in which containers can be stacked at most four tiers high.

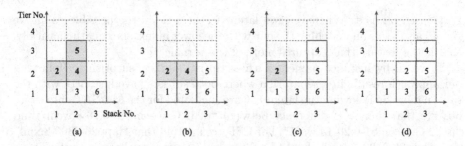

Fig. 9. An example solution to the CPMP with mis-overlays highlighted. (Reproduced from [46]).

Each container is represented by a box with its corresponding group.[3] This is not an ideal layout as the containers with groups 2, 4 and 5 will need to be relocated in order to retrieve the containers with higher groups (1 and 3). That is, containers with groups 2, 4 and 5 are mis-overlaid. Consider a container movement (f, t) defining the relocation of the container on top of the stack f to the top position of the stack t. The containers in the initial layout of Fig. 9 can reach the final layout (d) with three relocation moves: $(2, 3)$ reaching layout (b), $(2, 3)$ reaching layout (c) and $(1, 2)$ reaching layout (d) where no mis-overlays occur.

Pre-marshalling is important both in terms of operational and tactical goals at a container terminal. In particular, effective pre-marshalling of containers can help reduce delays moving containers from the terminal yard onto vessels, as well as from the yard onto trucks or trains. Consider [46] for more information and a discussion of related work.

The features used in our dataset are given in Fig. 10, split into three categories. Features 1 through 16 were designed before performing latent feature analysis. Features 17 through 20 were created based on our first iteration of latent feature analysis, and features 21 and 22 using our second iteration.

Original features are created in the standard way for algorithm selection benchmarks, based on domain knowledge. The first 5 features address the problem size and density of containers. Feature 6 counts the number of mis-overlaid containers, a naive lower bound to the problem, whereas Feature 7 counts how many stacks contain mis-overlaid containers. Feature 8 provides the lower bound from [6], analyzing indirect container movements in addition to the mis-overlays present in feature 7. Features 9 through 12 offer information on how many containers belong to each group. Features 12 through 15 attempt to uncover the structure of the groups of the top non-mis-overlaid container on each stack.

LFA features are constructed based on the suggestions of the latent features. Feature 17 is the density of containers on the "left" side of the instance. We

---

[3] We note that multiple containers may have the same group, but in order to make containers easily identifiable, in this example we have assigned a different group to each container.

1. Number of stacks
2. Number of tiers
3. Tiers/stacks ratio
4. Container density
5. Empty stack percentage
6,7. Percent of all {slots, stacks} that are mis-overlaid
8. Bortfeldt & Forster lower bound
9–12. Min/max/mean/stdev container group counts
13–16. Min/max/mean/stdev group of top non-mis-overlaid container in each stack

---

17. Container density in stacks 1 through #*Stacks*/3
18. Tier-weighted groups
19. Largest group L1 distance from top left (average)
20. Pct. contiguous empty space including one empty stack

---

21. Mis-overlaid stack (≥ 2 containers) percentage
22. Low-group containers near stack tops (percentage)

Fig. 10. Features for the CPMP.

note that this feature is likely "overtuned" to the algorithms in our benchmark. Feature 18 measures whether containers with high group values are on high or low tiers by multiplying the tier of a container by its group, summing these values together and dividing by the maximum this value could take (namely if the highest group container was in each slot). Feature 19 measures the L1 (manhattan) distance from the top left of a problem to each container in the latest exit time, averaging these distances if there are multiple containers in the latest exit group. The final feature from iteration 1 computes the percentage of empty space in the instance in which an area of contiguous empty space includes at least one empty stack. Features 21 and 22 come from LFA iteration 2. Feature 21 counts how many stacks with more than two containers are mis-overlaid, and Feature 22 counts "low" ($\leq max\text{-}group/4$) valued containers on the top of stacks.

Using the four available solvers to tackle the pre-marshaling problem, we evaluate the feature sets using a typical portfolio approach. Table 10 provides the performances of a portfolio when trained on the three datasets versus the best single solver (BSS) and the virtual best solver (VBS), which is a portfolio that always picks the correct solver. As is typical, using just the initial arbitrary features the portfolio already performs significantly better than the BSS, indicating even the original features have descriptive value.

When a portfolio is trained on the first iteration of features, the performance improves not only in the number of instances solved, but also on the average time taken to solve each instance. This shows that by utilizing the latent feature analysis, a researcher is able to develop a richer set of features to describe the

Table 10. Performance of a portfolio trained on the three feature sets.

Solver	Avg	PAR-10	Solved
BSS	78.6	5,923	458
Original features	51.6	3,469	495
LFA iteration 1 features	46.6	2,741	506
LFA iteration 2 features	45.4	2,543	509
VBS	12.8	12.8	547

instances. Furthermore, the process can be repeated, as is evidenced by the performance of the portfolio on the second iteration of features. Note that the overall performance is again improved not only in the number of instances solved, but the time taken to solve them on average. Thus, multiple iterations of the latent feature analysis process can lead to even better features, although there are clearly diminishing returns.

3.4   Conclusions

We have presented a number of advanced portfolio techniques. Specifically, one portfolio approach that does not rely on a single problem representation or set of solvers, but leverages our ability to convert between problem representations to increase the space of possible solving approaches. In doing so, the contrasting performance among solvers on different representations of the same problem can be exploited. The overall performance can be improved significantly compared to restricting the portfolio to a single problem representation. We demonstrated empirically the significant performance improvements Proteus can achieve on a large set of diverse benchmarks using a portfolio based on a range of different state-of-the-art solvers.

Furthermore, we have presented novel algorithm portfolio techniques that again help to improve performance by building the portfolio iteratively and dynamically adapting it. Finally, we have investigated methods for constructing enhanced instance features for algorithm selection.

Acknowledgements. This work is supported by Science Foundation Ireland (SFI) Grant 10/IN.1/I3032 and FP7 FET-Open Grant 284715. The Insight Centre for Data Analytics is supported by SFI Grant SFI/12/RC/2289.

References

1. CSP Solver Competition Benchmarks. http://www.cril.univ-artois.fr/~lecoutre/benchmarks.html (2009)
2. Ansótegui, C., Sellmann, M., Tierney, K.: A gender-based genetic algorithm for the automatic configuration of algorithms. In: Gent, I.P. (ed.) CP 2009. LNCS, vol. 5732, pp. 142–157. Springer, Heidelberg (2009). doi:10.1007/978-3-642-04244-7_14

3. Argelich, J., Li, C., Manyà, F., Planes, J.: Maxsat evaluations (2012). www.maxsat.udl.cat
4. Audemard, G., Simon, L.: Glucose 2.3 in the SAT 2013 competition. In: Proceedings of SAT Competition 2013, p. 42 (2013)
5. Biere, A.: Lingeling, plingeling and treengeling entering the SAT competition 2013. In: Proceedings of SAT Competition 2013 (2013)
6. Bortfeldt, A., Forster, F.: A tree search procedure for the container pre-marshalling problem. Eur. J. Oper. Res. **217**(3), 531–540 (2012)
7. Carlo, H., Vis, I., Roodbergen, K.: Storage yard operations in container terminals: literature overview, trends, and research directions. Eur. J. Oper. Res. **235**(2), 412–430 (2014)
8. Data, S.: (2011). http://www.cs.ubc.ca/labs/beta/Projects/SATzilla/
9. Een, N., Sörensson, N.: Minisat 2.2 (2013). http://minisat.se
10. Gebser, M., Kaufmann, B., Neumann, A., Schaub, T.: *clasp*: a conflict-driven answer set solver. In: Baral, C., Brewka, G., Schlipf, J. (eds.) LPNMR 2007. LNCS (LNAI), vol. 4483, pp. 260–265. Springer, Heidelberg (2007). doi:10.1007/978-3-540-72200-7_23
11. Gecode Team: Gecode: Generic Constraint Development Environment (2006). http://www.gecode.org
12. Ghahramani, Z., Griffiths, T.L., Sollich, P.: Bayesian nonparametric latent feature models. In: World Meeting on Bayesian Statistics (2006)
13. Hall, M., Frank, E., Holmes, G., Pfahringer, B., Reutemann, P., Witten, I.H.: The WEKA data mining software: an update. SIGKDD Explor. Newsl. **11**(1), 10–18 (2009)
14. Hebrard, E.: Mistral, a constraint satisfaction library. In: Proceedings of the Third International CSP Solver Competition (2008)
15. Hebrard, E., O'Mahony, E., O'Sullivan, B.: Constraint programming and combinatorial optimisation in numberjack. In: Lodi, A., Milano, M., Toth, P. (eds.) CPAIOR 2010. LNCS, vol. 6140, pp. 181–185. Springer, Heidelberg (2010). doi:10.1007/978-3-642-13520-0_22
16. Helmert, M., Röger, G., Karpas, E.: Fast downward stone soup: a baseline for building planner portfolios. In: ICAPS (2011)
17. Hoos, H.: Adaptive novelty+: novelty+ with adaptive noise. In: AAAI (2002)
18. Hubert, L., Arabie, P.: Comparing partitions. J. Classif. **2**(1), 193–218 (1985). http://dx.doi.org/10.1007/BF01908075
19. Hutter, F., Tompkins, D., Hoos, H.: Rsaps: reactive scaling and probabilistic smoothing. In: CP (2002)
20. Kadioglu, S., Malitsky, Y., Sabharwal, A., Samulowitz, H., Sellmann, M.: Algorithm selection and scheduling. In: Lee, J. (ed.) CP 2011. LNCS, vol. 6876, pp. 454–469. Springer, Heidelberg (2011). doi:10.1007/978-3-642-23786-7_35
21. Kadioglu, S., Malitsky, Y., Sellmann, M., Tierney, K.: ISAC - instance-specific algorithm configuration. In: ECAI, pp. 751–756 (2010)
22. Kotthoff, L.: LLAMA: leveraging learning to automatically manage algorithms. Technical report, June 2013. arXiv:1306.1031, http://arxiv.org/abs/1306.1031
23. Le Berre, D., Lynce, I.: CSP2SAT4J: a simple CSP to SAT translator. In: Proceedings of the Second International CSP Solver Competition (2008)
24. Lecoutre, C., Tabary, S.: Abscon 112, toward more robustness. In: Proceedings of the Third International CSP Solver Competition (2008)
25. Lee, Y., Hsu, N.: An optimization model for the container pre-marshalling problem. Comput. Oper. Res. **34**(11), 3295–3313 (2007)

26. Lehnfeld, J., Knust, S.: Loading, unloading and premarshalling of stacks in storage areas: survey and classification. Eur. J. Oper. Res. **239**(2), 297–312 (2014)
27. Li, C., Huang, W.: G2wsat: gradient-based greedy walksat. SAT **3569**, 158–172 (2005)
28. Malitsky, Y., Sellmann, M.: Instance-specific algorithm configuration as a method for non-model-based portfolio generation. In: Beldiceanu, N., Jussien, N., Pinson, É. (eds.) CPAIOR 2012. LNCS, vol. 7298, pp. 244–259. Springer, Heidelberg (2012). doi:10.1007/978-3-642-29828-8_16
29. Malitsky, Y., Sabharwal, A., Samulowitz, H., Sellmann, M.: Algorithm portfolios based on cost-sensitive hierarchical clustering. In: IJCAI (2013)
30. Manthey, N.: The SAT solver RISS3G at SC 2013. In: Proceedings of SAT Competition 2013, p. 72 (2013)
31. Martin, C., Porter, M.: The extraordinary SVD. Math. Assoc. Am. **119**(10), 838–851 (2012)
32. Nudelman, E., Leyton-Brown, K., Hoos, H.H., Devkar, A., Shoham, Y.: Understanding random SAT: beyond the clauses-to-variables ratio. In: Wallace, M. (ed.) CP 2004. LNCS, vol. 3258, pp. 438–452. Springer, Heidelberg (2004). doi:10.1007/978-3-540-30201-8_33
33. Pham, D., Anbulagan: ranov. Solver description. SAT Competition (2007)
34. Pham, D., Gretton, C.: gnovelty+. Solver description. SAT Competition (2007)
35. Prasantha, H.: Image compression using SVD. In: Conference on Computational Intelligence and Multimedia Applications, pp. 143–145 (2007)
36. Prestwich, S.: Vw: Variable weighting scheme. SAT (2005)
37. Rand, W.: Objective criteria for the evaluation of clustering methods. J. Am. Statist. Assoc. **66**(336), 846–850 (1971)
38. Roussel, O., Lecoutre, C.: XML Representation of Constraint Networks: Format XCSP 2.1. CoRR abs/0902.2362 (2009)
39. Rutz, O.J., Bucklin, R.E., Sonnier, G.P.: A latent instrumental variables approach to modeling keyword conversion in paid search advertising. J. Mark. Res. **49**, 306–319 (2012)
40. Soos, M.: Cryptominisat 2.9.0 (2011)
41. Stahlbock, R., Voß, S.: Operations research at container terminals: a literature update. OR Spectr. **30**(1), 1–52 (2008)
42. Tamura, N., Tanjo, T., Banbara, M.: System description of a SAT-based CSP solver sugar. In: Proceedings of the Third International CSP Solver Competition, pp. 71–75 (2009)
43. Tanjo, T., Tamura, N., Banbara, M.: Azucar: a SAT-based CSP solver using compact order encoding. In: Cimatti, A., Sebastiani, R. (eds.) SAT 2012. LNCS, vol. 7317, pp. 456–462. Springer, Heidelberg (2012). doi:10.1007/978-3-642-31612-8_37
44. Choco team: choco: an Open Source Java Constraint Programming Library (2008)
45. Thornton, J., Pham, D., Bain, S., Ferreira, V.: Additive versus multiplicative clause weighting for SAT. In: PRICAI, pp. 405–416 (2008)
46. Tierney, K., Pacino, D., Voß, S.: Solving the pre-marshalling problem to optimality with A* and IDA*. Technical report, WP#1401, DS&OR Lab, University of Paderborn (2014)
47. Tompkins, D., Hutter, F., Hoos, H.: saps. Solver description. SAT Competition(2007)
48. Wei, W., Li, C., Zhang, H.: adaptg2wsatp. Solver description. SAT Competition(2007)
49. Wei, W., Li, C., Zhang, H.: Combining adaptive noise and promising decreasing variables in local search for SAT. Solver description. SAT Competition(2007)

50. Wei, W., Li, C., Zhang, H.: Deterministic and random selection of variables in local search for sat. Solver description. SAT Competition (2007)
51. Xu, L., Hoos, H., Leyton-Brown, K.: Hydra: automatically configuring algorithms for portfolio-based selection. In: AAAI (2010)
52. Xu, L., Hutter, F., Shen, J., Hoos, H., Leyton-Brown, K.: SATzilla 2012: improved algorithm selection based on cost-sensitive classification models. SAT Competition (2012)
53. Xu, L., Hutter, F., Hoos, H.H., Leyton-Brown, K.: SATzilla: portfolio-based algorithm selection for SAT. J. Artif. Intell. Res. **32**, 565–606 (2008)
54. Yang, W., Yi, D., Xie, Y., Tian, F.: Statistical identification of syndromes feature and structure of disease of western medicine based on general latent structure mode. Chin. J. Integr. Med. **18**, 850–861 (2012)

Adapting Consistency in Constraint Solving

Amine Balafrej[1]([⊠]), Christian Bessiere[2], Anastasia Paparrizou[2],
and Gilles Trombettoni[2]

[1] TASC (INRIA/CNRS), Mines Nantes, Nantes, France
amine.balafrej@mines-nantes.fr
[2] CNRS, University of Montpellier, Montpellier, France
bessiere@lirmm.fr

Abstract. State-of-the-art constraint solvers uniformly maintain the
same level of local consistency (usually arc consistency) on all the
instances. We propose two approaches to adjust the level of consistency depending on the instance and on which part of the instance we
propagate. The first approach, parameterized local consistency, uses as
parameter the *stability* of values, which is a feature computed by arc
consistency algorithms during their execution. Parameterized local consistencies choose to enforce arc consistency or a higher level of local
consistency to a value depending on whether the stability of the value
is above or below a given threshold. In the adaptive version, the parameter is dynamically adapted during search, and so is the level of local
consistency. In the second approach, we focus on partition-one-AC, a
singleton-based consistency. We propose adaptive variants of partition-
one-AC that do not necessarily run until having proved the fixpoint. The
pruning can be weaker than the full version, but the computational effort
can be significantly reduced. Our experiments show that adaptive parameterized maxRPC and adaptive partition-one-AC can obtain significant
speed-ups over arc consistency and over the full versions of maxRPC and
partition-one-AC.

1 Introduction

Enforcing local consistency by applying constraint propagation during search is
one of the strengths of constraint programming (CP). It allows the constraint
solver to remove locally inconsistent values. This leads to a reduction of the
search space. Arc consistency is the oldest and most well-known way of propagating constraints [Bes06]. It has the nice feature that it does not modify the structure of the constraint network. It just prunes infeasible values. Arc consistency is
the standard level of consistency maintained in constraint solvers. Several other
local consistencies pruning only values and stronger than arc consistency have
been proposed, such as max restricted path consistency or singleton arc consistency [DB97]. These local consistencies are seldom used in practice because
of the high computational cost of maintaining them during search.However, on

The results contained in this chapter have been presented in [BBCB13] and
[BBBT14]. This work has been funded by the EU project ICON (FP7-284715).

© Springer International Publishing AG 2016
C. Bessiere et al. (Eds.): Data Mining and Constraint Programming, LNAI 10101, pp. 226–253, 2016.
DOI: 10.1007/978-3-319-50137-6_9

some instances of problems, maintaining arc consistency is not a good choice because of the high number of ineffective revisions of constraints that penalize the CPU time. For instance, Stergiou observed that when solving the scen11, an instance from the radio link frequency assignment problem (RLFAP) class, with an algorithm maintaining arc consistency, only 27 out of the 4103 constraints of the problem were identified as causing a domain wipe-out and 1921 constraints did not prune any value [Ste09].

Choosing the right level of local consistency for solving a problem requires finding a good trade-off between the ability of this local consistency to remove inconsistent values, and the cost of the algorithm that enforces it. The works of [Ste08] and [PS12] suggest to take advantage of the power of strong propagation algorithms to reduce the search space while avoiding the high cost of maintaining them in the whole network. These methods result in a heuristic approach based on the monitoring of propagation events to dynamically adapt the level of local consistency (arc consistency or max restricted path consistency) to individual constraints. This prunes more values than arc consistency and less than max restricted path consistency. The level of propagation obtained is not character-ized by a local consistency property. Depending on the order of propagation, we can converge on different closures. In other work, a high level of consistency is applied in a non exhaustive way, because it is very expensive when applied exhaustively everywhere in the network during the whole search. In [SS09], a preprocessing phase learns which level of consistency to apply on which parts of the instance. When dealing with global constraints, some authors propose to weaken arc consistency instead of strengthening it. In [KVH06], Katriel et al. proposed a randomized filtering scheme for AllDifferent and Global Cardi-nality Constraint. In [Sel03], Sellmann introduced the concept of approximated consistency for optimization constraints and provided filtering algorithms for Knapsack Constraints based on bounds with guaranteed accuracy.

In this chapter, we propose two approaches for adapting automatically the level of consistency during search. Our first approach is based on the notion of stability of values. This is an original notion independent of the characteristics of the instance to be solved, but based on the state of the arc consistency algorithm during its propagation. Based on this notion, we propose *parameterized consis-tencies*, an original approach to adjust the level of consistency inside a given instance. The intuition is that if a value is hard to prove arc consistent (i.e., the value is not stable for arc consistency), this value will perhaps be pruned by a stronger local consistency. The parameter p specifies the threshold of stability of a value v below which we will enforce a stronger consistency to v. A para-meterized consistency p-LC is thus an intermediate level of consistency between arc consistency and another consistency LC, stronger than arc consistency. The strength of p-LC depends on the parameter p. This approach allows us to find a trade-off between the pruning power of local consistency and the computational cost of the algorithm that achieves it. We apply p-LC to the case where LC is max restricted path consistency. We describe the algorithm p-maxRPC3 (based on maxRPC3 [BPSW11]) that achieves p-max restricted path consistency. Then, we propose ap-LC, an adaptive variant of p-LC that uses the number of failures in which variables or constraints are involved to assess the difficulty of the different parts of the problem during search. ap-LC dynamically and locally adapts the level p of local consistency to apply depending on this difficulty.

Our second approach is inspired by singleton-based consistencies. They have been shown extremely efficient to solve some classes of hard problems [BCDL11]. Singleton-based consistencies apply the *singleton test* principle, which consists of assigning a value to a variable and trying to refute it by enforcing a given level of consistency. If a contradiction occurs during this singleton test, the value is removed from its domain. The first example of such a local consistency is Singleton Arc Consistency (SAC), introduced in [DB97]. In SAC, the singleton test enforces arc consistency. By definition, SAC can only prune values in the variable domain on which it currently performs singleton tests. In [BA01], Partition-One-AC (which we call POAC) has been proposed. POAC is an extension of SAC that can prune values everywhere in the network as soon as a variable has been completely singleton tested. As a consequence, the fixpoint in terms of filtering is often quickly reached in practice. This observation has already been made on numerical constraint problems. In [TC07,NT13], a consistency called Constructive Interval Disjunction (CID), close to POAC in its principle, gave good results by simply calling the main procedure once on each variable or by adapting during search the number of times it is called. Based on these observations, we propose an adaptive version of POAC, called APOAC, where the number of times variables are processed for singleton tests on their values is dynamically and automatically adapted during search. A sequence of singleton tests on all values of one variable is called a `varPOAC` call. The number k of times `varPOAC` is called will depend on how effective POAC is or not in pruning values. This number k of `varPOAC` calls will be learned during a sequence of nodes of the search tree (learning nodes) by measuring stagnation in the amount of pruned values. This amount k of `varPOAC` calls will be applied at each node during a sequence of nodes (called exploitation nodes) before we enter a new learning phase to adapt k again. Observe that if the number of `varPOAC` calls learned is 0, then adaptive POAC will mimic AC.

The aim of both of the proposed adaptive approaches (i.e., ap-LC and APOAC) is to adapt the level of consistency automatically and dynamically during search. ap-LC uses failure information to learn what are the most difficult parts of the problem and it increases locally and dynamically the parameter p on those difficult parts. APOAC measures a stagnation in number of inconsistent values removed for k calls of `varPOAC`. APOAC then uses this information to stop enforcing POAC. APOAC avoids the cost of the last calls to `varPOAC` that delete very few values or no value at all. We thus see that both ap-LC and APOAC learn some information during search to adapt the level of consistency. This allows them to benefit from the pruning power of a high level of consistency while avoiding the prohibitive time cost of fully maintaining this high level.

The rest of the paper is organized as follows. Section 2 contains the necessary formal background. Section 3 describes the parameterized consistency approach and gives an algorithm for parameterized maxRPC. In Sect. 4, the adaptive variant of parameterized consistency is defined. Sections 5 and 6 are devoted to our study of singleton-based consistencies. In Sect. 5, we propose an efficient POAC algorithm that will be used as a basis for the adaptive versions of POAC. Section 6 presents different ways to learn the number of variables on which to perform singleton tests. All these sections contain experimental results that validate the different contributions. Section 7 concludes this work.

2   Background

A *constraint network* is defined as a set of n variables $X = \{x_1, \ldots, x_n\}$, a set of ordered domains $D = \{D(x_1), \ldots, D(x_n)\}$, and a set of e constraints $C = \{c_1, \ldots, c_e\}$. Each constraint c_k is defined by a pair $(var(c_k), sol(c_k))$, where $var(c_k)$ is an ordered subset of X, and $sol(c_k)$ is a set of combinations of values (tuples) satisfying c_k. In the following, we restrict ourselves to binary constraints, because the local consistency (maxRPC) we use here to instantiate our approach is defined on the binary case only. However, the notions we introduce can be extended to non-binary constraints, by using maxRPWC for instance [BSW08]. A binary constraint c between x_i and x_j will be denoted by c_{ij}, and $\Gamma(x_i)$ will denote the set of variables x_j involved in a constraint with x_i.

A value $v_j \in D(x_j)$ is called an *arc consistent support (AC support)* for $v_i \in D(x_i)$ on c_{ij} if $(v_i, v_j) \in sol(c_{ij})$. A value $v_i \in D(x_i)$ is *arc consistent (AC)* if and only if for all $x_j \in \Gamma(x_i)$ v_i has an AC support $v_j \in D(x_j)$ on c_{ij}. A domain $D(x_i)$ is arc consistent if it is non empty and all values in $D(x_i)$ are arc consistent. A network is arc consistent if all domains in D are arc consistent. If enforcing arc consistency on a network N leads to a domain wipe out, we say that N is arc inconsistent.

A tuple $(v_i, v_j) \in D(x_i) \times D(x_j)$ is *path consistent (PC)* if and only if for any third variable x_k there exists a value $v_k \in D(x_k)$ such that v_k is an AC support for both v_i and v_j. In such a case, v_k is called *witness* for the path consistency of (v_i, v_j).

A value $v_j \in D(x_j)$ is a *max restricted path consistent (maxRPC)* support for $v_i \in D(x_i)$ on c_{ij} if and only if it is an AC support and the tuple (v_i, v_j) is path consistent. A value $v_i \in D(x_i)$ is max restricted path consistent on a constraint c_{ij} if and only if there exist $v_j \in D(x_j)$ maxRPC support for v_i on c_{ij}. A value $v_i \in D(x_i)$ is max restricted path consistent if and only if for all $x_j \in \Gamma(x_i)$ v_i has a maxRPC support $v_j \in D(x_j)$ on c_{ij}. A variable x_i is maxRPC if its domain $D(x_i)$ is non empty and all values in $D(x_i)$ are maxRPC. A network is maxRPC if all domains in D are maxRPC.

A value $v_i \in D(x_i)$ is *singleton arc consistent* (SAC) if and only if the network $N|_{x_i=v_i}$ where $D(x_i)$ is reduced to the singleton $\{v_i\}$ is not arc inconsistent. A variable x_i is SAC if $D(x_i) \neq \emptyset$ and all values in $D(x_i)$ are SAC. A network is SAC if all its variables are SAC.

A variable x_i is partition-one-AC (POAC) if and only if $D(x_i) \neq \emptyset$, all values in $D(x_i)$ are SAC, and $\forall j \in 1 \ldots n, j \neq i, \forall v_j \in D(x_j), \exists v_i \in D(x_i)$ such that $v_j \in AC(N|_{x_i=v_i})$. A constraint network $N = (X, D, C)$ is POAC if and only if all its variables are POAC. Observe that POAC, as opposed to SAC, is able to prune values from all variable domains when being enforced on a given variable.

Following [DB97], we say that a local consistency LC_1 is stronger than a local consistency LC_2 ($LC_2 \preceq LC_1$) if LC_2 holds on any constraint network on which LC_1 holds. It has been shown in [BA01] that POAC is strictly stronger than SAC. Hence, SAC holds on any constraint network on which POAC holds and there exist constraint networks on which SAC holds but not POAC.

The problem of deciding whether a constraint network has solutions is called the *constraint satisfaction problem (CSP)*, and it is NP-complete. Solving a CSP is mainly done by backtrack search that maintains some level of consistency between each branching step.

3   Parameterized Consistency

In this section, we present an original approach to parameterize a level of consistency LC stronger than arc consistency so that it degenerates to arc consistency when the parameter equals 0, to LC when the parameters equals 1, and to levels in between when the parameter is between 0 and 1. The idea behind this is to be able to adjust the level of consistency to the instance to be solved, hoping that such an adapted level of consistency will prune significantly more values than arc consistency while being less time consuming than LC.

Parameterized consistency is based on the concept of stability of values. We first need to define the 'distance to end' of a value in a domain. This captures how far a value is from the last in its domain. In the following, $rank(v, S)$ is the position of value v in the ordered set of values S.

Definition 1 (Distance to end of a value). *The* distance to end *of a value* $v_i \in D(x_i)$ *is the ratio*

$$\Delta(x_i, v_i) = (|D_o(x_i)| - rank(v_i, D_o(x_i)))/|D_o(x_i)|,$$

where $D_o(x_i)$ *is the initial domain of* x_i.

We see that the first value in $D_o(x_i)$ has distance $(|D_o(x_i)| - 1)/|D_o(x_i)|$ and the last one has distance 0. Thus, $\forall v_i \in D(x_i), 0 \leq \Delta(x_i, v_i) < 1$.

We can now give the definition of what we call the parameterized stability of a value for arc consistency. The idea is to define stability for values based on the distance to the end of their AC supports. For instance, consider the constraint $x_1 \leq x_2$ with the domains $D(x_1) = D(x_2) = \{1, 2, 3, 4\}$ (see Fig. 1). $\Delta(x_2, 1) = (4 - 1)/4 = 0.75$, $\Delta(x_2, 2) = 0.5$, $\Delta(x_2, 3) = 0.25$ and $\Delta(x_2, 4) = 0$. If $p = 0.2$, the value $(x_1, 4)$ is not p-stable for AC, because the first and only AC support of $(x_1, 4)$ in the ordering used to look for supports, that is $(x_2, 4)$, has a distance to end smaller than the threshold p. Proving that the pair $(4, 4)$ is inconsistent (by a stronger consistency) could lead to the pruning of $(x_1, 4)$. In other words, applying a stronger consistency on $(x_1, 4)$ has a higher chance to lead to its removal than applying it to for instance $(x_1, 1)$, which had no difficulty to find its first AC support (distance to end of $(x_2, 1)$ is 0.75).

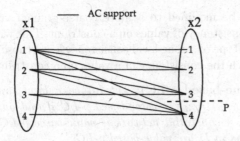

Fig. 1. Stability of supports on the example of the constraint $x_1 \leq x_2$ with the domains $D(x_1) = D(x_2) = \{1, 2, 3, 4\}$. $(x_1, 4)$ is not p-stable for AC.

At this point, we want to emphasize that the ordering of values used to look for supports in the domains is not related to the order in which values are selected by the branching heuristic used by the backtrack search procedure. That is, we can use a given order of values for looking for supports and another one for exploring the search tree.

Definition 2 (p-stability for AC). *A value* $v_i \in D(x_i)$ *is p-stable for AC on* c_{ij} *iff* v_i *has an AC support* $v_j \in D(x_j)$ *on* c_{ij} *such that* $\Delta(x_j, v_j) \geq p$. *A value* $v_i \in D(x_i)$ *is p-stable for AC iff* $\forall x_j \in \Gamma(x_i)$, v_i *is p-stable for AC on* c_{ij}.

We are now ready to give the first definition of parameterized local consistency. This first definition can be applied to any local consistency LC for which the consistency of a value on a constraint is well defined. This is the case for instance for all triangle-based consistencies [DB01, Bes06].

Definition 3 (Constraint-based p-LC). *Let LC be a local consistency stronger than AC for which the LC consistency of a value on a constraint is defined. A value* $v_i \in D(x_i)$ *is constraint-based p-LC on* c_{ij} *iff it is p-stable for AC on* c_{ij}, *or it is LC on* c_{ij}. *A value* $v_i \in D(x_i)$ *is constraint-based p-LC iff* $\forall c_{ij}$, v_i *is constraint-based p-LC on* c_{ij}. *A constraint network is constraint-based p-LC iff all values in all domains in D are constraint-based p-LC.*

Theorem 1. *Let LC be a local consistency stronger than AC for which the LC consistency of a value on a constraint is defined. Let* p_1 *and* p_2 *be two parameters in* $[0..1]$. *If* $p_1 < p_2$, *then* $AC \preceq$ *constraint-based* p_1-LC $\preceq$ *constraint-based* p_2-LC $\preceq$ LC.

Proof. Suppose that there exist two parameters p_1, p_2 such that $0 \leq p_1 < p_2 \leq 1$, and suppose that there exists a p_2-LC constraint network N that contains a p_2-LC value (x_i, v_i) that is p_1-LC inconsistent. Let c_{ij} be the constraint on which (x_i, v_i) is p_1-LC inconsistent. Then, $\nexists v_j \in D(x_j)$ that is an AC support for (x_i, v_i) on c_{ij} such that $\Delta(x_j, v_j) \geq p_1$. Thus, v_i is not p_2-stable for AC on c_{ij}. In addition, v_i is not LC on c_{ij}. Therefore, v_i is not p_2-LC, and N is not p_2-LC. ∎

Definition 3 can be modified to a more coarse-grained version that is not dependent on the consistency of values on a constraint. This will have the advantage to apply to any type of strong local consistency, even those, like singleton arc consistency, for which the consistency of a value on a constraint is not defined.◆

Definition 4 (Value-based p-LC). *Let LC be a local consistency stronger than AC. A value $v_i \in D(x_i)$ is value-based p-LC if and only if it is p-stable for AC or it is LC. A constraint network is value-based p-LC if and only if all values in all domains in D are value-based p-LC.*

Theorem 2. *Let LC be a local consistency stronger than AC. Let p_1 and p_2 be two parameters in $[0..1]$. If $p_1 < p_2$ then $AC \preceq$ value-based p_1-LC $\preceq$ value-based p_2-LC $\preceq LC$.*

Proof. Suppose that there exist two parameters p_1, p_2 such that $0 \leq p_1 < p_2 \leq 1$, and suppose that there exists a p_2-LC constraint network N that contains a p_2-LC value (x_i, v_i) that is p_1-LC-inconsistent. v_i is p_1-LC-inconsistent means that:

1. v_i is not p_1-stable for AC: $\exists c_{ij}$ on which v_i is not p_1-stable for AC. Then $\nexists v_j \in D(x_j)$ that is an AC support for (x_i, v_i) on c_{ij} such that $\Delta(x_j, v_j) \geq p_1$. Therefore, v_i is not p_2-stable for AC on c_{ij}, then v_i is not p_2-stable for AC.
2. v_i is LC inconsistent.

(1) and (2) imply that v_i is not p_2-LC and N is not p_2-LC.   ∎

For both types of definitions of p-LC, we have the following property on the extreme cases $(p = 0, p = 1)$.

Corollary 1. *Let LC_1 and LC_2 be two local consistencies stronger than AC. We have: value-based 0-$LC_2 = AC$ and value-based 1-$LC_2 = LC$. If the LC_1 consistency of a value on a constraint is defined, we also have: constraint-based 0-$LC_1 = AC$ and constraint-based 1-$LC_1 = LC$.*

3.1 Parameterized MaxRPC: p-maxRPC

To illustrate the benefit of our approach, we apply *parameterized consistency* to maxRPC to obtain the p-maxRPC level of consistency that achieves a consistency level between AC and maxRPC.

Definition 5 (p-maxRPC). *A value is p-maxRPC if and only if it is constraint-based p-maxRPC. A network is p-maxRPC if and only if it is constraint-based p-maxRPC.*

From Theorem 1 and Corollary 1 we derive the following corollary.

Corollary 2. *For any two parameters p_1, $p_2, 0 \leq p_1 < p_2 \leq 1$, $AC \preceq p_1$-maxRPC $\preceq p_2$-maxRPC $\preceq$ maxRPC. 0-maxRPC $= AC$ and 1-maxRPC $=$ maxRPC.*

Algorithm 1. Initialization(X, D, C, Q)

```
 1 begin
 2 │   foreach x_i ∈ X do
 3 │   │   foreach v_i ∈ D(x_i) do
 4 │   │   │   foreach x_j ∈ Γ(x_i) do
 5 │   │   │   │   p-support ← false;
 6 │   │   │   │   foreach v_j ∈ D(x_j) do
 7 │   │   │   │   │   if (v_i, v_j) ∈ c_ij then
 8 │   │   │   │   │   │   LastAC_{x_i,v_i,x_j} ← v_j;
 9 │   │   │   │   │   │   if Δ(x_j, v_j) ≥ p then
10 │   │   │   │   │   │   │   p-support ← true;
11 │   │   │   │   │   │   │   LastPC_{x_i,v_i,x_j} ← v_j;
12 │   │   │   │   │   │   │   break;
13 │   │   │   │   │   │   if searchPCwit(v_i, v_j) then
14 │   │   │   │   │   │   │   p-support ← true;
15 │   │   │   │   │   │   │   LastPC_{x_i,v_i,x_j} ← v_j;
16 │   │   │   │   │   │   │   break;

17 │   │   │   if ¬p-support then
18 │   │   │   │   remove v_i from D(x_i);
19 │   │   │   │   Q ← Q ∪ {x_i};
20 │   │   │   │   break;

21 │   │   if D(x_i) = ∅ then return false;
22 │   return true;
```

We propose an algorithm for p-maxRPC, based on maxRPC3, the best existing maxRPC algorithm. We do not describe maxRPC3 in full detail, as it can be found in [BPSW11]. We only describe procedures where changes to maxRPC3 are necessary to design p-maxRPC3, a coarse grained algorithm that performs p-maxRPC. We use light grey to emphasize the modified parts of the original maxRPC3 algorithm.

maxRPC3 uses a propagation list Q where it inserts the variables whose domains have changed. It also uses two other data structures: LastAC and LastPC. For each value (x_i, v_i), LastAC$_{x_i, v_i, x_j}$ stores the smallest AC support for (x_i, v_i) on c_{ij} and LastPC$_{x_i, v_i, x_j}$ stores the smallest PC support for (x_i, v_i) on c_{ij} (i.e., the smallest AC support (x_j, v_j) for (x_i, v_i) on c_{ij} such that (v_i, v_j) is PC). This algorithm comprises two phases: initialization and propagation.

In the initialization phase (Algorithm 1) maxRPC3 checks if each value (x_i, v_i) has a maxRPC-support (x_j, v_j) on each constraint c_{ij}. If not, it removes v_i from $D(x_i)$ and inserts x_i in Q. To check if a value (x_i, v_i) has a maxRPC-support on a constraint c_{ij}, maxRPC3 looks first for an AC-support (x_j, v_j) for (x_i, v_i) on c_{ij}, then it checks if (v_i, v_j) is PC. In this last step, changes were

---

Algorithm 2. checkPCsupLoss(v_j, x_i)

---

1 **begin**
2 | **if** $LastAC_{x_j,v_j,x_i} \in D(x_i)$ **then**
3 | | $b_i \leftarrow max(LastPC_{x_j,v_j,x_i}+1, LastAC_{x_j,v_j,x_i})$;
4 | **else**
5 | | $b_i \leftarrow max(LastPC_{x_j,v_j,x_i}+1, LastAC_{x_j,v_j,x_i}+1)$;
6 | **foreach** $v_i \in D(x_i), v_i \geq b_i$ **do**
7 | | **if** $(v_j, v_i) \in c_{ji}$ **then**
8 | | | **if** $LastAC_{x_j,v_j,x_i} \notin D(x_i)$ & $LastAC_{x_j,v_j,x_i} > LastPC_{x_j,v_j,x_i}$ **then**
9 | | | | $LastAC_{x_j,v_j,x_i} \leftarrow v_i$;
10 | | | **if** $\Delta(x_i, v_i) \geq p$ **then**
11 | | | | $LastPC_{x_j,v_j,x_i} \leftarrow v_i$;
12 | | | | **return** $true$;
13 | | | **if** searchPCwit(v_j, v_i) **then**
14 | | | | $LastPC_{x_j,v_j,x_i} \leftarrow v_i$;
15 | | | | **return** $true$;
16 | **return** $false$;

---

necessary to obtain p-maxRPC3 (lines 9–12). We check if (v_i, v_j) is PC (line 13) only if $\Delta(x_j, v_j)$ is smaller than the parameter p (line 9).

The propagation phase of maxRPC3 involves propagating the effect of deletions. While Q is non empty, maxRPC3 extracts a variable x_i from Q and checks for each value (x_j, v_j) of each neighboring variable $x_j \in \Gamma(x_i)$ if it is not maxRPC because of deletions of values in $D(x_i)$. A value (x_j, v_j) becomes maxRPC inconsistent in two cases: if its unique PC-support (x_i, v_i) on c_{ij} has been deleted, or if we deleted the unique witness (x_i, v_i) for a pair (v_j, v_k) such that (x_k, v_k) is the unique PC-support for (x_j, v_j) on c_{jk}. So, to propagate deletions, maxRPC3 checks if the last maxRPC support (last known support) of (x_j, v_j) on c_{ij} still belongs to the domain of x_i, otherwise it looks for the next support (Algorithm 2). If such a support does not exist, it removes the value v_j and adds the variable x_j to Q. Then if (x_j, v_j) has not been removed in the previous step, maxRPC3 checks (Algorithm 3) whether there is still a witness for each pair (v_j, v_k) such that (x_k, v_k) is the PC support for (x_j, v_j) on c_{jk}. If not, it looks for the next maxRPC support for (x_j, v_j) on c_{jk}. If such a support does not exist, it removes v_j from $D(x_j)$ and adds the variable x_j to Q.

In the propagation phase, we also modified maxRPC3 to check if the values are still p-maxRPC instead of checking if they are maxRPC. In p-maxRPC3, the last p-maxRPC support for (x_j, v_j) on c_{ij} is the last AC support if (x_j, v_j) is p-stable for AC on c_{ij}. If not, it is the last PC support. Thus, p-maxRPC3 checks if the last p-maxRPC support (last known support) of (x_j, v_j) on c_{ij} still belongs to the domain of x_i. If not, it looks (Algorithm 2) for the next AC support (x_i, v_i) on c_{ij}, and checks if (v_i, v_j) is PC (line 13) only when $\Delta(x_i, v_i) < p$

Algorithm 3. checkPCwitLoss(x_j, v_j, x_i)

---

1 **begin**
2   **foreach** $x_k \in \Gamma(x_j) \cap \Gamma(x_i)$ **do**
3   witness $\leftarrow$ *false*;
4   **if** $v_k \leftarrow LastPC_{x_j,v_j,x_k} \in D(x_k)$ **then**
5   **if** $\Delta(x_k, v_k) \geq p$ **then**
6   witness $\leftarrow$ *true*;
7   **else**
8   **if** $LastAC_{x_j,v_j,x_i} \in D(x_i)$ & $LastAC_{x_j,v_j,x_i} = LastAC_{x_k,v_k,x_i}$
9   **OR** $LastAC_{x_j,v_j,x_i} \in D(x_i)$ & ($LastAC_{x_j,v_j,x_i}, v_k) \in c_{ik}$
10   **OR** $LastAC_{x_k,v_k,x_i} \in D(x_i)$ & ($LastAC_{x_k,v_k,x_i}, v_j) \in c_{ij}$
11   **then** witness $\leftarrow$ *true* ;
12   **else**
13   **if** searchACsup(x_j, v_j, x_i) & searchACsup(x_k, v_k, x_i) **then**
14   **foreach**
     $v_i \in D(x_i), v_i \geq max(LastAC_{x_j,v_j,x_i}, LastAC_{x_k,v_k,x_i})$
   do
15   **if** $(v_j, v_i) \in c_{ji}$ & $(v_k, v_i) \in c_{ki}$ **then**
16   witness $\leftarrow$ *true*;
17   break;

18   **if** ¬witness & ¬checkPCsupLoss(v_j, x_k) **then return** *false* ;
19   **return** *true*;

(line 10). If no p-maxRPC support exists, p-maxRPC3 removes the value and adds the variable x_j to Q. If the value (x_j, v_j) has not been removed in the previous phase, p-maxRPC3 checks (Algorithm 3) whether there is still a witness for each pair (v_j, v_k) such that (x_k, v_k) is the p-maxRPC support for v_j on c_{jk} and $\Delta(x_k, v_k) < p$. If not, it looks for the next p-maxRPC support for v_j on c_{jk}. If such a support does not exist, it removes v_j from $D(x_j)$ and adds the variable x_j to Q.

p-maxRPC3 uses the data structure *LastPC* to store the last p-maxRPC support (i.e., the latest AC support for the p-stable values and the latest PC support for the others). Algorithms 1 and 2 update the data structure *LastPC* of maxRPC3 to be *LastAC* for all the values that are p-stable for AC (line 11 of Algorithm 1 and line 11 of Algorithm 2) and avoid seeking witnesses for those values. Algorithm 3 avoids checking the loss of witnesses for the p-stable values by setting the flag **witness** to *true* (line 6). Correctness of p-maxRPC3 directly comes from maxRPC3: The removed values are necessarily p-maxRPC-inconsistent and all the values that are p-maxRPC-inconsistent are removed.

3.2   Experimental Validation of p-maxRPC

To validate the approach of parameterized local consistency, we conducted a first basic experiment. The purpose of this experiment is to see if there exist instances on which a given level of p-maxRPC, with a value p that is uniform (i.e., identical for the entire constraint network) and static (i.e., constant through the entire search process), is more efficient than AC or maxRPC, or both.

We have implemented the algorithms that achieve p-maxRPC as described in the previous section in our own binary constraint solver, in addition to maxRPC (maxRPC3 version [BPSW11]) and AC (AC2001 version [BRYZ05]). All the algorithms are implemented in our JAVA CSP solver. We tested these algorithms on several classes of CSP instances from the International Constraint Solver Competition 09[1]. We have only selected instances involving binary constraints. To isolate the effect of propagation, we used the lexicographic ordering for variables and values. We set the CPU timeout to one hour. Our experiments were conducted on a 12-core Genuine Intel machine with 16 Gb of RAM running at 2.92 GHz.

On each instance of our experiment, we ran AC, max-RPC, and p-maxRPC for all values of p in $\{0.1, 0.2, \ldots, 0.9\}$. Performance has been measured in terms of CPU time in seconds, the number of visited nodes (NODE) and the number of constraint checks (CCK). Results are presented as "CPU time (p)", where p is the parameter for which p-maxRPC gives the best result.

Table 1 reports the performance of AC, maxRPC, and p-maxRPC for the value of p producing the best CPU time, on instances from Radio Link Frequency Assignment Problems (RLFAPs), geom problems, and queens knights problems. The CPU time of the best algorithm is bold-faced. On RLFAP and geom, we observe the existence of a parameter p for which p-maxRPC is faster than *both* AC and maxRPC for most instances of these two classes of problems. On the queens-knight problem, however, AC is always the best algorithm. In Figs. 2 and 3, we try to understand more closely what makes p-maxRPC better or worse than AC and maxRPC. Figures 2 and 3 plot the performance (CPU, NODE and CCK) of p-maxRPC for all values of p from 0 to 1 by steps of 0.1 against performance of AC and maxRPC. Figure 2 shows an instance where p-maxRPC solves the problem faster than AC and maxRPC for values of p in the range [0.3..0.8]. We observe that p-maxRPC is faster than AC and maxRPC when it reduces the size of the search space as much as maxRPC (same number of nodes visited) with a number of CCK closer to the number of CCK produced by AC. Figure 3 shows an instance where the CPU time for p-maxRPC is never better than *both* AC and maxRPC, whatever the value of p. We see that p-maxRPC is two to three times faster than maxRPC. But p-maxRPC fails to improve AC because the number of constraint checks performed by p-maxRPC is much higher than the number of constraint checks performed by AC, whereas the number of nodes visited by p-maxRPC is not significantly reduced compared to the number of nodes visited by AC. From these observations, it thus seems that p-maxRPC

---

[1] http://cpai.ucc.ie/09/.

Table 1. Performance (CPU time, nodes and constraint checks) of AC, p-maxRPC, and maxRPC on various instances.

		AC	p-maxRPC	p	maxRPC
scen1-f8	CPU	>3600	**1.39**	(0.2)	6.10
	#nodes	–	927		917
	#ccks	–	1,397,440		26,932,990
scen2-f24	CPU	>3600	**0.13**	(0.3)	0.65
	#nodes	–	201		201
	#ccks	–	296,974		3,462,070
scen3-f10	CPU .	>3600	**0.89**	(0.5)	2.80
	#nodes	–	469		408
	#ccks	–	874,930		13,311,797
geo50-20-d4-75-26	CPU	111.48	17.80	(1.0)	**15.07**
	#nodes	477,696	3,768		3,768
	#ccks	96,192,822	40,784,017		40,784,017
geo50-20-d4-75-43	CPU	1,671.35	**1,264.36**	(0.5)	1,530.02
	#nodes	4,118,134	555,259		279,130
	#ccks	1,160,664,461	1,801,402,535		3,898,964,831
geo50-20-d4-75-46	CPU	1,732.22	**371.30**	(0.6)	517.35
	#nodes	3,682,394	125,151		64,138
	#ccks	1,516,856,615	584,743,023		1,287,674,430
geo50-20-d4-75-84	CPU	404.63	**0.44**	(0.6)	0.56
	#nodes	2,581,794	513		333
	#ccks	293,092,144	800,657		1,606,047
queensKnights10-5-add	CPU	**27.14**	30.79	(0.2)	98.44
	#nodes	82,208	81,033		78,498
	#ccks	131,098,933	148,919,686		954,982,880
queensKnights10-5-mul	CPU	**43.89**	83.27	(0.1)	300.74
	#nodes	74,968	74,414		70,474
	#ccks	104,376,698	140,309,576		1,128,564,278

outperforms AC and maxRPC when it finds a compromise between the number of nodes visited (the power of maxRPC) and the number of CCK needed to maintain (the light cost of AC).

In Figs. 2 and 3 we can see that the CPU time for 1-maxRPC (respectively 0-maxRPC) is greater than the CPU time for maxRPC (respectively AC), although the two consistencies are equivalent. The reason is that p-maxRPC performs tests on the distances. For $p = 0$, we also explain this difference by the fact that p-maxRPC maintains data structures that AC does not use.

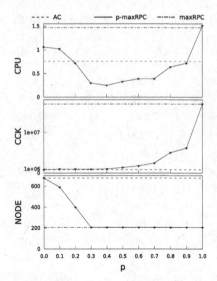
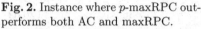

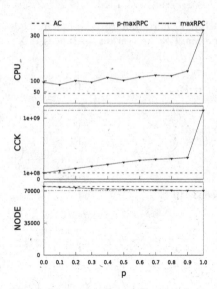

Fig. 2. Instance where p-maxRPC outperforms both AC and maxRPC.

Fig. 3. Instance where AC outperforms p-maxRPC.

4   Adaptative Parameterized Consistency: ap-maxRPC

In the previous section, we have defined p-maxRPC, a version of parameterized consistency where the strong local consistency is maxRPC. We have performed some initial experiments where p has the same value during the whole search and everywhere in the constraint network. However, the algorithm we proposed to enforce p-maxRPC does not specify how p is chosen. In this section, we propose two possible ways to dynamically and locally adapt the parameter p in order to solve the problem faster than both AC and maxRPC. Instead of using a single value for p during the whole search and for the whole constraint network, we propose to use several local parameters and to adapt the level of local consistency by dynamically adjusting the value of the different local parameters during search. The idea is to concentrate the effort of propagation by increasing the level of consistency in the most difficult parts of the given instance. We can determine these difficult parts using heuristics based on conflicts in the same vein as the weight of a constraint or the weighted degree of a variable in [BHLS04].

4.1   Constraint-Based ap-maxRPC: apc-maxRPC

The first technique we propose, called constraint-based ap-maxRPC, assigns a parameter $p(c_k)$ to each constraint c_k in C. We define this parameter to be correlated to the *weight* of the constraint. The idea is to apply a higher level of consistency in parts of the problem where the constraints are the most active.

Definition 6 (The weight of a constraint [BHLS04]). *The weight $w(c_k)$ of a constraint $c_k \in C$ is an integer that is incremented every time a domain wipe-out occurs while performing propagation on this constraint.*

We define the adaptive parameter $p(c_k)$ local to constraint c_k in such a way that it is greater when the weight $w(c_k)$ is higher w.r.t. other constraints.

$$\forall c_k \in C, p(c_k) = \frac{w(c_k) - min_{c \in C}(w(c))}{max_{c \in C}(w(c)) - min_{c \in C}(w(c))} \tag{1}$$

Equation 1 is normalized so that we are guaranteed that $0 \leq p(c_k) \leq 1$ for all $c_k \in C$ and that there exists c_{k_1} with $p(c_{k_1}) = 0$ (the constraint with lowest weight) and c_{k_2} with $p(c_{k_2}) = 1$ (the constraint with highest weight).

We are now ready to define adaptive parameterized consistency based on constraints.

Definition 7 (constraint-based ap-maxRPC). *A value $v_i \in D(x_i)$ is constraint-based ap-maxRPC (or apc-maxRPC) on a constraint c_{ij} if and only if it is constraint-based $p(c_{ij})$-maxRPC. A value $v_i \in D(x_i)$ is apc-maxRPC iff $\forall c_{ij}$, v_i is apc-maxRPC on c_{ij}. A constraint network is apc-maxRPC iff all values in all domains in D are apc-maxRPC.*

4.2 Variable-Based ap-maxRPC: apx-maxRPC

The technique proposed in Sect. 4.1 can only be used on consistencies where the consistency of a value on a constraint is defined. We present a second technique which can be used on constraint-based or variable-based local consistencies indifferently. We instantiate our definitions to maxRPC but the extension to other consistencies is direct. We call this new technique variable-based ap-maxRPC. We need to define the weighted degree of a variable as the aggregation of the weights of all constraints involving it.

Definition 8 (The weighted degree of a variable [BHLS04]). *The weighted degree $wdeg(x_i)$ of a variable x_i is the sum of the weights of the constraints involving x_i and one other uninstantiated variable.*

We associate each variable with an adaptive local parameter based on its weighted degree.

$$\forall x_i \in X, p(x_i) = \frac{wdeg(x_i) - min_{x \in X}(wdeg(x))}{max_{x \in X}(wdeg(x)) - min_{x \in X}(wdeg(x))} \tag{2}$$

As in Eq. 1, we see that the local parameter is normalized so that we are guaranteed that $0 \leq p(x_i) \leq 1$ for all $x_i \in X$ and that there exists x_{k_1} with $p(x_{k_1}) = 0$ (the variable with lowest weighted degree) and x_{k_2} with $p(x_{k_2}) = 1$ (the variable with highest weighted degree).

Definition 9 (variable-based ap-maxRPC). *A value $v_i \in D(x_i)$ is variable-based ap-maxRPC (or apx-maxRPC) if and only if it is value-based $p(x_i)$-maxRPC. A constraint network is apx-maxRPC iff all values in all domains in D are apx-maxRPC.*

4.3   Experimental Evaluation of ap-maxRPC

In Sect. 3.2 we have shown that maintaining a static form of p-maxRPC during the entire search can lead to a promising trade-off between computational effort and pruning when all algorithms follow the same static variable ordering. In this section, we want to put our contributions in the real context of a solver using the best known variable ordering heuristic, $dom/wdeg$, though it is known that this heuristic is so good that it substantially reduces the differences in performance that other features of the solver could provide. We have compared the two variants of adaptive parameterized consistency, namely apc-maxRPC and apx-maxRPC, to AC and maxRPC. We ran the four algorithms on instances of radio link frequency assignment problems, geom problems, and queens knights problems.

Table 2 reports some representative results. A first observation is that, thanks to the $dom/wdeg$ heuristic, we were able to solve more instances before the cutoff of one hour, especially the scen11 variants of RLFAP. A second observation is that apc-maxRPC and apx-maxRPC are both faster than at least one of the two extreme consistencies (AC and maxRPC) on all instances except scen7-w1-f4 and geo50-20-d4-75-30. Third, when apx-maxRPC and/or apc-maxRPC are faster than both AC and maxRPC (scen1-f9, scen2-f25, scen11-f9, scen11-f10 and scen11-f11), we observe that the gap in performance in terms of nodes and CCKs between AC and maxRPC is significant. Except for scen7-w1-f4, the number of nodes visited by AC is three to five times greater than the number of nodes visited by maxRPC and the number of constraint checks performed by maxRPC is twelve to sixteen times greater than the number of constraint checks performed by AC. For the geom instances the CPU time of the ap-maxRPC algorithms is between AC and maxRPC, and it is never lower than the CPU time of AC. This probably means that when solving these instances with the $dom/wdeg$ heuristic, there is no need for sophisticated local consistencies. In general we see that the ap-maxRPC algorithms fail to improve both the two extreme consistencies simultaneously for the instances where the performance gap between AC and maxRPC is low.

If we compare apx-maxRPC to apc-maxRPC, we observe that although apx-maxRPC is coarser in its design than apc-maxRPC, apx-maxRPC is often faster than apc-maxRPC. We can explain this by the fact that the constraints initially all have the same weight equal to 1. Hence, all local parameters $ap(c_k)$ initially have the same value 0, so that apc-maxRPC starts resolution by applying AC everywhere. It will start enforcing some amount of maxRPC only after the first wipe-out occurred. On the contrary, in apx-maxRPC, when constraints all have the same weight, the local parameter $p(x_i)$ is correlated to the degree of the variable x_i. As a result, apx-maxRPC benefits from the filtering power of maxRPC even before the first wipe-out.

In Table 2, we reported only the results on a few representative instances. Table 3 summarizes the entire set of experiments. It shows the average CPU time for each algorithm on all instances of the different classes of problems tested. We considered only the instances solved before the cutoff of one hour by at least one

Table 2. Performance (CPU time, nodes and constraint checks) of AC, variable-based ap-maxRPC (apx-maxRPC), constraint-based ap-maxRPC (apc-maxRPC), and maxRPC on various instances.

		AC	apx-maxRPC	apc-maxRPC	maxRPC
scen1-f9	CPU	90.34	**31.17**	33.40	41.56
	#nodes	2,291	1,080	1,241	726
	#ccks	3,740,502	3,567,369	2,340,417	50,045,838
scen2-f25	CPU	70.57	46.40	**27.22**	81.40
	#nodes	12,591	4,688	3,928	3,002
	#ccks	15,116,992	38,239,829	8,796,638	194,909,585
scen6-w2	CPU	7.30	1.25	2.63	**0.01**
	#nodes	2,045	249	610	0
	#ccks	2,401,057	1,708,812	1,914,113	85,769
scen7-w1-f4	CPU	0.28	**0.17**	0.54	0.30
	#nodes	567	430	523	424
	#ccks	608,040	623,258	584,308	1,345,473
scen11-f9	CPU	2,718.65	**1,110.80**	1,552.20	2,005.61
	#nodes	103,506	40,413	61,292	32,882
	#ccks	227,751,301	399,396,873	123,984,968	3,637,652,122
scen11-f10	CPU	225.29	**83.89**	134.46	112.18
	#nodes	9,511	3,510	4,642	2,298
	#ccks	12,972,427	17,778,458	6,717,485	156,005,235
scen11-f11	CPU	156.76	**39.39**	93.69	76.95
	#nodes	7,050	2,154	3,431	1,337
	#ccks	7,840,552	10,006,821	5,143,592	91,518,348
scen11-f12	CPU	139.91	69.50	88.76	**61.92**
	#nodes	7,050	2,597	3,424	1,337
	#ccks	7,827,974	11,327,536	5,144,835	91,288,023
geo50-20d4-75-19	CPU	**242.13**	553.53	657.72	982.34
	#nodes	195,058	114,065	160,826	71,896
	#ccks	224,671,319	594,514,132	507,131,322	2,669,750,690
geo50-20d4-75-30	CPU	**0.84**	1.01	1.07	1.02
	#nodes	359	115	278	98
	#ccks	261,029	432,705	313,168	1,880,927
geo50-20d4-75-84	CPU	**0.02**	0.09	0.05	0.29
	#nodes	59	54	59	52
	#ccks	33,876	80,626	32,878	697,706
queensK20-5-mul	CPU	787.35	2,345.43	**709.45**	>3600
	#nodes	55,596	40,606	41,743	–
	#ccks	347,596,389	6,875,941,876	379,826,516	–
queensK15-5-add	CPU	24.69	17.01	**14.98**	35.05
	#nodes	24,639	12,905	12,677	11,595
	#ccks	90,439,795	91,562,150	58,225,434	394,073,525

Table 3. Average CPU time of AC, variable-based ap-maxRPC (apx-maxRPC), constraint-based ap-maxRPC (apc-maxRPC), and maxRPC on all instances of each class of problems tested, when the local parameters are updated at each node

class (#instances)		AC	apx-maxRPC	apc-maxRPC	maxRPC
geom (10)	#solved	**10**	10	10	10
	average CPU	**69.28**	180.57	191.03	279.30
scen (10)	#solved	10	10	**10**	10
	average CPU	18.95	9.63	**8.30**	13.94
scen11 (10)	#solved	4	**4**	4	4
	average CPU	810.15	**325.90**	467.28	564.17
queensK (11)	#solved	6	6	**6**	5
	average CPU	135.95	395.41	**121.75**	>610.51

Table 4. Average CPU time of AC, variable-based ap-maxRPC (apx-maxRPC), constraint-based ap-maxRPC (apc-maxRPC), and maxRPC on all instances of each class of problems tested, when the local parameters are updated every 10 nodes

class (#instances)		AC	apx-maxRPC	apc-maxRPC	maxRPC
geom (10)	#solved	**10**	10	10	10
	average CPU	**69.28**	147.20	189.42	279.30
scen (10)	#solved	10	10	**10**	10
	average CPU	18.95	**7.40**	8.86	13.94
scen11 (10)	#solved	4	**4**	4	4
	average CPU	810.15	**311.74**	417.97	564.17
queensK (11)	#solved	6	6	**6**	5
	average CPU	135.95	269.51	**117.18**	>610.52

of the four algorithms. To compute the average CPU time of an algorithm on a class of instances, we add the CPU time needed to solve each instance solved before the cutoff of one hour, and for the instances not solved before the cutoff, we add one hour. We observe that the adaptive approach is, on average, faster than the two extreme consistencies AC and maxRPC, except on the geom class.

In apx-maxRPC and apc-maxRPC, we update the local parameters $p(x_i)$ or $p(c_k)$ at each node in the search tree. We could wonder if such a frequent update does not produce too much overhead. To answer this question we performed a simple experiment in which we update the local parameters every 10 nodes only. We re-ran the whole set of experiments with this new setting. Table 4 reports the average CPU time for these results. We observe that when the local parameters are updated every 10 nodes, the gain for the adaptive approach is, on average, greater than when the local parameters are updated at each node. This gives room for improvement, by trying to adapt the frequency of update of these parameters.

5   Partition-One-Arc-Consistency

In this section, we describe our second approach, which is inspired from singleton-based consistencies. Singleton Arc Consistency (SAC) [DB97] makes a singleton test by enforcing arc consistency and can only prune values in the variable domain on which it currently performs singleton tests. Partition-One-AC (POAC) [BA01] is an extension of SAC, which, as observed in [BD08], combines singleton tests and *constructive disjunction* [VSD98]. POAC can prune values everywhere in the network as soon as a variable has been completely singleton tested.

We propose an adaptive version of POAC, where the number of times variables are processed for singleton tests on their values is dynamically and automatically adapted during search. Before moving to adaptive partition-one-AC, we first propose an efficient algorithm enforcing POAC and we compare its behaviour to SAC.

5.1   The Algorithm

The efficiency of our POAC algorithm, POAC1, is based on the use of counters associated with each value (x_j, v_j) in the constraint network. These counters are used to count how many times a value v_j from a variable x_j is pruned during the sequence of POAC tests on all the values of another variable x_i (the varPOAC call to x_i). If v_j is pruned $|D(x_i)|$ times, this means that it is not POAC and can be removed from $D(x_j)$.

POAC1 (Algorithm 4) starts by enforcing arc consistency on the network (line 2). Then it puts all variables in the ordered cyclic list S using any total ordering on X (line 3). varPOAC iterates on all variables from S (line 7) to make them POAC until the fixpoint is reached (line 12) or a domain wipe-out occurs (line 8). The counter FPP (FixPoint Proof) counts how many calls to varPOAC have been processed in a row without any change in any domain (line 9).

Algorithm 4. POAC1(X, D, C)

---

1 **begin**
2 | **if** ¬EnforceAC(X, D, C) **then return** *false* ;
3 | $S \leftarrow CyclicList(Ordering(X))$;
4 | FPP $\leftarrow 0$;
5 | $x_i \leftarrow first(S)$;
6 | **while** FPP $< |X|$ **do**
7 | | **if** ¬varPOAC$(x_i, X, D, C, \text{CHANGE})$ **then**
8 | | | **return** *false*;
9 | | **if** CHANGE **then** FPP $\leftarrow 1$;
10 | | **else** FPP++;
11 | | $x_i \leftarrow NextElement(x_i, S)$;
12 | **return** *true*;

Algorithm 5. varPOAC(x_i, X, D, C, CHANGE)

---

1 **begin**
2   SIZE $\leftarrow |D(x_i)|$; CHANGE $\leftarrow$ *false*;
3   **foreach** $v_i \in D(x_i)$ **do**
4   **if** $\neg$TestAC($X, D, C \cup \{x_i = v_i\}$) **then**
5   remove v_i from $D(x_i)$;
6   **if** $\neg$EnforceAC(X, D, C, x_i) **then return** *false* ;
7   **if** $D(x_i) = \emptyset$ **then return** *false*;
8   **if** SIZE $\neq |D(x_i)|$ **then** CHANGE $\leftarrow$ *true*;
9   **foreach** $x_j \in X \backslash \{x_i\}$ **do**
10   SIZE $\leftarrow |D(x_j)|$;
11   **foreach** $v_j \in D(x_j)$ **do**
12   **if** $counter(x_j, v_j) = |D(x_i)|$ **then** remove v_j from $D(x_j)$;
13             $counter(x_j, v_j) \leftarrow 0$;
14   **if** $D(x_j) = \emptyset$ **then return** *false*;
15   **if** SIZE $\neq |D(x_j)|$ **then** CHANGE $\leftarrow$ *true*;
16   **return** *true*

Algorithm 6. TestAC($X, D, C \cup \{x_i = v_i\}$)

---

1 **begin**
2      $Q \leftarrow \{(x_j, c_k) \mid c_k \in \Gamma(x_i), x_j \in var(c_k), x_j \neq x_i\}$;
3      $L \leftarrow \emptyset$;
4   **while** $Q \neq \emptyset$ **do**
5   pick and delete (x_j, c_k) from Q ;
6   SIZE $\leftarrow |D(x_j)|$;
7   **foreach** $v_j \in D(x_j)$ **do**
8   **if** $\neg$HasSupport(x_j, v_j, c_k) **then**
9   remove v_j from $D(x_j)$;
10                 $L \leftarrow L \cup (x_j, v_j)$;
11   **if** $D(x_j) = \emptyset$ **then**
12   RestoreDomains(L, *false*) ;
13   **return** *false* ;
14   **if** $|D(x_j)| <$ SIZE **then**
15             $Q \leftarrow Q \cup \{(x_{j'}, c_{k'}) | c_{k'} \in \Gamma(x_j), x_{j'} \in var(c_{k'}), x_{j'} \neq x_j, c_{k'} \neq c_k\}$;
16   RestoreDomains(L, *true*) ;
17   **return** *true* ;

The procedure varPOAC (Algorithm 5) is called to establish POAC w.r.t. a variable x_i. It works in two steps. The first step enforces arc consistency in each sub-network $N = (X, D, C \cup \{x_i = v_i\})$ (line 4) and removes v_i from $D(x_i)$ (line 5) if the sub-network is arc-inconsistent. Otherwise, the procedure TestAC (Algorithm 6) increments the counter associated with every arc inconsistent value

---

Algorithm 7. RestoreDomains(L, UPDATE)

---

1 **begin**
2   **if** UPDATE **then**
3   **foreach** $(x_j, v_j) \in L$ **do**
4          $D(x_j) \leftarrow D(x_j) \cup \{v_j\}$;
5   counter$(x_j, v_j) \leftarrow$ counter$(x_j, v_j) + 1$;
6   **else**
7   **foreach** $(x_j, v_j) \in L$ **do**
8          $D(x_j) \leftarrow D(x_j) \cup \{v_j\}$;

$(x_j, v_j), j \neq i$ in the sub-network $N = (X, D, C \cup \{x_i = v_i\})$. (Lines 6 and 7 have been added for improving the performance in practice but are not necessary for reaching the required level of consistency.) In line 8 the Boolean CHANGE is set to *true* if $D(x_i)$ has changed. The second step deletes all the values $(x_j, v_j), j \neq i$ with a counter equal to $|D(x_i)|$ and sets back the counter of each value to 0 (lines 12–13). Whenever a domain change occurs in $D(x_j)$, if the domain is empty, varPOAC returns failure (line 14); otherwise it sets the Boolean CHANGE to *true* (line 15).

Enforcing arc consistency on the sub-networks $N = (X, D, C \cup \{x_i = v_i\})$ is done by calling the procedure TestAC (Algorithm 6). TestAC just checks whether arc consistency on the sub-network $N = (X, D, C \cup \{x_i = v_i\})$ leads to a domain wipe-out or not. It is an instrumented AC algorithm that increments a counter for all removed values and restores them all at the end. In addition to the standard propagation queue Q, TestAC uses a list L to store all the removed values. After the initialisation of Q and L (lines 2–3), TestAC revises each arc (x_j, c_k) in Q and adds each removed value (x_j, v_j) to L (lines 5–10). If a domain wipe-out occurs (line 11), TestAC restores all removed values (line 12) without incrementing the counters (call to RestoreDomains with UPDATE $= false$) and it returns failure (line 13). Otherwise, if values have been pruned from the revised variable (line 14) it puts in Q the neighbouring arcs to be revised. At the end, removed values are restored (line 16) and their counters are incremented (call to RestoreDomains with UPDATE $= true$) before returning success (line 17).

Proposition 1. POAC1 *has a worst-case time complexity in $O(n^2 d^2 (T + n))$, where T is the time complexity of the arc-consistency algorithm used for singleton tests, n is the number of variables, and d is the number of values in the largest domain.*

Proof. The cost of calling varPOAC on a single variable is $O(dT + nd)$ because varPOAC runs AC on d values and updates nd counters. In the worst case, each of the nd value removals trigger n calls to varPOAC. Therefore POAC1 has a time complexity in $O(n^2 d^2 (T + n))$.   □

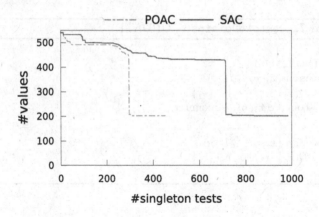

Fig. 4. The convergence speed of POAC and SAC.

5.2   Comparison of POAC and SAC Behaviors

Although POAC has a worst-case time complexity greater than SAC, we observed in practice that maintaining POAC during search is often faster than maintaining SAC. This behavior occurs even when POAC cannot remove more values than SAC, i.e. when the same number of nodes is visited with the same static variable ordering. This is due to what we call the *(filtering) convergence speed*: when both POAC and SAC reach the same fixpoint, POAC reaches the fixpoint with fewer singleton tests than SAC.

Figure 4 compares the convergence speed of POAC and SAC on an CSP instance where they have the same fixpoint. We observe that POAC is able to reduce the domains, to reach the fixpoint, and to prove the fixpoint, all in fewer singleton tests than SAC. This pattern has been observed on most of the instances and whatever ordering was used in the list S. The reason is that each time POAC applies varPOAC to a variable x_i, it is able to remove inconsistent values from $D(x_i)$ (like SAC), but also from any other variable domain (unlike SAC).

The fact that SAC cannot remove values in variables other than the one on which the singleton test is performed makes it a poor candidate for adapting the number of singleton tests. A SAC-inconsistent variable/value pair never singleton tested has no chance to be pruned by such a technique.

6   Adaptive POAC

This section presents an adaptive version of POAC that approximates POAC by monitoring the number of variables on which to perform singleton tests.

To achieve POAC, POAC1 calls the procedure varPOAC until it has proved that the fixpoint is reached. This means that, when the fixpoint is reached, POAC1 needs to call n (additional) times the procedure varPOAC without any pruning to prove that the fixpoint was reached. Furthermore, we experimentally observed that in most cases there is a long sequence of calls to `varPOAC` that prune very few values, even before the fixpoint has been reached (see Fig. 4 as an example). The goal of *Adaptive POAC* (APOAC) is to stop iterating on `varPOAC` as soon as possible. We want to benefit from strong propagation of singleton tests while avoiding the cost of the last calls to `varPOAC` that delete very few values or no value at all.

6.1   Principle

The APOAC approach alternates between two phases during search: a short *learning* phase and a longer *exploitation* phase. One of the two phases is executed on a sequence of nodes before switching to the other phase for another sequence of nodes. The search starts with a learning phase. The total length of a pair of sequences learning + exploitation is fixed to the parameter LE.

Before providing a more detailed description, let us define the (log_2 of the) *volume* of a constraint network $N = (X, D, C)$, used to approximate the size of the search space:

$$V = log_2 \prod_{i=1}^{n} |D(x_i)|$$

We use the logarithm of the volume instead of the volume itself, because of the large integers the volume generates. We also could have used the perimeter (i.e., $\sum_i |D(x_i)|$) for approximating the search space size, as done in [NT13]. However, experiments have confirmed that the volume is a more precise and effective criterion for adaptive POAC.

The ith learning phase is applied to a sequence of $L = \frac{1}{10} \cdot LE$ consecutive nodes. During that phase, we learn a cutoff value k_i, which is the maximum number of calls to the procedure `varPOAC` that each node of the next (ith) exploitation phase will be allowed to perform. A good cutoff k_i is such that `varPOAC` removes many inconsistent values (that is, obtains a significant volume reduction in the network) while avoiding calls to `varPOAC` that delete very few values or no value at all. During the ith exploitation phase, applied to a sequence of $\frac{9}{10} \cdot LE$ consecutive nodes, the procedure `varPOAC` is called at each node until fixpoint is proved or the cutoff limit of k_i calls to `varPOAC` is reached.

The ith learning phase works as follows. Let k_{i-1} be the cutoff learned at the previous learning phase. We initialize $maxK$ to $max(2 \cdot k_{i-1}, 2)$. At each node n_j in the new learning sequence $n_1, n_2, \ldots n_L$, APOAC is used with a cutoff $maxK$ on the number of calls to the procedure `varPOAC`. APOAC stores the sequence of volumes $(V_1, \ldots, V_{last})$, where V_p is the volume resulting from the pth call to `varPOAC` and *last* is the smallest among $maxK$ and the number of calls needed to prove fixpoint. Once the fixpoint is proved or the $maxK$th call to `varPOAC` performed, APOAC computes $k_i(j)$, the number of `varPOAC` calls that are enough to *sufficiently* reduce the volume while avoiding the extra cost of the last calls that remove few or no value. (The criteria to decide what 'sufficiently' means are described in Sect. 6.2.) Then, to make the learning phase more adaptive, $maxK$ is updated before starting node n_{j+1}. If $k_i(j)$ is close to

$maxK$, that is, greater than $\frac{3}{4} \cdot maxK$, we increase $maxK$ by 20%. If $k_i(j)$ is less than $\frac{1}{2} \cdot maxK$, we reduce $maxK$ by 20%. Otherwise, $maxK$ is unchanged. Once the learning phase ends, APOAC computes the cutoff k_i that will be applied to the next exploitation phase. k_i is an aggregation of the $k_i(j)$ values, $j = 1, \ldots, L$, computed using one of the aggregation techniques presented in Sect. 6.3.

6.2   Computing $k_i(j)$

We implemented APOAC using two different techniques to compute $k_i(j)$ at a node n_j of the learning phase:

- LR (*Last Reduction*) $k_i(j)$ is the rank of the last call to `varPOAC` that reduced the volume of the constraint network.
- LD (*Last Drop*) $k_i(j)$ is the rank of the last call to `varPOAC` that has produced a *significant* drop of the volume. The significance of a drop is captured by a ratio $\beta \in [0,1]$. More formally, $k_i(j) = max\{p \mid V_p \leq (1 - \beta)V_{p-1}\}$.

6.3   Aggregation of the $k_i(j)$ Values

Once the ith learning phase is complete, APOAC aggregates the $k_i(j)$ values computed during that phase to generate k_i, the new cutoff value on the number of calls to the procedure `varPOAC` allowed at each node of the ith exploitation phase. We propose two techniques to aggregate the $k_i(j)$ values into k_i.

- Med k_i is the median of the $k_i(j), j \in 1..L$.
- q-PER This technique generalizes the previous one. Instead of taking the median, we use any percentile. That is, k_i is equal to the smallest value among $k_i(1), \ldots, k_i(L)$ such that $q\%$ of the values among $k_i(1), \ldots, k_i(L)$ are less than or equal to k_i.

Several variants of APOAC can be proposed, depending on how we compute the $k_i(j)$ values in the learning phase and how we aggregate the different $k_i(j)$ values. In the next section, we give an experimental comparison of the different variants we tested.

6.4   Experimental Evaluation of (A)POAC

This section presents experiments that compare the performance of maintaining AC, POAC, or adaptive variants of POAC during search. For the adaptive variants we use two techniques to determine $k_i(j)$: the last reduction (LR) and the last drop (LD) with $\beta = 5\%$ (see Sect. 6.2). We also use two techniques to aggregate these $k_i(j)$ values: the median (Med) and the qth percentile (q-PER) with $q = 70\%$ (see Sect. 6.3). In experiments not presented in this paper we tested the performance of APOAC using the 10th to 90th percentiles. The 70th percentile showed the best behavior. We have performed experiments for the four variants obtained by combining two by two the parameters LR vs LD and Med vs 70-PER. For each variant we compared three initial values for the $maxK$ used by the first learning phase: $maxK \in \{2, n, \infty\}$, where n is the number of variable in the instance to be solved. These three versions are denoted by APOAC-2, APOAC-n and APOAC-fp respectively.

We compare these search algorithms on instances available from Lecoultre's webpage.[2] We selected four binary classes containing at least one difficult instance for MAC (>10 s): mug, K-insertions, myciel and Qwh-20. We also selected all the n-ary classes in extension: the traveling-salesman problem (TSP-20, TSP-25), the Renault Megane configuration problem (Renault) and the Cril instances (Cril). These eight problem classes contain instances with 11 to 1406 variables, domains of size 3 to 1600 and 20 to 9695 constraints.

For the search algorithm maintaining AC, the algorithm AC2001 (resp. GAC2001) [BRYZ05] is used for the binary (resp. non-binary) problems. The

Table 5. Total number of instances solved by AC, several variants of APOAC, and POAC.

$k_i(j)$	k_i		AC	APOAC-2	APOAC-n	APOAC-fp	POAC
LR	70-PER	#solved	115	116	**119**	118	115
	Med	#solved	115	114	**118**	**118**	115
LD	70-PER	#solved	115	117	**121**	120	115
	Med	#solved	115	116	**119**	**119**	115

Table 6. CPU time for AC, APOAC-2, APOAC-n, APOAC-fp and POAC on the eight problem classes.

class (#instances)		AC	APOAC-2	APOAC-n	APOAC-fp	POAC
Tsp-20 (15)	#solved	**15**	15	15	15	15
	sum CPU	**1,596.38**	3,215.07	4,830.10	7,768.33	18,878.81
Tsp-25 (15)	#solved	15	14	**15**	15	11
	sum CPU	20,260.08	>37,160.63	**16,408.35**	33,546.10	>100,947.01
renault (50)	#solved	50	50	50	50	50
	sum CPU	**837.72**	2,885.66	11,488.61	15,673.81	18,660.01
cril (8)	#solved	4	5	**7**	7	7
	sum CPU	>45,332.55	>42,436.17	**747.05**	876.57	1,882.88
mug (8)	#solved	5	6	6	6	**6**
	sum CPU	>29,931.45	12,267.39	12,491.38	12,475.66	**2,758.10**
K-insertions (10)	#solved	4	5	**6**	5	5
	sum CPU	>30,614.45	>29,229.71	**27,775.40**	>29,839.39	>20,790.69
myciel (15)	#solved	**12**	12	12	12	11
	sum CPU	**1,737.12**	2,490.15	2,688.80	2,695.32	>20,399.70
Qwh-20 (10)	#solved	10	10	**10**	10	10
	sum CPU	16,489.63	12,588.54	**11,791.27**	12,333.89	27,033.73
Sum of CPU times		>146,799	>142,273	**88,221**	>115,209	>211,351
Sum of average CPU times per class		>18,484	>14,717	**8,773**	>9,467	>10,229

---

[2] www.cril.univ-artois.fr/~lecoutre/benchmarks.html.

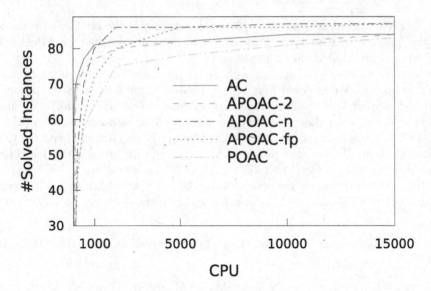

Fig. 5. Number of instances solved when the time allowed increases.

same AC algorithms are used as refutation procedure for POAC and APOAC algorithms. The *dom/wdeg* heuristic [BHLS04] is used both to order variables in the *Ordering(X)* function (see line 3 of Algorithm 4) and to order variables during search for all the search algorithms. The results presented involve all the instances solved before the cutoff of 15,000 s by at least one algorithm.

Table 5 compares all the competitors and shows the number of instances (#solved) solved before the cutoff. We observe that, on the set of instances tested, adaptive versions of POAC are better than AC and POAC. All of them, except APOAC-2+LR+Med, solve more instances than AC and POAC. All the versions using the last drop (LD) technique to determine the $k_i(j)$ values in the learning phase are better than those using last reduction (LR). We also see that the versions that use the 70th percentile (70-PER) to aggregate the $k_i(j)$ values are better than those using the median (Med). This suggests that the best combination is LD+70-PER. This is the only combination we will consider in the following.

Table 6 focuses on the performance of the three variants of APOAC (APOAC-2, APOAC-n and APOAC-fp), all with the combination (LD+70-PER). When a competitor cannot solve an instance before the cutoff, we count 15,000 s for that instance and we write '>' in front of the corresponding sum of CPU times. The last two rows of the table give the sum of CPU times and the sum of average CPU times per class. For each class taken separately, the three versions of APOAC are never worse than AC and POAC at the same time. APOAC-n solves all the instances solved by AC and POAC, and for four of the eight problem classes it outperforms both AC and POAC. However, there remain a few classes, such as Tsp-20 and renault, where even the first learning phase of APOAC is too costly

Table 7. Performance of APOAC-n compared to AC and POAC on n-ary problems.

	AC	APOAC-n	POAC
#solved	84/87	**87/87**	83/87
sum CPU	>68,027	**33,474**	>140,369
gain w.r.t. AC	–	*>51%*	–
gain w.r.t. POAC	–	*>76%*	–

to compete with AC despite our agile auto-adaptation policy that limits the number of calls to `varPOAC` during learning (see Sect. 6.1). Table 6 also shows that maintaining a high level of consistency, such as POAC, throughout the entire network generally produces a significant overhead.

Table 7 and Fig. 5 sum up the performance results obtained on all the instances with n-ary constraints. The binary classes are not included in the table and figure, because they have not been exhaustively tested. Figure 5 gives the performance profile for each algorithm presented in Table 6: AC, APOAC-2, APOAC-n, APOAC-fp and POAC. Each point (t, i) on a curve indicates the number i of instances that an algorithm can solve in less than t seconds. The performance profile underlines that AC and APOAC are better than POAC: whatever the time given, they solve more instances than POAC. The comparison between AC and APOAC highlights two phases: A first phase (for easy instances), during which AC is better than APOAC, and a second phase, where APOAC becomes better than AC. Among the adaptive versions, APOAC-n is the variant with the shortest first phase (it adapts quite well to easy instances), and it remains the best even when time increases.

Finally, Table 7 compares the best APOAC version (APOAC-n) to AC and POAC on n-ary problems. The first row of the table gives the number of solved instances by each algorithm before the cutoff. We observe that APOAC-n solves more instances than AC and POAC. The second row of the table gives the sum of CPU time required to solve all the instances. Again, when an instance cannot be solved before the cutoff of 15,000 s, we count 15,000 s for that instance. We observe that APOAC-n significantly outperforms both AC and POAC. The last two rows of the table give the gain of APOAC-n w.r.t. AC and w.r.t. POAC. We see that APOAC-n has a positive total gain greater than 51% compared to AC and greater than 76% compared to POAC.

7 Conclusion

We have proposed two approaches to adjust the level of consistency automatically during search. For the parameterized local consistency approach, we introduced the notion of stability of values for arc consistency, a notion based on the depth of their supports in their respective domain. This approach us allows us to define levels of local consistency of increasing strength between arc consistency and a given strong local consistency. We have introduced two techniques which allow us to make the parameter adaptable dynamically and locally during search. As a second approach, we proposed POAC1, an algorithm that enforces partition-one-AC efficiently in practice. We have also proposed an adaptive version of POAC that monitors the number of variables on which to perform singleton tests. Our experiments show that in both approaches, adapting the level of local consistency during search can outperform both MAC and maintaining a chosen local consistency stronger than AC.

Our approaches concentrate on adapting the level of consistency between the standard arc consistency and a chosen higher level. There are many constraints (especially global constraints) on which arc consistency is already a (too) high level of consistency and on which the standard consistency is bound consistency or some simple propagation rules. In these cases, an approach to that chosen in this paper could allow us to adapt automatically between arc consistency and the given lower level.

References

[BA01] Bennaceur, H., Affane, M.-S.: Partition-k-AC: an efficient filtering technique combining domain partition and arc consistency. In: Walsh, T. (ed.) CP 2001. LNCS, vol. 2239, pp. 560–564. Springer, Heidelberg (2001). doi:10.1007/3-540-45578-7_39

[BBBT14] Balafrej, A., Bessiere, C., Bouyakhf, E.H., Trombettoni, G.: Adaptive singleton-based consistencies. In: Proceedings of the Twenty-Eighth AAAI Conference on Artificial Intelligence (AAAI 2014), Quebec City, Canada, pp. 2601–2607 (2014)

[BBCB13] Balafrej, A., Bessiere, C., Coletta, R., Bouyakhf, E.H.: Adaptive parameterized consistency. In: Schulte, C. (ed.) CP 2013. LNCS, vol. 8124, pp. 143–158. Springer, Heidelberg (2013). doi:10.1007/978-3-642-40627-0_14

[BCDL11] Bessiere, C., Cardon, S., Debruyne, R., Lecoutre, C.: Efficient algorithms for singleton arc consistency. Constraints 16(1), 25–53 (2011)

[BD08] Bessiere, C., Debruyne, R.: Theoretical analysis of singleton arc consistency and its extensions. Artif. Intell. 172(1), 29–41 (2008)

[Bes06] Bessiere, C.: Constraint propagation. In: Rossi, F., van Beek, P., Walsh, T. (eds.) Handbook of Constraint Programming, chap. 3. Elsevier, Amsterdam (2006)

[BHLS04] Boussemart, F., Hemery, F., Lecoutre, C., Sais, L.: Boosting systematic search by weighting constraints. In: Proceedings of the 16th Eureopean Conference on Artificial Intelligence (ECAI 2004), Valencia, Spain, pp. 146–150. IOS Press (2004)

[BPSW11] Balafoutis, T., Paparrizou, A., Stergiou, K., Walsh, T.: New algorithms for max restricted path consistency. Constraints 16(4), 372–406 (2011)

[BRYZ05] Bessiere, C., Régin, J.-C., Yap, R.H.C., Zhang, Y.: An optimal coarse-grained arc consistency algorithm. Artif. Intell. 165(2), 165–185 (2005)

[BSW08] Bessiere, C., Stergiou, K., Walsh, T.: Domain filtering consistencies for non-binary constraints. Artif. Intell. 172(6–7), 800–822 (2008)

[DB97] Debruyne, R., Bessiere, C.: Some practicable filtering techniques for the constraint satisfaction problem. In: Proceedings of the Fifteenth International Joint Conference on Artificial Intelligence (IJCAI 1997), Nagoya, Japan, pp. 412–417 (1997)

[DB01] Debruyne, R., Bessiere, C.: Domain filtering consistencies. J. Artif. Intell. Res. **14**, 205–230 (2001)

[KVH06] Katriel, I., Van Hentenryck, P.: Randomized filtering algorithms. Technical report CS-06-09, Brown University, June 2006

[NT13] Neveu, B., Trombettoni, G.: Adaptive constructive interval disjunction. In: Proceedings of the 25th IEEE International Conference on Tools for Artificial Intelligence (IEEE-ICTAI 2013), Washington D.C., USA, pp. 900–906 (2013)

[PS12] Paparrizou, A., Stergiou, K.: Evaluating simple fully automated heuristics for adaptive constraint propagation. In: Proceedings of the 24th IEEE International Conference on Tools for Artificial Intelligence (IEEE-ICTAI 2012), Athens, Greece, pp. 880–885 (2012)

[Sel03] Sellmann, M.: Approximated consistency for Knapsack constraints. In: Rossi, F. (ed.) CP 2003. LNCS, vol. 2833, pp. 679–693. Springer, Heidelberg (2003). doi:10.1007/978-3-540-45193-8_46

[SS09] Stamatatos, E., Stergiou, K.: Learning how to propagate using random probing. In: Hoeve, W.-J., Hooker, J.N. (eds.) CPAIOR 2009. LNCS, vol. 5547, pp. 263–278. Springer, Heidelberg (2009). doi:10.1007/978-3-642-01929-6_20

[Ste08] Stergiou, K.: Heuristics for dynamically adapting propagation. In: Proceedings of the Eighteenth European Conference on Artificial Intelligence (ECAI 2008), Patras, Greece, pp. 485–489 (2008)

[Ste09] Stergiou, K.: Heuristics for dynamically adapting propagation in constraint satisfaction problems. AI Commun. **22**, 125–141 (2009)

[TC07] Trombettoni, G., Chabert, G.: Constructive interval disjunction. In: Bessière, C. (ed.) CP 2007. LNCS, vol. 4741, pp. 635–650. Springer, Heidelberg (2007). doi:10.1007/978-3-540-74970-7_45

[VSD98] Van Hentenryck, P., Saraswat, V.A., Deville, Y.: Design, implementation, and evaluation of the constraint language cc(FD). J. Log. Program. **37** (1–3), 139–164 (1998)

Constraint Programming for Data Mining

Modeling in MiningZinc

Anton Dries[1], Tias Guns[1], Siegfried Nijssen[1,2], Behrouz Babaki[1],
Thanh Le Van[1], Benjamin Negrevergne[1], Sergey Paramonov[1],
and Luc De Raedt[1(✉)]

[1] DTAI, KU Leuven, Leuven, Belgium
luc.deraedt@cs.kuleuven.be
[2] LIACS, Universiteit Leiden, Leiden, The Netherlands

Abstract. MiningZinc offers a framework for modeling and solving
constraint-based mining problems. The language used is MiniZinc, a
high-level declarative language for modeling combinatorial (optimisation) problems. This language is augmented with a library of functions
and predicates that help modeling data mining problems and facilities
for interfacing with databases. We show how MiningZinc can be used
to model constraint-based itemset mining problems, for which it was
originally designed, as well as sequence mining, Bayesian pattern mining, linear regression, clustering data factorization and ranked tiling.
The underlying framework can use any existing MiniZinc solver. We also
showcase how the framework and modeling capabilities can be integrated
into an imperative language, for example as part of a greedy algorithm.

1 Introduction

The traditional approach to data mining is to develop specialized algorithms
for specific tasks. This has led to specialized algorithms for many tasks, among
which classification, clustering and association rule discovery [19,30,35]. In many
cases these algorithms support certain kinds of constraints as well; in particular
constraint-based clustering and constraint-based pattern mining are established
research areas [2,5,6]. Even though successful for specific applications, the down-
side of specialized algorithms is that it is hard to adapt them to novel tasks.

In recent years, researchers have explored the idea of using generic solvers to
tackle data mining problems such as itemset mining [16,22], sequence mining [7,31]
and clustering [11,13]. These approaches start from the insight that many data
mining problems can be formalized as either a constraint satisfaction problem, or
a constrained optimization problem. The advantage is that, just as in constraint
programming, new tasks can be addressed by changing the constraint specification.

Siegfried Nijssen can currently be reached at the Institute of Information and
Communication Technologies, Electronics and Applied Mathematics, UC Louvain,
Belgium.
Tias Guns can currently be reached at the Vrije Universiteit Brussel.

© Springer International Publishing AG 2016
C. Bessiere et al. (Eds.): Data Mining and Constraint Programming, LNAI 10101, pp. 257–281, 2016.
DOI: 10.1007/978-3-319-50137-6_10

Although these techniques allow flexibility in modeling new tasks, they are often tied to one particular solver. To address this shortcoming we introduced MiningZinc [15], a solver-independent language and framework for modeling constraint-based data mining tasks. That work focussed on constraint-based itemset mining problems and the solving capabilities of the framework. In this work, we focus on the modeling aspects of the MiningZinc language and we show for a wide range of data mining tasks how they can be modeled in MiningZinc, namely sequence mining, Bayesian pattern mining, linear regression, clustering data factorization and ranked tiling. We end with a discussion of related work and conclusion.

2   Language

Ideally, a language for mining allows one to express the problems in a natural way, while at the same time being generic enough to express additional constraints or different optimization criteria. We choose to build on the MiniZinc constraint programming language for this purpose [32]. It is a subset of Zinc [25] restricted to the built-in types *bool, int, set* and *float*, and user-defined predicates and functions [34].

MiningZinc uses the MiniZinc language, that is, all models written for MiningZinc are compatible with the standard MiniZinc compiler and toolchain. However, MiningZinc offers additional functionality aimed towards modeling data mining problems:

- a library of predicates and functions that are commonly used in data mining
- extensions for reading data from different sources (e.g. a database)
- integration with Python for easy implementation of greedy algorithms

2.1   MiniZinc

MiniZinc is a language designed for specifying constraint problems. Listing 1 shows an example of a MiniZinc model for the well-known "Send+More=Money" problem. In this problem the goal is to assign distinct digits to the letters such that the formula holds.

This model starts with declaring the decision variables with their domains (Lines 1 and 2). The problem specification states that the variables S and M should not be zero which we encode in their domains. Next, we specify the constraints on these decision variables. On Line 4 we specify that all variables should take a different value. For this we use the all_different global constraint which is defined in MiniZinc's standard library. In order to use this constraint we include that library (Line 3). On Line 5 we specify that the numbers formed by the digits "SEND" and "MORE" should sum up to the number formed by the digits "MONEY". For the translation between the list of digits and the number they represent, we define a helper function on Line 6; it first creates a local parameter max_i that represents the largest index, and then sums over each variable

Listing 1. An example MiniZinc model

```
1  var 1..9: S; var 0..9: E; var 0..9: N; var 0..9: D;
2  var 1..9: M; var 0..9: O; var 0..9: R; var 0..9: Y;

3  include "globals.mzn";
4  constraint all_different([S,E,N,D,M,O,R,Y]);

5  constraint number([S,E,N,D]) + number([M,O,R,E]) =
                                     number([M,O,N,E,Y]);

6  function var int: number(array[int] of var int: digits) =
     let { int: max_i = max(index_set(digits)) } in
     sum(i in index_set(digits))
          (pow(10,max_i-i) * digits[i]);

7  solve satisfy;

8  output [show([S,E,N,D]), "+", show([M,O,R,E]),
                            "=", show([M,O,N,E,Y])];
```

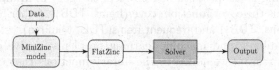

Fig. 1. Overview of the MiniZinc toolchain

multiplied by $1, 10, 100, \ldots$ depending on its position in the array (for example, number([S,E,N,D]) = 1000*S+100*E+10*N+1*D). Line 7 states that this is a constraint satisfaction problem. MiniZinc also supports optimization problems in which case this statement would be replaced by solve minimize <variable expression>, or likewise with maximize. Finally, Line 8 defines the output of the model.

Apart from the functionality demonstrated in the example, MiniZinc models can be parameterized, for example, to include external data. The values of these parameters can be loaded from an external file, or passed in through the command line.

MiniZinc is *solver-independent*, that is, MiniZinc models can be translated to the lower level language FlatZinc which is understood by a wide range of solvers. The MiniZinc toolchain does however support solver-specific optimizations through the use of solver-specific versions of the standard library and annotations. A schematic overview of the MiniZinc toolchain is shown in Fig. 1.

In the following we describe how we extended MiniZinc.

2.2 Library

MiniZinc offers the ability to add additional libraries of commonly used predicates and functions. As part of MiningZinc, we created a minimal library for help with specifying mining and learning problems. It has two purposes: (1) to simplify modeling for the user and (2) to simplify the model analysis by the MiningZinc solver.

There are four categories of functions:

- generic helper functions
- itemset mining functions
- norms and distance functions
- extra solver annotations

There are two generic helper functions that we feel are missing in MiniZinc, namely a direct bool2float conversion function var float: b2f(var bool: B) and an explicit weighted sum:

function var int: weighted_sum(array[int] of var int W, array[int] of var int X).

Itemset mining is the best studied problem category in MiningZinc, and the key abstraction is the cover function: cover(Items, TDB)). Other helper functions are coverInv(Trans, TDB)) and frequent_items(TDB, MinFreq).

Many data mining and machine learning problems involve computing distances. For this reason, we added to the library functions that compute the l_1, l_2 and l_∞ norms, the Manhattan, Euclidean and Chebyshev distance as well as the sum of squared errors and mean squared error:

Listing 2. "norms"

```
1 function var float: norm1(array[int] of var float: W) =
2   sum(j in index_set(W))( abs(W[j]) );
3 function var float: norm2sq(array[int] of var float: W) =
4   sum(j in index_set(W))( W[j]*W[j] );
5 function var float: norm2(array[int] of var float: W) =
6   sqrt(norm2sq(W));
7 function var float: normInf(array[int] of var float: W) =
8   max(j in index_set(W))( W[j] );
```

Listing 3. "distances"

```
1  function var float: manhDist(array[int] of var float: A,
2                               array[int] of var float: B) =
3    norm1([ A[d] - B[d] | d in index_set(A) ]);
4  function var float: euclDist(array[int] of var float: A,
5                               array[int] of var float: B) =
6    norm2([ A[d] - B[d] | d in index_set(A) ]);
7  function var float: chebDist(array[int] of var float: A,
8                               array[int] of var float: B) =
9    normInf([ A[d] - B[d] | d in index_set(A) ]);
10 function var float: sumSqErr(array[int] of var float: A,
11                              array[int] of var float: B) =
12   norm2sq([ A[d] - B[d] | d in index_set(A) ]);
13 function var float: meanSqErr(array[int] of var float: A,
14                               array[int] of var float: B) =
15   sumSqErr(A,B)/length(A);
```

Finally, the library also declares a few annotations that provide additional information to the solver such as load_data and vartype, which are discussed in the next section.

The MiningZinc library can be used by adding the following statement.

```
1  include" miningzinc.mzn";
```

The library is written in MiniZinc and is fully compatible with the standard MiniZinc toolchain.

2.3   Facilities for Loading Data

The MiningZinc library also declares a few annotations that allow us to extend the functionality of MiniZinc with respect to loading data from external sources. This consists of two components: (1) accessing data from an external data source and (2) translating it to a data structure supported by MiniZinc.

When using standard MiniZinc, if one wants to use external data, the workflow would be as follows:

1. Determine the relevant information from the database
2. Use SQL to extract this information
3. Translate the data to numeric identifiers using a script
4. Write out the data into MiniZinc format (dzn) using a script
5. Execute MiniZinc
6. Translate the results' identifiers back to the original data using a script
7 Analyze the results and, if necessary, repeat the process

Using the data loading facilities available in MiningZinc, the workflow becomes:

1. Determine the relevant information from the database
2. Update the MiningZinc model with data sources pointing to the relevant information
3. Execute MiningZinc
4. Analyze the results and, if necessary, repeat the process

Reading Data. MiningZinc facilitates loading data from external sources through the load_data(specification) annotation. The specification is a string describing the data source. By default, MiningZinc supports the following specifications:

sqlite;<filename>;<SQL query> Retrieve data from an SQLite database based on an SQL query.
arff;<filename> Retrieve data from a file in Weka's ARFF format [18].
csv;<filename>;<field separator> Retrieve data from a CSV file.

The use of these annotations is illustrated in Listing 4.

Listing 4. Examples of external data loading

```
1  array [int] of set of int: TDB
       :: load_data("sqlite;data/uci.db;SELECT * FROM zoo;");

2  array [int] of set of int: TDB
       :: load_data("arff;data/zoo.arff;");

3  string: datasource;
4  array [int] of set of int: TDB :: load_data(datasource);
```

The translation process is determined based on the structure of the input data and the target type in MiniZinc. For example, given an input table with two columns and the output type array[int] of set of int, the first column is interpreted as the index of the array and the second column as an element of the set. This is illustrated in Fig. 2.

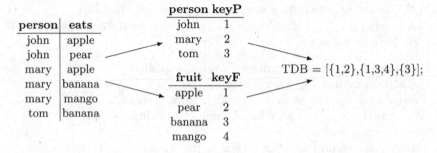

Fig. 2. Default translation to an array[int] of set of int from a table with two columns.

Automatic Translations. The previous example shows that during the loading of the data, we need to translate some of the data to an integer range. MiningZinc performs these translations automatically. The user can guide this translation process by adding type annotations to variable definitions. This can be done using the vartype annotation as illustrated in Listing 5. The additional information allows MiningZinc to translate the solutions back to the original values.

Listing 5. Examples of type annotations

```
1  array [int] of set of int: TDB
       :: load_data("sqlite;data/uci.db;SELECT * FROM zoo;")
       :: vartype("Animals","Features")

2  var set of Items: Items :: vartype("Features");
3  var set of int: Trans :: vartype("Animals") = cover(Items,TDB);
4  constraint card(cover(Items, TDB)) >= MinFreq;

5. solve satisfy;

6  output [show(Items), show(Trans)];
```

2.4   Python Integration

MiniZinc is a language that is specifically designed for expressing constrained optimization problems. It is therefore not very suitable to write complete systems, but should be seen as a domain specific language instead. To facilitate the use of MiningZinc we provide an interface with Python through the mngzn module. This interface allows the user to parse a MiningZinc model from a Python string, provide parameters using Python's dictionaries and query its results as a Python data structure. The main interface is demonstrated in Listing 6.

Listing 6. "Python interface"

```
1   import mngzn
2   modelstr = """
    int: sum; int: max;
    var  0..max: a; var 0..max: b;
    constraint a+b == sum;
    solve satisfy;
    output [show(a), show(b)];
    """

3   params = {'sum': 3, 'max': 2}
4   model = mngzn.parseModel(modelstr, params)
5   solutions = model.solve()
6   for sol in solutions:
7       print model.format_solution(sol)
```

First, we load the MiningZinc package (Line 1) and we define a model as a string of MiniZinc code (Line 2). The model takes two parameters sum and max and finds all pairs of integers up to max that sum to sum. On Line 3 we set the values of these parameters in a Python dictionary. Next, we parse the model string together with the parameters to obtain a solvable model (Line 4). On Line 5 we solve the model and obtain the solutions. This returns a sequence of Python dictionaries containing the output variables of the model, in this case [{'a': 1, 'b': 2}, {'a': 2, 'b': 1}]. Finally, we format and print out each solution (Line 7).

In Sect. 3.7 (Listing 17) we show an example of how this interface can be used to implement a greedy algorithm.

3   Modeling Data Mining Problems

We show how to model a variety of data mining problems, including constraint-based itemset mining, sequence mining, clustering, linear regression, ranked tiling and more.

In each case, the problem is modelled as a standard constraint satisfaction or optimisation problem, and it is modelled using the primitives available in MiniZinc, as well some common functions and predicates that we have added to the MiningZinc library.

3.1   Itemset Mining

The MiningZinc work originates from our work on using constraint programming (solvers) for constraint-based itemset mining [16].

Problem Statement. Itemset mining was introduced by Agrawal et al. [1] and can be defined as follows. The input consists of a set of transactions, each of which contains a set of *items*. Transactions are identified by identifiers $\mathcal{S} = \{1, \ldots, n\}$; the set of all items is $\mathcal{I} = \{1, \ldots, m\}$. An itemset database $\mathcal{D}$ maps transaction identifiers to sets of items: $\mathcal{D}(t) \subseteq \mathcal{I}$. The frequent itemset mining problem is then defined mathematically as follows.

Definition 1 (Frequent Itemset Mining). *Given an itemset database $\mathcal{D}$ and a threshold α, the frequent itemset mining problem consists of finding all itemsets $I \subseteq \mathcal{I}$ such that $|\phi_{\mathcal{D}}(I)| > \alpha$, where $\phi_{\mathcal{D}}(I) = \{t \mid I \subseteq \mathcal{D}(t)\}$.*

The set $\phi_{\mathcal{D}}(I)$ is called the *cover* of the itemset, and the threshold α the *minimum frequency* threshold. An itemset I which has $|\phi_{\mathcal{D}}(I)| > \alpha$ is called a *frequent itemset*.

Listing 7. "Frequent Itemset mining"

```
1  % Data
2  int: NrI; int: NrT; int: Freq; array[1..NrT] of
3  set of 1..NrI: TDB;

4  % Pattern
5  var set of 1..NrI: Items;

6  % Min. frequency constraint
7  constraint card(cover(Items,TDB)) >= Freq;

8  solve satisfy;
```

Listing 8. "Cover function for itemsets"

```
1  function var set of int: cover(var set of int: Items,
2                             array[int] of set of int: D) =
3    let {
4      var set of index_set(D): CoverSet;
5      constraint forall (t in index_set(D))
6         ( t in CoverSet <-> Items subset D[t] );
7    } in CoverSet;
```

MiningZinc Model. Listing 7 shows the frequent itemset mining problem in MiningZinc. Lines 2 and 3 are parameters and data, which a user can provide separate from the actual model or through load_data statements. The model represents the items and transaction identifiers in $\mathcal{I}$ and $\mathcal{S}$ by natural numbers (from 1 to NrI and 1 to NrT respectively) and the dataset $\mathcal{D}$ by the array TDB, mapping each transaction identifier to the corresponding set of items. The set of items we are looking for is represented on line 5 as a *set variable* with elements between value 1 and Nrl. The minimum frequency constraint is posted on line 7; it naturally corresponds to the formal notation $|\phi_{\mathcal{D}}(I)| \geq \alpha$. The *cover* relation used on line 7 and shown in Listing 8 is part of the MiningZinc library and implements $\phi_{\mathcal{D}}(I) = \{t | I \subseteq \mathcal{D}(t)\}$; note that this constraint is *not* a hard-coded constraint in the solver, such as in other systems, but is implemented in the MiningZinc language itself, and can hence be changed if this is desired. Finally, line 8 states that it is a satisfaction problem. Enumerating all solutions that satisfy the constraints corresponds to enumerating all frequent itemsets.

This example demonstrates the appeal of using a modeling language like MiniZinc for pattern mining: The formulation is high-level, declarative and close to the mathematical notation of the problem. Furthermore, the use of user-defined functions allows us to abstract away concepts that are common when expressing constraint-based mining problems.

Constraint-Based Mining. In constraint-based mining the idea is to incorporate additional user-constraints into the mining process. Such constraints can be motivated by an overwhelming number of (redundant) results otherwise found, or by application-specific constraints such as searching for patterns with high profit margins in sales data.

Listing 9 shows an example constraint-based mining setting. Compared to Listing 7, two constraints have been added: a *closure* constraint on line 6, which avoids non-closed patterns in the output, and a minimum cost constraint 10, requiring that the sum of the costs of the individual items is above a threshold. Other constraints could be added and combined in a similar way. See [16] for the range of constraints that has been studied in a constraint programming setting.

Listing 9. "Constraint-based itemset mining"

```
1   int: NrI; int: NrT; int: Freq; array[1..NrT] of
2   set of 1..NrI: TDB;

3   var set of 1..NrI: Items;

4   constraint card(cover(Items,TDB)) >= Freq;

5   % Closure
6   constraint Items = cover_inv(cover(Items,TDB),TDB);
7

8   % Minimum cost
9   array[1..NrI] of int: item_c; int: Cost;
10  constraint sum(i in Items)(item_c[i]) >= Cost;

11  solve satisfy :: enumerate;
```

3.2   Sequence Mining

Sequence mining [1] can be seen as a variation of the itemset mining problem discussed above. Whereas in itemset mining each transaction is a set of items, in sequence mining both transactions and patterns are ordered, (i.e. they are sequences instead of sets) and symbols can be repeated. For example, $\langle b, a, c, b \rangle$ and $\langle a, c, c, b, b \rangle$ are two sequences, and the sequence $\langle a, b \rangle$ is one possible pattern included in both.

Problem Statement. Two key concepts in any pattern mining setting are the structure of the pattern, and the *cover* relation that defines when a pattern covers a transaction.

In sequence mining, a transaction is covered by a pattern if there exists an embedding of the sequence pattern in the transaction; where an embedding is a mapping of every symbol in the pattern to the same symbol in the transaction such that the order is respected.

Definition 2 (Embedding in a sequence). *Let $S = \langle s_1, \ldots, s_m \rangle$ and $S' = \langle s'_1, \ldots, s'_n \rangle$ be two sequences of size m and n respectively with $m \leq n$. The tuple of integers $e = (e_1, \ldots, e_m)$ is an **embedding** of S in S' (denoted $S \sqsubseteq_e S'$) if and only if:*

$$S \sqsubseteq_e S' \leftrightarrow e_1 < \ldots < e_m \text{ and } \forall i \in 1, \ldots, m : s_i = s'_{e_i} \tag{1}$$

For example, let $S = \langle a, b \rangle$ be a pattern, then $(2, 4)$ is an embedding of S in $\langle b, a, c, b \rangle$ and $(1, 4), (1, 5)$ are both embeddings of S in $\langle a, c, c, b, b \rangle$.

Given an alphabet Σ of symbols, a *sequential* database D is a set of transactions where each transaction is a sequences defined over symbols in Σ. As in itemset mining, let $\mathcal{D}$ be a mapping from transaction identifiers to transactions. The frequent sequence mining problem is then defined as follows:

Definition 3 (Frequent Sequence Mining). *Given a sequential database $\mathcal{D}$ with alphabet Σ and a threshold α, the frequent sequence mining problem consists of finding all sequences S over alphabet Σ such that $|\psi_{\mathcal{D}}(S)| > \alpha$, where $\psi_{\mathcal{D}}(S) = \{t \mid \exists e \text{ s.t. } S \sqsubseteq_e \mathcal{D}(t)\}$.*

The set $\psi_{\mathcal{D}}(S)$ is the *cover* of the sequence, similar to the cover of an itemset.

MiningZinc Model. Modeling this in MiningZinc is somewhat more complex than itemset mining, as for itemsets we could reuse the *set* variable type, while sequences and the embedding relation need to be *encoded*. Each symbol is given an identifier (offset 1), and a transaction is represented as an array of symbol identifiers. The data is hence represented by a two dimensional array, and all sequences are padded with the identifier 0 such that they have the same length (MiniZinc does not support an array of arrays of different length). This data is given in lines (2)–(7) in Listing 10.

The pattern itself is also an array of integers, representing symbol identifiers. The 0 identifier can be used as padding at the end of the pattern, so that patterns of any size can be represented. Line (9) represents the array of integer variables while line (11)–(13) enforce the *padding* meaning of the 0 identifier.

To encode the cover relation we can not quantify over all possible embeddings e explicitly, as there can be an exponential number of them. Instead, we add one array of variables for every transaction that will represent the embedding of the pattern in that transaction, if one exists (line 15). Furthermore, we add one Boolean variable for every transaction, which will indicate whether the embedding is valid, e.g. whether the transaction is covered (line 17). Using these variables, we can encode the cover relation (line 19), explained below, as well as that the number of covered transactions must be larger than the minimum frequency threshold (line 21).

Listing 10. "Frequent sequence mining"

```
1   % Data
2   int: NrS; % number of distinct symbols (symbol identifiers)
3   int: NrPos; % number of positions = maximum transaction size
4   int: NrT;
5   int: Freq;
6   % dataset: 2D array of symbols
7   array[1..NrT,1..NrPos] of 1..NrS: data;

8   % Pattern (0 means 'end of sequence')
9   array[1..NrPos] of var 0..NrS: Seq;

10  % enforce meaning of '0'
11  constraint Seq[1] != 0;
12  constraint forall(j in 1..NrPos-1) (
13        (Seq[j] == 0) -> (Seq[j+1] == 0) );

14  % Helper variables for embeddings (0 means 'no match')
15  array[1..NrT, 1..NrPos] of var 0..NrPos: Emb;

16  % Helper variables for Boolean representation of cover set
17  array[1..NrT] of var bool: Cov;

18  % Constrain cover relation
19  constraint sequence_cover(Seq, Emb, Cov);

20  % Min. frequency constraint
21  constraint sum(i in 1..NrT) (bool2int(Cov[i])) >= Freq;

22  solve satisfy;
```

The actual formulation of the cover relation is shown in Listing 11; it could be made available as a function returning a *set* variable too. The formulation consists of three parts. In the first part (line 6) we constrain that for each transaction, the jth embedding variable must point to a position in the transaction that matches the symbol of the jth symbol in the pattern. Note that if no match is possible then the embedding variable will only have symbol 0 in its domain. The second part (line 9) requires that embedding variables must be increasing (except when 0). Finally, on line 12 we state that a transaction is covered if for every non-0 valued position in the pattern there is a matching embedding variable.

Listing 11. "Cover relation for sequences"

```
1   predicate sequence_cover(array[int] of var int: Seq,
2                            array[int,int] of var int: Emb,
3                            array[int] of var bool: Cov) =
4
5     % Individual positions should match (else: 0)
6     forall(i in 1..NrT, j,x in 1..NrPos) (
        (Emb[i,j] == x) -> (Seq[j] == data[i,x]) ) /\
7
8     % Positions increase (except when 0)
9     forall(i in 1..NrT, j in 1..NrPos-1, x in 1..NrPos) (
        (Emb[i,j+1] == x) -> (Emb[i,j] < x) ) /\
10
11    % Covered if all its positions match
12    forall(i in 1..NrT) (
13      Cov[i] <-> forall(j in 1..NrT) ( (Seq[j] != 0) ->
                                         (Emb[i,j] != 0) ));
```

As in sequence mining, extra constraints can be added to extract fewer, but more relevant or interesting patterns. An overview of sequence mining constraints that have been studied in a sequence mining setting is available in [31].

3.3 Constraint-Based Pattern Mining in Bayesian Networks

Just as one can mine patterns in data, it is also possible to mine patterns in Bayesian Networks (BNs). These patterns can help in understanding the knowledge that resides in the network. Extracting such knowledge can be useful when trying to better understand a network, for example when presented with a new network, in case of large and complex networks or when it is updated frequently.

Problem Statement. A Bayesian Network $\mathcal{G}$ defines a probability distribution over a set of random variables $\mathcal{X}$. Each variable has a set of values it can take, called its *domain*. We define a Bayesian network pattern as an assignment to a subset of the variables:

Definition 4 (BN pattern). *A pattern A over a Bayesian network $\mathcal{G}$ is a partial assignment, that is, an assignment to a subset of the variables in $\mathcal{G}$:* $A = \{(X_1 = x_1), \ldots, (X_m = x_m)\}$, *where each X_i is a different variables and x_i is a possible value in its domain.*

A BN pattern can be seen as an itemset, where each item is an assignment of a variable to a value. One can compute the (relative) *frequency* of an itemset in a database; related, for a BN pattern one can compute the *probability* of the pattern in the Bayesian network. The probability of a pattern A, denoted by $P_{\mathcal{G}}(A)$, is $P((X_1 = x_1), \ldots, (X_m = x_m))$, that is, the probability of the partial assignment marginalized over the unassigned variables.

Given this problem setting, one can define a range of constraint-based pattern mining tasks over Bayesian Networks, similar to constraint-based mining tasks over itemsets or sequences. In line with frequent itemset mining, the following defines the probable BN pattern mining problem:

Definition 5 (Probable BN pattern Mining). *Given a Bayesian network $\mathcal{G}$ over variables $\mathcal{X}$ and a threshold α, the probable BN pattern mining problem consists of finding all BN patterns A over $\mathcal{X}$ such that $P_{\mathcal{G}}(A) > \alpha$.*

MiningZinc Model. We encode a BN pattern with an array of integer CP variables, as many as there are random variables in the BN. Each CP variable has $m+1$ possible values, where m is the number of values in the domain of the corresponding random variable: value 0 represents that the variable is not part of the partial assignment, e.g. it should be marginalized over when computing the probability. The other values each correspond to a value the domain of the random variable.

The main issue is then to encode the probability computation. As this computation will be performed many times during search, we choose to first compile the BN into an *arithmetic circuit*. Computing the probability over the circuit is polynomial in its size (which may be worst-case exponential to the size of the origin network) [8]. Nevertheless, using ACs is generally recognized as one of the most effective techniques for exact computation of probabilities, especially when doing so repeatedly.

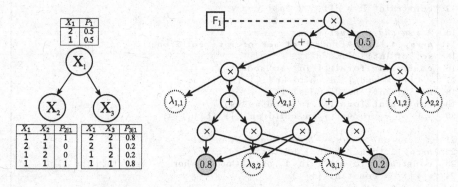

Fig. 3. Left: The Bayesian network and CPTs for a distribution with 3 variables. The domain for all variables is $\{1, 2\}$. Right: The compiled arithmetic circuit for this BN. A λ_{ij} leaf of in the AC corresponds to assignment $X_i = j$ in the BN.

Figure 3 shows an example AC. It consists of product nodes, sum nodes, constants and *indicator variables* λ. The Boolean indicator variables $\lambda_{i,j}$ indicate whether $(X_i = j)$. An assignment sets the corresponding indicator variable to 1 and the other indicator variables of the random variable to 0. To marginalize a random variable away, all indicator variables of that variable must be set to 1. The value obtained in the root node after evaluating the arithmetic circuit corresponds to the probability of the pattern given the assignment to the indicator variable nodes.

To encode this in CP, we will create a float variable for every node in the AC. Listing 12 shows how to encode an AC in CP. We assume given the set of product and sum node identifiers (root node has identifier 1), an array of sets representing the 'children' relation, an array with the constants, and a 2D array that maps the indicator variables to nodes of the tree. The last two constraints in the listing provide a mapping from the Q variables to indicator variables, such that they must all be set to 1 (=marginalize) or exclusively have one indicator set to 1 (=assignment).

Listing 12. Probable BN pattern mining

```
1   int: num_vars;
2   array[variables] of int: num_values;
3   float: min_prob;

4   array[1..num_vars] of int: Q;
5   var 0.0..1.0: P;

6   constraint P > min_prob;

7   constraint forall(i in 1..num_vars) (
8     0 <= Q[i] /\ Q[i] <= num_values[i]+1 );

9   % encode AC
10  int: num_ACnodes;
11  array[1..num_ACnodes] of var 0.0..1.0: F;
12  constraint P = F[1]; % root node

13  % sum and product nodes
14  array[1..num_ACnodes] of set of int: children;
15  set of int: sum_nodes;
16  constraint forall(i in sum_nodes) (
17    F[i] = sum(j in children[i]) (F[j]) );
18  set of int: prod_nodes;
19  constraint forall(i in prod_nodes) (
20    F[i] = product(j in children[i]) (F[j]) );

21  % constant nodes, -1 means non-constant
22  array[1..num_ACnodes] of int: constants;
23  constraint forall(i in 1..num_ACnodes where constants[i]!=-1) (
24    F[i] = constants[i] );

25  % Q to indicator nodes (which must take either 0 or 1)
26  array[1..num_vars,int] of int: mapQ;
27  constraint forall(i in 1..num_vars) (
28    Q[i]=0 -> forall(j in 1..num_vals[i]) (F[mapQ[i,j]] == 1) );
29  constraint forall(i in 1..num_vars, k in 1..num_vals[i]) (
30    Q[i]=k -> (F[mapQ[i,k]] == 1) /\
31            forall(j in 1..num_vals[i] where j!=k)
32              (F[mapQ[i,j]] == 0) );

33  solve satisfy;
```

Many constraints from the itemset mining literature have a counterpart over BN patterns and can be formulated as well, such as size constraints, closed/maximal/free constraints and constraints to discriminate results from two networks from each other, or to discriminate a probability in a network to relative frequency in a dataset. This is currently work in progress.

3.4   Linear Regression

A common problem studied in data mining is regression, where the goal is to estimate the value of a variable based on the values of dependent variables.

Problem Statement. In simple regression, the goal is to predict the value of a *target* attribute given the value of an M-dimensional vector of *input* variables $x = (x_1, ..., x_M)$.

We are given a set of N observations $\mathbf{X}$ and their corresponding target values y. The goal is to find a function $\hat{y}(x)$ that approximates the target values y for the given observations. In linear regression [3] we assume $\hat{y}(x)$ to be a linear function. Mathematically, such a function can be formulated as

$$\hat{y}(x) = w_1 x_1 + ... + w_M x_M + w_{M+1}$$

where $w = (w_1, ..., w_{M+1})$ is a vector of weights that minimizes a given error function. This error function is typically defined as the sum of squared errors

$$sumSqErr(\hat{y}, y) = \|\mathbf{X}w - y\|_2^2,$$

where $\mathbf{X}$ is an $N \times (M+1)$ matrix where each row corresponds to a given observation (extended with the constant 1), and y is a vector of length N containing the corresponding target values. The vector $\mathbf{X}w$ contains the result of computing $\hat{y}$ for each observation. The goal is then to find the vector of weights w that minimizes this error, that is,

$$\arg\min_w \|\mathbf{X}w - y\|_2^2.$$

MiningZinc Model. We can formulate this problem as an optimization problem as shown in Listing 13. The model starts with defining the input data (Lines 2 and 3) and its dimensions (Lines 5–6). The input data can be specified in an external file. Line 9 specifies the weight vector that needs to be found. Based on this weight vector and the input variable x, we specify the estimate $(\hat{y}(x))$ of the linear model in Line 11. Finally, on Line 13 we specify the optimization criterion, i.e. to minimize the sum of squared errors. The function sumSqErr is defined in the MiningZinc library (Listing 3).

Listing 13. Model for min-squared-error linear regression

```
1   % Observations
2   array[int, int] of float: X;
3   array[int] of float: y;
4   % Data dimensions
5   set of int: Points = indexset_1of2(X);
6   set of int: Dimensions = indexset_2of2(X);
7   int: NumWeights = max(Dimensions)+1;
8   % Weights to find
9   array[1..NumWeights] of var float: w;

10  % Estimate for each data point
11  array[Points] of var float: yh =
```

```
      [ sum(j in Dimensions) (w[j]*X[i,j]) + w[NumWeights]
                                          | i in Points ];
12  % Optimization criterium
13  solve minimize sumSqErr(y, yh);
```

By replacing the error function we can easily model other linear regression tasks, for example linear regression with *elastic net* regularization [38] where the optimization criterium is defined as

$$\arg\min_w \frac{1}{2n_{samples}} \|\mathbf{X}w - y\|_2^2 + \alpha\rho\|w\|_1 + \frac{\alpha(1-\rho)}{2}\|w\|_2^2$$

with α and ρ parameters that determine the trade-off between L1 and L2 regularization. Listing 14 shows the implementation of this scoring function in MiniZinc.

Listing 14. Elastic net error function for linear regression

```
1  function var float: elastic_net(
              array[int] of float: Y,
              array[int] of var float: E,
              array[int] of var float: W,
              float: Alpha, float: Rho) =
    (0.5 / int2float(length(Y))) *
      norm2([ Est[i] − Y[i] | i in indexset(Y)])
    + (Alpha*Rho) * norm2( W )
    + (0.5*Alpha*(1.0−Rho)) * norm1( W );
```

3.5 Clustering

The task of clustering is discussed in the next chapter. We would like to point out however that the clustering problems explained there can be modeled in MiningZinc too.

Problem Statement. Let us consider the minimum sum of squared error clustering problem (which k-means provides an approximation of), where the goal is to group all examples into k non-overlapping groups [21]. The objective to minimize is the 'error' of each cluster, that is, the distance of each point in the cluster to the mean (centroid) of that cluster.

The *centroid* of a cluster can be computed by computing the mean of the data points that belong to it:

$$z_C = \text{mean}(C) = \frac{\sum_{p \in C} p}{|C|} \tag{2}$$

The error is then measured as the sum of squared errors of the clusters:

$$\sum_{C \in \mathcal{C}} \sum_{p \in C} d^2(p, z_C) \tag{3}$$

MiningZinc Model. The model below shows a MiningZinc specification of this problem. As variables, it uses an array of integers, one integer variable for every example. This variable will indicate which cluster the example belongs too. The optimisation criterion is specified over all clusters and examples; the b2f(Belongs[j])==i) part converts the Boolean valuation of whether point j belongs to cluster i into a float variable, such that this indicator variable can be multiplied by the sum of squared errors of point j to cluster i.

The functions b2f() (bool 2 float) and sumSqErr() (sum of squared errors) are part of the MiningZinc library, see Sect. 2.2. The definition of mean() is shown below and follows the mathematical definition above.

```
1   % Data
2   int : NrDim; % number of dimensions
3   int : NrE; % number of examples
4   int : K; % number of clusters
5   array [1..NrE, 1..NrDim] of float : Data;

6   % Clustering (each point belongs to one cluster)
7   array [1..NrE] of var 1..K: Belongs;

8   solve minimize sum( i in 1..K, j in 1..NrE) (
9                   b2f ( Belongs [ j ] == i )*
10                  sumSqErr ( Data [ j ] , mean ( Data , Belongs , i ))
11                  );

12  function array [ int ] of var float : mean (
13                      array [ int , int ] of var float : Data ,
14                      array [ int ] of var int : Belongs ,
15                      int : c ) =
16      let {
17        set of int : Exs = index_set_1of2 ( Data ) ,
18        set of int : Dims = index_set_2of2 ( Data ) ,
19        array [ Dims ] of var float : Mean ,
20        constraint forall ( d in Dims ) (
21          Mean [ d ] =
22            sum( i in Exs )( b2f ( Belongs [ i ] == c ) * Data [ i , d ] ) /
23            sum( i in Exs )( b2f ( Belongs [ i ] == c ) )
24        )
25      } in Mean ;
```

More clustering problems and how to model them for use with constraint solvers can be found in the next chapter.

3.6   Relational Data Factorization

Motivated by an analogy with Boolean matrix factorization [29] (cf. Fig. 4), [9] introduces the problem of factorizing a relation in a database. In matrix factorization, one is given a matrix and has to factorize it as a product of other matrices.

Problem Statement. In relational data factorization (RDF), the task is to factorize a given relation as a conjunctive query over other relations, i.e., as a combination of natural join operations. The problem is then to compute the

$$\begin{matrix} \mathbf{A} & & \mathbf{B} & & \mathbf{C} \end{matrix}$$

$$\begin{vmatrix} 1\,1\,0 \\ 1\,1\,1 \\ 1\,0\,1 \end{vmatrix} = \begin{vmatrix} 1\,0 \\ 1\,1 \\ 0\,1 \end{vmatrix} \times \begin{vmatrix} 1\,1\,0 \\ 0\,1\,1 \end{vmatrix}$$

Fig. 4. Boolean matrix factorization.

extensions of these relations starting from the input relation and a conjunctive query. Thus relational data factorization is a form of both *inverse* querying and *abduction*, as one has to compute the relations in the query from the result of the query. The result of relational data factorization is not necessarily unique, constraints on the desired factorization can be imposed and a scoring function is used to determine the quality of a factorization. Relational data factorization is thus a constraint satisfaction and optimization problem.

More specifically, relational data factorization is a generalization of abductive reasoning [10]:

1. instead of working with a single observation f, we now assume a set of facts D for a unique target predicate db is given;
2. instead of assuming any definition for the target predicate, we assume a single definite rule defines db in terms of a set of abducibles A, the conjunctive query;
3. instead of assuming that a minimal set of facts be abduced, we score the different solutions based on the observed error.

Formally we can specify the relational factorization problem as follows:
Given:

- a dataset D (of ground facts for a target predicate db);
- a factorization shape Q: $db(\bar{T}) \leftarrow q_1(\bar{T}_1), \ldots, q_k(\bar{T}_k)$, where some of the q_i are abducibles;
- a set of rules and constraints P;
- a scoring function *opt*.

Find: the set of ground facts F, the extensions of relation Q, that scores best w.r.t. $opt(D, approx(P, Q, F))$ and for which $Q \cup P \cup F$ is consistent.

MiningZinc Model. Listing 15 shows the model for a relational factorization problem with a ternary conjunctive query, using the sum of absolute errors as scoring function and without additional constraints.

Listing 15. Relational decomposition

```
1   % Input data
2   array[int, int, int] of int: paper;
3   % index set of authors
4   set of int: Authors = index_set_1of3(paper);
5   % index set of universities
6   set of int: Universities = index_set_2of3(paper);
7   % index set of venues
8   set of int: Venues = index_set_3of3(paper);
```

```
 9    % Search for
10    array[Authors, Universities] of var bool: worksAt;
11    array[Authors, Venues] of var bool: publishesAt;
12    array[Universities, Venues] of var bool: knownAt;

13    solve minimize
         sum(a in Authors, u in Universities, v in Venues) (
            abs(paper[a,u,v] - bool2int(
               worksAt[a,u] /\ publishesAt[a,v] /\ knownAt[u,v])) );

14    output [ show(worksAt), show(publishesAt), show(knownAt) ];
```

3.7 Ranked Tiling

Ranked tiling was introduced in [23] to find interesting areas in ranked data. In this data, each transaction defines a complete ranking of the columns. Ranked data occurs naturally in applications like sports or other competitions. It is also a useful abstraction when dealing with numeric data in which the rows are incomparable.

Problem Statement. Ranked tiling discovers regions that have high average rank scores in rank matrices. These regions are called *ranked tiles*. Formally, a rank tile is defined by the following optimization problem:

Problem 1 (Maximal ranked tile mining). *Given a rank matrix* $\mathcal{M} \in \sigma^{m \times n}$, $\sigma \in \{1, \ldots n\}$ *and a threshold* θ, *find the ranked tile* $B = (R^*, C^*)$, *with* $R^* \subseteq \{1 \ldots m\}$ *and* $C^* \subseteq \{1 \ldots n\}$, *such that:*

$$B = (R^*, C^*) = \operatorname*{argmax}_{R,C} \sum_{r \in R, c \in C} (\mathcal{M}_{r,c} - \theta). \tag{4}$$

where θ *is an absolute-valued threshold.*

Example 1. Figure 5 depicts a rank matrix containing five rows and ten columns. When $\theta = 5$, the maximal ranked tile is defined by $R = \{1, 2, 3, 5\}$ and $C = \{1, 2, 3\}$. The score obtained by this tile is 37, and no more columns or rows can be added without decreasing the score.

The maximal ranked tiling problem aims to find a single tile, but we are also interested in finding a set of such tiles. This maximizes the amount of information we can get from the data. In other words, we would like to discover a ranked tiling.

Problem 2 (Ranked tiling). *Given a rank matrix* $\mathcal{M}$, *a number* k, *a threshold* θ, *and a penalty term* P, *the ranked tiling problem is to find a set of ranked tiles* $B_i = (R_i, C_i)$, $i = 1 \ldots k$, *such that they together maximize the following objective function:*

$$\operatorname*{argmax}_{R_i, C_i} \sum_{r \in \mathcal{R}, c \in \mathcal{C}} \mathbb{1}_{(t_{r,c} \geq 1)}((\mathcal{M}_{r,c} - \theta) - (t_{r,c} - 1)P) \tag{5}$$

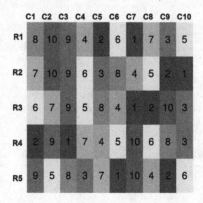

Fig. 5. Example rank matrix, with maximal ranked tile $B = (\{R1, R2, R3, R5\}, \{C1, C2, C3\})$.

where $t_{r,c} = |\{i \in \{1, \ldots, k\} \mid r \in R_i, c \in C_i\}|$ indicates the number of tiles that cover a cell, and $\mathbb{1}_\varphi$ is an indicator function that returns 1 if the test φ is true, and 0 otherwise. P indicates a penalty that is assigned when tiles overlap.

To solve Problem 1 efficiently, we introduce two Boolean decision vectors: $T = (T_1, T_2, \ldots, T_m)$, with $T_i \in \{0, 1\}$, for rows and $I = (I_1, I_2, \ldots, I_n)$, with $I_i \in \{0, 1\}$, for columns. An assignment to the Boolean vectors T and I corresponds to an indication of rows and columns belonging to a tile. Then, the maximal ranked tile can be found by solving the following equivalent problem:

$$\operatorname*{argmax}_{T,I} \sum_{t \in \mathcal{R}} T_t * \left(\sum_{i \in \mathcal{C}} (\mathcal{M}_{t,i} - \theta) * I_i \right) \tag{6}$$

subject to

$$\forall t \in \mathcal{R} : T_t = 1 \leftrightarrow \sum_{i \in \mathcal{C}} (\mathcal{M}_{t,i} - \theta) * I_i \geq 0 \tag{7}$$

$$\forall i \in \mathcal{C} : I_i = 1 \leftrightarrow \sum_{t \in \mathcal{R}} (\mathcal{M}_{t,i} - \theta) * T_t \geq 0 \tag{8}$$

where redundant constraints (7), (8) are introduced to boost the search.

MiningZinc Model for Finding a Single Tile. This problem specification translates directly into the MiningZinc model shown in Listing 16 where Eqs. 7 and 8 correspond to lines 7 and 8, respectively, and the optimization criterion of Eq. 6 corresponds to line 9.

Listing 16. MiniZinc model for finding a single best tile

```
1  array[int, int] of int: TDB;
2  int: th;  % Theta
3  set of int: Rows = index_set_1of2(TDB);
4  set of int: Cols = index_set_2of2(TDB);

5  array [Cols] of var bool: I;
6  array [Rows] of var bool: T;

7  constraint forall (r in Rows) (
     T[r] == (sum(c in Cols)((TDB[r,c]-th)*bool2int(I[c])) >= 0)
   );

8  constraint forall (c in Cols) (
     I[c] == (sum(r in Rows)((TDB[r,c]-th)*bool2int(T[r]))>=0)
   );

9  solve maximize
     sum(r in Rows)(
        bool2int(I[r])*sum(c in Cols)((TDB[r,c]-th)*bool2int(T[c])
     )
   );

10 output [ show(T), "\n", show(I), "\n"];
```

To solve Problem 2, Le Van et al. [23] propose to approximate the optimal solution by using a greedy approach, as is common for this type of pattern set mining problem. The first tile is found by solving the optimization problem in Listing 16. Next, we remove that tile by setting all cells in the matrix that are covered to the lowest rank (or another value, depending on parameter P). Then, we search in the resulting matrix for the second tile. This process is repeated until the number of desired tiles is found. The sum of the scores of all discovered tiles will correspond to the score of Eq. 5 for this solution. However, as the search is greedy, the solution is not necessarily optimal.

Python Wrapper for Greedy tile Mining. The greedy approach cannot be modelled directly in the MiningZinc language. However, the MiningZinc framework allows direct access to the solving infrastructure from Python. The complete algorithm is shown in Listing 17. The interaction with the MiningZinc module (**mngzn**) occurs on line 7 where the model is initialized, and on line 9 where one solution of the model is retrieved. In lines 13 through 19 the obtained tile is removed from the original matrix (by setting its entries to 0). The process is repeated until the tile no longer covers a new entry of the matrix.

Listing 17. Wrapper for finding all tiles (in Python)

```
1  TDB = ...    # Load matrix mznmodel = ...    # See Listing
   reflst:rankedtilingspsmzn params = { 'TDB': TDB, 'th': 5}
2  tiles = [] stop = False while not stop :
3      model = mngzn.parseModel(mznmodel, params)
4      # Solve the model to find one tile
5      solution = next(model.solve())
6      tiles.append(solution)
7      stop = True
8      # Update the ranking matrix => zero out values
```

```
 9      for i,r in enumerate(solution['T']) :
10          for j,c in enumerate(solution['I']) :
11              if r and c :
12                  # Stop unless the new tile covers a new item
13                  if TDB[i][j] > 0 :
14                      TDB[i][j] = 0
15                      stop = False
16  return tiles
```

4   Related Work

We have shown how MiningZinc can be used to model a wide variation of data mining and machine learning tasks in a high-level and declarative way. Our modeling language is based on MiniZinc [32] because it is a well-developed existing language with wide support in the CP community, it supports user-defined constraint, and is solver-independent. Other modeling languages such as Essence [12], Comet [37] and OPL [36] have no, or only limited, support for building libraries of user-defined constraints, and/or are tied to a specific solver.

Integrating declarative modeling and data mining has been studied before in the case of itemset mining [16,22], clustering [11,27] and sequence mining [31]. However, these approaches were *low-level* and *solver dependent*. The use of higher-level modeling languages and primitives has been studied before [17,28], though again tied to one particular solving technology.

The idea of combining multiple types of data mining and machine learning techniques also lies at the basis of machine learning packages such as WEKA [18] and scikit-learn [33]. However, these packages do not offer a unified declarative language and they do not support going beyond the capabilities of the algorithms offered.

In data mining, our work is related to that on inductive databases [24]; these are databases in which both data and patterns can be queried. Most inductive query languages, e.g., [20,26], extend SQL with primitives for pattern mining. They have only a restricted language for expressing mining problems, and are usually tied to one mining algorithm. A more advanced development is that of mining views [4], which provides lazy access to patterns through a virtual table. Standard SQL can be used for querying, and the implementation will only materialize those patterns in the table that are relevant for the query. This is realized using a traditional mining algorithm. In MiningZinc we support the integration of data from an external database through the use of SQL queries directly.

5   Solving

This chapter does not expand on solving, but the MiningZinc framework [14] supports three types of solving: (1) to use an existing MiniZinc solver; (2) to detect that the specified tasks is a standard known task and to use a specialised algorithm to solve it; and (3) a hybrid solving approach that uses both specialised algorithms and generic constraint solvers, for example by solving a master problem and subproblem with different technology, or to incorporate specialised algorithms inside *global* constraint propagators. The first approach is typically least efficient but most flexible towards adding extra constraints. The second approach is least flexible but typically most scalable. The third, hybrid, approach offers a trade-off between generality and efficiency, but requires modifications to the solving process, which is hence beyond what can be expressed in a modeling language like Mini(ng)Zinc.

6   Conclusion

In this chapter we showed how a wide range of data mining problems can be modeled in MiningZinc. Only a minimal library of extra predicates and functions was needed to express these problems, meaning that standard MiniZinc is often sufficient to model such problem. Two additions are the ability to load data from a database, and a library of distance functions, which are often used in data mining.

The key feature of MiningZinc as a language for expressing data mining problems is the ability to add and modify constraints and objective functions. Hence constraint-based mining problems are those where the language and framework has most to offer, such as in constraint-based pattern mining and constrained clustering. Another valuable use is for prototyping new data mining problems, as was done for relational data factorization and ranked tiling. Many other problem settings are yet unexplored.

References

1. Agrawal, R., Imielinski, T., Swami, A.N.: Mining association rules between sets of items in large databases. In: Proceedings of the 1993 ACM SIGMOD International Conference on Management of Data, pp. 207–216. ACM Press (1993)
2. Basu, S., Davidson, I., Wagstaff, K.: Constrained Clustering: Advances in Algorithms, Theory, and Applications. Chapman & Hall/CRC Data Mining and Knowledge Discovery Series. CRC Press, Boca Raton (2008)
3. Bishop, C.M.: Pattern Recognition and Machine Learning. Springer, Heidelberg (2006)
4. Blockeel, H., Calders, T., Fromont, É., Goethals, B., Prado, A., Robardet, C.: An inductive database system based on virtual mining views. Data Min. Knowl. Discov. **24**(1), 247–287 (2012)
5. Boulicaut, J.F., Dzeroski, S. (eds.): Proceedings of the Second International Workshop on Inductive Databases, 22 September, Cavtat-Dubrovnik, Croatia. Rudjer Boskovic Institute, Zagreb (2003)
6. Boulicaut, J.-F., Raedt, L., Mannila, H. (eds.): Constraint-Based Mining and Inductive Databases. LNCS (LNAI), vol. 3848. Springer, Heidelberg (2006). doi:10.1007/11615576
7. Coquery, E., Jabbour, S., Sais, L., Salhi, Y., et al.: A SAT-based approach for discovering frequent, closed and maximal patterns in a sequence. In: European Conference on Artificial Intelligence (ECAI), vol. 242, pp. 258–263 (2012)

8. Darwiche, A.: A differential approach to inference in bayesian networks. J. ACM **50**(3), 280–305 (2003). http://doi.acm.org/10.1145/765568.765570
9. De Raedt, L., Paramonov, S., van Leeuwen, M.: Relational decomposition using answer set programming. In: Online Preprints 23rd International Conference on Inductive Logic Programming, International Conference on Inductive Logic Programming, Rio de Janeiro, 28–30 August 2013, August 2013. https://lirias.kuleuven.be/handle/123456789/439287
10. Denecker, M., Kakas, A.: Abduction in logic programming. In: Kakas, A.C., Sadri, F. (eds.) Computational Logic: Logic Programming and Beyond. LNCS (LNAI), vol. 2407, pp. 402–436. Springer, Heidelberg (2002). doi:10.1007/3-540-45628-7_16
11. Dao, T.-B.-H., Duong, K.-C., Vrain, C.: A declarative framework for constrained clustering. In: Blockeel, H., Kersting, K., Nijssen, S., Železný, F. (eds.) ECML PKDD 2013. LNCS (LNAI), vol. 8190, pp. 419–434. Springer, Heidelberg (2013). doi:10.1007/978-3-642-40994-3_27
12. Frisch, A., Harvey, W., Jefferson, C., Hernández, B.M., Miguel, I.: Essence: a constraint language for specifying combinatorial problems. Constraints **13**(3), 268–306 (2008)
13. Gilpin, S., Davidson, I.N.: Incorporating SAT solvers into hierarchical clustering algorithms: an efficient and flexible approach. In: Proceedings of the 17th ACM SIGKDD International Conference on Knowledge Discovery and Data Mining, San Diego, CA, USA, 21–24 August 2011, pp. 1136–1144 (2011)
14. Guns, T., Dries, A., Tack, G., Nijssen, S., De Raedt, L.: MiningZinc: a modeling language for constraint-based mining. In: Proceedings of the Twenty-Third International Joint Conference on Artificial Intelligence, pp. 1365–1372. AAAI Press, August 2013
15. Guns, T., Dries, A., Tack, G., Nijssen, S., Raedt, L.D.: Miningzinc: a language for constraint-based mining. In: International Joint Conference on Artificial Intelligence (2013)
16. Guns, T., Nijssen, S., De Raedt, L.: Itemset mining: a constraint programming perspective. Artif. Intell. **175**(12–13), 1951–1983 (2011)
17. Guns, T., Nijssen, S., De Raedt, L.: k-Pattern set mining under constraints. IEEE Trans. Knowl. Data Eng. **25**(2), 402–418 (2013)
18. Hall, M., Frank, E., Holmes, G., Pfahringer, B., Reutemann, P., Witten, I.H.: The weka data mining software: an update. SIGKDD Explor. **11**(1), 10–18 (2009)
19. Han, J., Kamber, M.: Data Mining: Concepts and Techniques. Morgan Kaufmann, Burlington (2000)
20. Imielinski, T., Virmani, A.: MSQL: a query language for database mining. Data Min. Knowl. Disc. **3**, 373–408 (1999)
21. Jain, A.K., Murty, M.N., Flynn, P.J.: Data clustering: a review. ACM Comput. Surv. **31**(3), 264–323 (1999). http://doi.acm.org/10.1145/331499.331504
22. Järvisalo, M.: Itemset mining as a challenge application for answer set enumeration. In: Delgrande, J.P., Faber, W. (eds.) LPNMR 2011. LNCS (LNAI), vol. 6645, pp. 304–310. Springer, Heidelberg (2011). doi:10.1007/978-3-642-20895-9_35
23. Van, T., Leeuwen, M., Nijssen, S., Fierro, A.C., Marchal, K., Raedt, L.: Ranked tiling. In: Calders, T., Esposito, F., Hüllermeier, E., Meo, R. (eds.) ECML PKDD 2014. LNCS (LNAI), vol. 8725, pp. 98–113. Springer, Heidelberg (2014). doi:10.1007/978-3-662-44851-9_7. https://lirias.kuleuven.be/handle/123456789/457022
24. Mannila, H.: Inductive databases and condensed representations for data mining. In: ILPS, pp. 21–30 (1997)
25. Marriott, K., Nethercote, N., Rafeh, R., Stuckey, P.J., De La Banda, M.G., Wallace, M.: The design of the Zinc modelling language. Constraints **13**(3), 229–267 (2008)

26. Meo, R., Psaila, G., Ceri, S.: A new SQL-like operator for mining association rules. In: VLDB, pp. 122–133 (1996)
27. Métivier, J.-P., Boizumault, P., Crémilleux, B., Khiari, M., Loudni, S.: Constrained clustering using SAT. In: Hollmén, J., Klawonn, F., Tucker, A. (eds.) IDA 2012. LNCS, vol. 7619, pp. 207–218. Springer, Heidelberg (2012). doi:10.1007/978-3-642-34156-4_20
28. Métivier, J.P., Boizumault, P., Crémilleux, B., Khiari, M., Loudni, S.: A constraint language for declarative pattern discovery. In: SAC 2012, pp. 119–125. ACM (2012). http://doi.acm.org/10.1145/2245276.2245302
29. Miettinen, P., Mielikäinen, T., Gionis, A., Das, G., Mannila, H.: The discrete basis problem. IEEE Trans. Knowl. Data Eng. 20(10), 1348–1362 (2008)
30. Mitchell, T.: Machine Learning, 1st edn. McGraw-Hill, New York (1997)
31. Negrevergne, B., Guns, T.: Constraint-based sequence mining using constraint programming. In: Michel, L. (ed.) CPAIOR 2015. LNCS, vol. 9075, pp. 288–305. Springer, Heidelberg (2015). doi:10.1007/978-3-319-18008-3_20
32. Nethercote, N., Stuckey, P.J., Becket, R., Brand, S., Duck, G.J., Tack, G.: MiniZinc: towards a standard CP modelling language. In: Bessière, C. (ed.) CP 2007. LNCS, vol. 4741, pp. 529–543. Springer, Heidelberg (2007). doi:10.1007/978-3-540-74970-7_38
33. Pedregosa, F., Varoquaux, G., Gramfort, A., Michel, V., Thirion, B., Grisel, O., Blondel, M., Prettenhofer, P., Weiss, R., Dubourg, V., Vanderplas, J., Passos, A., Cournapeau, D., Brucher, M., Perrot, M., Duchesnay, E.: Scikit-learn: machine learning in Python. J. Mach. Learn. Res. 12, 2825–2830 (2011)
34. Stuckey, P.J., Tack, G.: MiniZinc with functions. In: Gomes, C., Sellmann, M. (eds.) CPAIOR 2013. LNCS, vol. 7874, pp. 268–283. Springer, Heidelberg (2013). doi:10.1007/978-3-642-38171-3_18
35. Tan, P.N., Steinbach, M., Kumar, V.: Introduction to Data Mining. Addison-Wesley, Boston (2005)
36. Van Hentenryck, P.: The OPL Optimization Programming Language. MIT Press, Cambridge (1999)
37. Van Hentenryck, P., Michel, L.: Constraint-Based Local Search. MIT Press, Cambridge (2005)
38. Zou, H., Hastie, T.: Regularization and variable selection via the elastic net. J. Roy. Stat. Soc. Series B 67, 301–320 (2005)

Partition-Based Clustering Using Constraint Optimization

Valerio Grossi[1], Tias Guns[3], Anna Monreale[1], Mirco Nanni[2],
and Siegfried Nijssen[3,4(✉)]

[1] University of Pisa, Pisa, Italy
[2] ISTI - CNR, Pisa, Italy
[3] DTAI, KU Leuven, Leuven, Belgium
[4] LIACS, Universiteit Leiden, Leiden, The Netherlands
siegfried.nijssen@uclouvain.be

Abstract. Partition-based clustering is the task of partitioning a
dataset in a number of groups of examples, such that examples in each
group are similar to each other. Many criteria for what constitutes a good
clustering have been identified in the literature; furthermore, the use of
additional constraints to find more useful clusterings has been proposed.
In this chapter, it will be shown that most of these clustering tasks can be
formalized using optimization criteria and constraints. We demonstrate
how a range of clustering tasks can be modelled in generic constraint
programming languages with these constraints and optimization criteria. Using the constraint-based modeling approach we also relate the
DBSCAN method for density-based clustering to the label propagation
technique for community discovery.

1 Introduction

Clustering [15] is the data analysis task of grouping sets of object. It is an unsupervised task, meaning that no information is known about the true grouping
of the objects. In general, the goal is to find clusters whose objects are similar
to each other while different from the objects in the other clusters. Clustering
can lead to better insights into data and to discoveries of previously unknown
groupings.

Many different clustering settings have been studied in the literature. The
focus of this chapter is on *partition-based clustering*. In partition-based clustering, the clustering must form a partition, that is, each object can only belong
to one cluster. This is in contrast to for instance hierarchical clustering, where
clusters form a tree in which one cluster can be a subset of another cluster.

An important aspect in partition-based clustering is the scoring function that
is used to determine the quality of a clustering. In the literature many different

---

Siegfried Nijssen can currently be reached at the Institute of Information and
Communication Technologies, Electronics and Applied Mathematics, UC Louvain,
Belgium.
Tias Guns can currently be reached at the Vrije Universiteit Brussel.

© Springer International Publishing AG 2016
C. Bessiere et al. (Eds.): Data Mining and Constraint Programming, LNAI 10101, pp. 282–299, 2016.
DOI: 10.1007/978-3-319-50137-6_11

methods for calculating the quality of a clustering have been proposed. A range of popular partition-based methods are based on the concept of a cluster *prototype*. Prototype-based techniques evaluate clusters based on the distance of points in the cluster to the prototype. These approaches provide clusters having spherical shapes. Other approaches consider the diameter of the clusters, or their distance to other clusters. Density-based techniques (e.g. *DBSCAN*) discover clusters of any shape, and are designed for discovering dense areas surrounded by areas with low density, typically formed by noise or outliers.

Another aspect of clustering methods is which constraints they support. Constraints can be used to specify additional requirements on the clusters that need to be found. The most well-known of such requirements are the *must-link* and *cannot-link* constraints, which specify that certain data points should or may not be clustered together [3,18].

In this chapter, we will show that many of these clustering problems can be formalized as generic constraint optimization problems. Consequently, generic constraint optimization solvers can be used to address a wide range of clustering problems. One motivation for the use of generic constraint optimization techniques in this context is the large number of choices that need to be made in defining a clustering setting. Central questions are here:

- how do we define the coherence of a cluster?
- how do we define the number of clusters that we wish to find?
- what other properties must the clusters satisfy?

While such constraints and optimization criteria may sometimes be added in specialized techniques, generic techniques that allow for the specification of such constraints and optimization criteria would be applicable more widely.

Within this chapter, we will distinguish two types of partitioning-based clustering settings: *direct* and *indirect* methods. These settings differ in how the number of clusters is determined. The *direct* methods require that a user specifies the number of clusters explicitly by setting a parameter k, which can be interpreted as a constraint on the number of clusters. For *indirect* methods, the number of clusters is indirectly specified through constraints on the coherence of a cluster; more clusters are created if a smaller number of clusters would not be sufficiently coherent [1].

We first discuss the direct approaches based on a parameter k (Sect. 2), followed by the indirect approaches (Sect. 3). Here, Sect. 2 first introduces several optimization criteria and then outlines common user-specified constraints in clustering. Different modeling choices are presented and demonstrated on a range of clustering problems.

The section on indirect approaches (Sect. 2) shows how clusters can also be modeled as separated regions of high data density. This corresponds to the principle behind the *DBSCAN* algorithm. Furthermore, we draw a link between this data clustering task and the mechanism of *Label Propagation* as used in community detection in graphs.

2   Direct Methods

Characteristic for direct methods is that users need to specify the number of clusters in advance by means of a parameter k. These methods will subsequently focus on finding a good clustering with this number of clusters.

Of crucial importance is then how to evaluate the quality of one cluster. Here, several approaches are possible.

The most studied and applied approaches are those in which a cluster prototype is identified. Every cluster is represented by a prototype called the *centroid* of the cluster. Two popular algorithms employing this approach are *K-means* and *K-medoids*. In K-means each centroid represents the average of all points in the cluster, while in K-medoids the centroid is the most representative actual point in the cluster.

Other approaches do not identify an explicit prototype, but evaluate all pairwise distances between points in the cluster, or evaluate the pairwise distances between points inside and outside the cluster.

From an algorithmic perspective, most algorithms for finding clusters are heuristic. K-means and K-medoids are good examples. Given a user-specified value k, these algorithms select k initial centroids. Successively each point is assigned to the closest centroid based on a distance measure. Finally, the centroids are updated iteratively based on the points assigned to the clusters. This process stops when centroids do not change.

In this chapter, we take a step back from this algorithmic view and look at the underlying optimisation problems that clustering methods are trying to solve. We first describe the different optimization criteria that can be used, followed by constraints that can be put on clusters or the entire clusterings.

In the following, we assume given a set of data points D of size n. Each point p is represented by an m-dimensional vector. A cluster C is a set of points: $C \subseteq D$, and a clustering $\mathscr{C}$ is a partitioning of the data into clusters: $\forall C \in \mathscr{C} : C \subseteq D, \bigcup_{C \in \mathscr{C}} C = D, \forall C_1, C_2 \in \mathscr{C} : C_1 \cap C_2 = \emptyset$. Note that we consider non-overlapping clusters here.

2.1   Optimization Criteria

Intuitively, a clustering consists of clusters that are coherent and whose data points are similar to each other; on the other hand we also expect the clusters (and data points therein) to not be similar to the other clusters [15].

There are many different ways to characterize how good a clustering is, by measuring the (dis)similarity of its clusters and data points. We identify a number of these measures below. Each measure can be used as an optimisation criterium to find a 'good' clustering according to this measure.

Sum of Squared Inter-cluster Distances. Given some distance function $d(\cdot, \cdot)$ over points, for example, the Euclidean distance, we can measure the sum of squared distances within each cluster as follows:

$$\sum_{C \in \mathscr{C}} \sum_{p,q \in C, p<q} d^2(p,q) \tag{1}$$

Here we assume that $p < q$ iff data point p is before point q in the database; this ensures that every pair of points is considered only once.

Sum of Squared Error to Centroid. A more common approach is to measure the "error" of each cluster, that is, the distance of each point in the cluster to the mean (centroid) of that cluster.

We compute the *centroid* of a cluster by computing the mean of the data points that belong to it:

$$z_C = \text{mean}(C) = \frac{\sum_{p \in C} p}{|C|} \tag{2}$$

Here, we assume that the points p are represented as vectors and traditional vector algebra is used. The sum of squared error is then measured as:

$$\sum_{C \in \mathscr{C}} \sum_{p \in C} d^2(p, z_C) \tag{3}$$

Note that this is identical to the sum of all pairwise distances between the points of a cluster, divided by the size of that cluster: $\sum_{C \in \mathscr{C}} \sum_{p,q \in C, p<q} d^2(p,q)/|C|$.

Sum of Squared Error to Medoids. Instead of using the mean (centroid) of the cluster, one can also use the medoid of the cluster, that is, the point that is most representative of the cluster. Let the medoid of a cluster be the point with smallest average distance to the other points:

$$y_C = \text{medoid}(C) = \arg\min_{y \in C} \sum_{p \in C} d^2(p, y). \tag{4}$$

The sum of squared error to the medoids is then measured as follows:

$$\sum_{C \in \mathscr{C}} \sum_{p \in C} d^2(p, y_C) \tag{5}$$

Cluster Diameter. Another measure of coherence is to measure the diameter of the largest cluster, where the diameter is defined as the largest distance between any two points of a cluster. This leads to the following measure of maximum cluster diameter:

$$\max_{C \in \mathscr{C}} \max_{p,q \in C, p<q} d(p,q) \tag{6}$$

One can imagine other variants such as the sum of diameters.

Inter-cluster Margin. The margin between two clusters is the minimal distance between any two points that belong to the different clusters. The margin gives an indication of how different the clusters are from each other (e.g. how far apart they are). This can be optimized using the following measure of minimum inter-cluster margin:

$$\min_{C,D \in \mathscr{C}} \min_{p \in C, q \in D} d(p,q) \quad \text{(7)}$$

2.2 Constraints

Using constraints for defining data mining tasks guarantees a high level of expressivity, since adding new constraints on the required output is quite easy and natural. Constraints typically specify background knowledge that the user has about the clustering. A famous example [18] is that clusters should group cars in lanes, and hence one can derive that some objects can certainly not be in the same cluster (e.g. when known to be driving side-by-side) while others certainly are (when driving in tandem).

The above example is an illustration of *instance-level* constraints, that is, constraints between specific points. *Must-link* constraints require that two points belong to the same cluster, while *Cannot-link* constraints require that two points belong to different clusters [18]. A *Must-link* constraint on two points p and q is expressed by: $\forall C \in \mathscr{C} : p \in C \leftrightarrow q \in C$; while a *Cannot-link* constraint is expressed by: $\forall C \in \mathscr{C} : p \in C \rightarrow q \notin C$.

Another type of constraints is *cluster-level* constraints [9]. The ϵ-constraint or maximal diameter constraint requires that the diameter of a cluster is at most ϵ, that is, each two points in a cluster are at most ϵ apart. This can also be formulated as requiring that each pair of points p and q that is further apart cannot be together in the same cluster: $\forall p, q : d(p,q) > \epsilon \rightarrow (\forall C \in \mathscr{C} : p \in C \rightarrow q \notin C)$. The δ-constraint or minimal margin constraint requires that two points belonging to different clusters have to be at least δ apart. Alternatively formulated: any two points that are closer than δ must belong to the same cluster: $\forall p, q : d(p,q) < \delta \rightarrow (\forall C \in \mathscr{C} : p \in C \leftrightarrow q \in C)$.

Other user-defined constraints can be expressed [7]. One can impose constraint on the clusters *size*, e.g. requiring clusters with a minimum or maximum number of points. Constraining the number of points to be minimum or maximum α is expressed as: $\forall C \in \mathscr{C} : |C| \geq \alpha$ and $\forall C \in \mathscr{C} : |C| \leq \alpha$.

Furthermore, any of the measures introduced in the previous section on optimization criteria can also be constrained to take a value within a certain interval. Other variants and combinations of these constraints can be employed as well, such as disjunctions of constraints or conditional must-link and cannot-link constraints. One can add constraints that certain individual clusters must be similar or different from predefined sets of points, or add *soft* constraints such as a bound on the number of points that can have a cannot-link constraint [2].

The complexity of adding constraints to (k-means) clustering has been studied in [10]. A general overview of constraint-based methods in clustering is available in the book "Constrained Clustering: Advances in Algorithms, Theory, and Applications" [3]. Furthermore, the chapter "*Data Mining & Constraints: an*

Overview" of this book provides several references to using constraints in data mining tasks also for clustering.

2.3   Modeling Clustering as Constraint Optimization

A constraint optimization problem $P = (V, D, X, f)$ consists of variables V, a domain D that lists the possible values the variables can take, a set of constraints X over V and an optimization function f over V that must be minimized or maximized.

Building on the primitives introduced earlier, many well-known clustering problems can now be modelled as follows, for a given optimization criterion $quality(\mathscr{C})$:

$$\text{maximize}_{\mathscr{C}} \quad quality(\mathscr{C}), \tag{8}$$

$$s.t.$$

$$C_1 \cap C_2 = \emptyset \qquad \forall C_1, C_2 \in \mathscr{C} \tag{9}$$

$$| \bigcup_{C \in \mathscr{C}} C | = n \tag{10}$$

$$|\mathscr{C}| = k \tag{11}$$

Here n is the total number of points. In this setting, the number of clusters to be found is fixed and has to be k.

Note that the model above uses a set notation for the clusters. Not all constraint solvers support sets; set variables may not always be the most efficient representation either. For these reasons, an *encoding* of the sets in variables of other types is sometimes necessary. There are various ways to model these sets, as well as different solving techniques that can be used on these models. We differentiate between three kinds of approaches:

- Constraint formulations that can directly be solved by most state-of-the-art constraint programming systems;
- Formulations that require an extension of a Constraint Programming system;
- Hybrid approaches with a specialized system that can deal with a range of clustering problems, but no other problems.

We will discuss these possibilities in more detail below.

2.3.1   Constraint Solving Formulations

To use constraint programming systems off-the-shelf, an important question that needs to be answered is how to encode a clustering in such systems. Next to a set-based notation, several representations have been proposed. We will use these representations to construct clustering models in the next section:

- a Boolean representation, in which a variable a_{it} with domain $\{0, 1\}$ takes value 1 iff point i (with $1 \leq i \leq n$) is in cluster t (with $1 \leq t \leq k$), and takes value 0 otherwise. Constraints

$$\sum_{t=1}^{k} a_{it} = 1$$

for all points i ensure that a point is in only one cluster [14];
- a Boolean representation, in which a variable a_t with domain $\{0, 1\}$ takes value 1 iff possible cluster t (with $1 \leq t \leq 2^n$, i.e. each possible cluster is given an index) is in the clustering; constraint $\sum_{t=1}^{2^n} a_t = k$ ensures exactly k possible clusters are selected and constraints

$$\sum_{t=1}^{2^n} [i \in C_t] a_t = 1 \quad.$$

ensure that every point i is in exactly one chosen cluster; here $[i \in C_t]$ is an indicator that takes value 1 iff point i is in possible cluster t [11, 16];
- an integer representation, in which a variable a_i with domain $\{1, \ldots, k\}$ indicates that point i (with $1 \leq i \leq n$) is in cluster a_i [8];
- an integer representation, in which a variable g_i with domain $\{1, \ldots, n\}$ identifies the point with the smallest index that is in the same cluster as point i; note that $g_i = i$ iff there is no point $j < i$ in the same cluster as point i [7].

An important benefit of the first Boolean representation is that it is easy to formalize additional constraints in this representation. A *Must-link* constraint on two points p_i and p_j is expressed by a set of $a_{it} = a_{jt}$ constraints, where $1 \leq t \leq k$; a *cannot-link* constraint is expressed by: $\forall t \in \{1, \ldots, k\} : a_{it} + a_{jt} \leq 1$. A *size* constraint can be expressed by:

$$\forall t \in \{1, \ldots, k\} : \sum_{i=1}^{n} a_{it} \geq \alpha$$

$$\forall t \in \{1, \ldots, k\} : \sum_{i=1}^{n} a_{it} \leq \alpha.$$

A drawback of this representation is that additional constraints are required to ensure that a point is not in two clusters; this is not necessary in the integer representations.

The second Boolean representation has as most important drawback that its number of variables is very large. One way to address this is to limit the number of possible clusters apriori; ideas for this were presented in [16].

The main difference between the integer representations is that in the second representation the indexes of representative points are used to identify clusters, while in the first cluster indexes are used. In the second representation the number of clusters is not fixed; to achieve a fixed number of clusters, additional variables c_j with domain $\{1, \ldots, n\}$, where $1 \leq j \leq k$, can be used, with the constraints:

- $g_{c_j} = c_j$ for all clusters j, i.e., variable c_j points to the identifying point for each cluster;
- $\sum_{j=1}^{k} [g_i = c_j] = 1$ for each point i; this ensures that each point i also belongs to one of the k clusters identified by c.

A remaining challenge is how to represent the optimization criterion. In many cases, additional variables are needed. This is illustrated for a number of cases below.

K-medoid Clustering. In k–medoid clustering, an important aspect is that we need to identify the cluster medoids. One approach is to represent these medoids using Boolean variables m_{ij}, where $1 \leq i \leq n$ and $1 \leq j \leq k$; these variables indicate whether a point is the medoid of a cluster or not. Constraints enforce that each cluster has only one medoid.

The optimization criterion then becomes:

$$\sum_{i=1}^{n} \sum_{j=k}^{n} a_{ij} \sum_{h=1}^{n} m_{hj} \, d(p_i, p_h)^2$$

This leads to the overall optimization problem below:

$$\underset{a,m}{\text{minimize}} \quad \sum_{i=1}^{n} \sum_{j=k}^{n} a_{ij} \sum_{h=1}^{n} m_{hj} \, d(p_i, p_h)^2 \tag{12}$$

$$s.t.$$

$$\sum_{j=1}^{k} a_{ij} = 1 \qquad \forall i \in \{1, \ldots, n\} \tag{13}$$

$$\sum_{i=1}^{N} m_{ij} = 1 \qquad \forall j \in \{1, \ldots, k\} \tag{14}$$

Hence, this model assumes that both the *assignment of points to clusters* and *the actual medoids* are discovered by the constraint programming system. Note that this model does not explicitly constrain the medoid to its cluster, as in an optimal solution, the chosen centers need to be medoids for their cluster in order to minimize the optimization criterion.

Furthermore, this model does not impose additional constraints. Constraints such as those discussed in Sect. 2.2 can be added to the model without modification.

Sum of Squared Inter-cluster Distances. This clustering setting has been modeled in constraint programming using the integer representation where each variable g_i points to the 'identifying' point of the cluster, e.g. its point with smallest index [7]. The sum of squared inter-cluster distances is then expressed as:

$$\sum_{i,j \in \{1, \ldots, n\}, i < j} [g_i = g_j] d^2(p_i, p_j)$$

Using variable c_j to represent the identifying point of cluster j, where the identifying point is the point with smallest index, this leads to the following constraint specification:

$$\underset{g,c}{\text{minimize}} \quad \sum_{i,j \in \{1,\dots,n\}, i<j} [g_i = g_j] d^2(p_i, p_j), \tag{15}$$

$$\text{s.t.}$$

$$g_i \leq i \qquad\qquad\qquad \forall i \in \{1,\dots,n\} \tag{16}$$

$$g_{c_j} = c_j \qquad\qquad\qquad \forall c \in \{1,\dots,k\} \tag{17}$$

$$\sum_{j=1}^{k} [g_i = c_j] = 1 \qquad\qquad\qquad \forall i \in \{1,\dots,n\} \tag{18}$$

$$c_j < c_{j'} \qquad\qquad\qquad \forall j,j' \in \{1,\dots,k\}, j < j' \tag{19}$$

$$c_1 = 1 \tag{20}$$

Equation 16 ensures that either it is the smallest (identifying) point of its cluster, or g_i points to another (smaller) identifying point. Equation 17 materializes the concept of identifying point in a variable c_j. The identifying point's index is the cluster identifier, so $g_{c_j} = c_j$. This constraint is known in the constraint solving literature as an element constraints. The last two constraints are symmetry breaking constraints.

Maximal Diameter and Minimal Margin. The same integer representation with identifying points has been used to model the problem of minimizing the maximal diameter and maximizing the minimal margin [7].

The main difference is the optimization criterion. Instead of computing the maximal diameter or minimal margin explicitly and optimising this, it is possible to constrain each pair of points individually. Let D be a new variable representing the maximum diameter, then each pair of points p_i, p_j that is further than d apart may not be in the same cluster: $d(p_i, p_j) > D \rightarrow (g_i \neq g_j)$. The model is shown below and shares a number of constraints with the previous model [11]:

$$\underset{D,g,c}{\text{minimize}} \quad D, \tag{21}$$

$$\text{s.t.}$$

$$d(p_i, p_j) > D \rightarrow (g_i \neq g_j) \qquad\qquad \forall i,j \in \{1,\dots,n\} \tag{22}$$

Equations 16...20 in the previous model

Maximizing the minimal margin is specified in a similar way [11]:

$$\underset{M,c,g}{\text{maximize}} \quad M, \tag{23}$$

$$\text{s.t.}$$

$$d(p_i, p_j) < M \rightarrow (g_i = g_j) \qquad\qquad \forall i,j \in \{1,\dots,n\} \tag{24}$$

Equations 16...20 in the above model

Squared Error to the Centroids (K-means). K-means aims to find non-overlapping clusters that minimize the sum of squared errors to the centroid of the cluster. As pointed out earlier, one formulation of the optimization criterion

is $\sum_{C \in \mathscr{C}} \sum_{i,j \in |C|, i<j} \frac{d^2(p_i, p_j)}{|C|}$. While we could model this with the Boolean or inte-
ger representations used so far, the division in this optimization criterion creates
a non-linearity that makes the problem a lot harder to solve.

Instead, we can use the approach in which we have 2^n Boolean variables a_t,
i.e., a variable for each possible cluster. Let m be an n by 2^n matrix of Boolean
values, where each column is a cluster with $m_{it} = 1$ if data point p_i is in cluster
t and $m_{it} = 0$ otherwise. For each cluster t, the squared error to the centroid
can then be precomputed as

$$c_t = \frac{\sum_{i=1}^{n} m_{it} \sum_{j=i+1}^{n} m_{jt} d^2(p_i, p_j)}{\sum_{i=1}^{n} m_{it}}$$

Using these costs, the problem can be formulated as follows [11]:

$$\underset{a}{\text{minimize}} \quad \sum_{t \in T} a_t c_t, \tag{25}$$

$$s.t.$$

$$\sum_{t \in T} a_t m_{it} = 1 \qquad\qquad \forall i \in \{1, \dots, n\} \tag{26}$$

$$\sum_{t \in T} a_t = k \tag{27}$$

where $T = \{1, \dots, 2^n\}$ denotes all possible clusters. Equation 25 is the sum of
the squared errors to the centroid of all clusters that are selected (e.g. $a_t = 1$).
Equation 26 states that each data point must be covered exactly once. Hence
it enforces both that the clusters are not overlapping and that all points are
covered. Equation 27 finally ensures that exactly k clusters are found.

2.3.2   Extending Constraint Solvers

The previous subsection introduced how to model many clustering problems
using generic constraint programming formulations. These formulations can be
decomposed into low-level constraints such as sum, (in)equality and implication.
Such constraints are supported by most CP systems.

While correct, however, these decompositions and the corresponding *propagation* of the low-level constraints is often not efficient. To improve the performance
one of the possible approaches is to add *global constraints* to these CP systems,
which implement specialized propagation methods for specific constraints.

For example, Thi-Bich-Hanh et al. [6] introduced a global constraint for the
sum of squared inter-cluster distances:

$$\sum_{i,j \in \{1, \dots, n\}, i<j} [g_i = g_j] d^2(p_i, p_j) \tag{28}$$

In a standard constraint solver, this constraint is decomposed by introducing
auxiliary variables $b_{ij} \leftrightarrow [g_i = g_j]$ and having a linear sum constraint over these
b_{ij}: $s = \sum_{i,j \in \{1, \dots, n\}, i<j} b_{ij} v_{ij}$, where $v_{ij} = d^2(p_i, p_j)$ are precomputed constants.

Instead, the authors introduce a global constraint for the entire Eq. 28, which can reason over the fact that each point can only belong to one cluster. In this way, a tighter lower bound on the sum can be computed than when using the standard decomposition.

Computing this tighter lower bound is achieved by splitting the sum into three distinct cases: (a) cases for which g_i and g_j are already assigned, (b) cases for which g_i or g_j is assigned, but not the other one, and (c) cases for which both g_i and g_j are not assigned. Case (a) can be deterministically computed. For case (b), for each point, the minimum value is chosen among all existing clusters to which this point could be added. For case (c) a clever heuristic is used to compute a lower bound based on the minimum number of possible connections that must still be added to obtain k clusters.

Apart from adding global constraints, efficiency improvements can typically also be obtained by adding redundant constraints or by breaking symmetries in the constraint formulation. Another important aspect is the order in which to search over the variables and their possible values. For example, one could use a furthest-point-first heuristic [7,13].

2.3.3 Hybrid Approach: Column Generation

Further problems of efficiency are posed by models that introduce an exponential number of variables, having one variable for each potential cluster. Global constraints can not solve this problem.

One approach to solve this challenge is to lazily add candidate clusters until the optimal subset of clusters is found. This is the idea behind *column generation* in Integer Linear Programming. This was first investigated for minimum sum of squared error clustering by DuMerle et al. [11] and later extended to support additional constraints by Babaki et al. [2].

The core idea is to only consider a subset of the clusters in the set T of the above model, and to *relax* the a_t variables such that they can take on real values instead of being Boolean. This problem is called the *restricted master problem*. Solving the restricted master problem can be done with standard (integer) linear programming solvers, and one obtains a real-valued solution to a_t. Then, using the dual values of this solution, one can search for the best cluster (column) to add, that is, the cluster that can best improve the objective function. This is called the *subproblem* and is typically done with a specialized method. If no such column can be found, the solution of the restricted master problem is also a solution of the original master problem [11].

A key observation in [2] is that most constraints considered in constraint-based clustering are constraints on individual clusters. Consequently, these constraints do not change the (restricted) master problem; they only affect the set T and hence the definition of the *subproblem*. Babaki et al. [2] have devised a method to solve the subproblem directly while taking must-link and cannot-link constraints into account, as well as other anti-monotone constraints such as cluster size and overlap constraints.

3   Indirect Methods

The approaches presented in Sect. 2 require to know in advance the number of clusters to be found. Moreover, they tend to provide clusters that are sphere-shaped. Unfortunately, in a number of real applications, the data points are grouped into non-spherical regions or regions that are quite dense surrounded by areas with low density, typically formed by noise. From this perspective, clusters can also be defined implicitly as regions of higher data density, separated from each other by regions of lower density. The price for this flexibility is a difficult interpretation of the obtained clusters. One of the most famous clustering algorithms based on the notion of density of regions is DBSCAN [12]. This algorithm does not rely on an optimization algorithm however, and in this chapter we present a constraint programming formulation (Sect. 3.1). Using this formulation, we show how re-defining this task as a community discovery problem in a network, this approach becomes very similar to the label propagation approach that finds clusters of nodes in networks [17].

3.1   Density-Based Clustering

Density-based clustering is based on measuring the data density at a certain region of the data space and then defining clusters as regions that exceed a certain density threshold. The final clusters are obtained by connecting neighboring dense regions. Figure 1 shows an illustrative example in a two-dimensional space. Four groups are recognized as clusters and they are separated by an area where the data density is low.

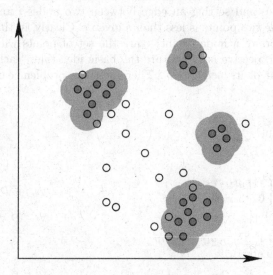

Fig. 1. Example of density-based clusters [4].

DBSCAN. DBSCAN [12] locates regions of high density that are separated from one another by regions of low density. The approach identifies three different classes of points:

Core points. These points are in the interior of a density-based cluster. A point is a core point if the number of points within a given neighborhood around the point as determined by the distance function and a user- specified distance parameter, ϵ, exceeds a certain threshold, $MinPts$, which is also a user-specified parameter.

Border points. These points are not core points, but fall within the neighborhood of a core point. A border point can fall within the neighborhoods of several core points.

Noise points. A noise point is any point that is neither a core point nor a border point.

The DBSCAN algorithm works as follows:

1. Label all points as core, border, or noise points.
2. Eliminate noise points.
3. Put an edge between all core points that are within ϵ of each other.
4. Make each group of connected core points into a separate cluster.
5. Assign each border point to one of the clusters of its associated core points.

Below we introduce the constraint programming model for this clustering problem.

Constraint Programming Model for Density Based Clustering. We reformulate the problem in the context of networks by considering the set of points D as nodes and setting an edge between two nodes i and j if the distance between the two points is less than a given ϵ. Clearly, in this way we have that the neighbors of a node (point) i are the set of points within a distance ϵ. Our intended objective is to capture the basic idea that "each node has the *same* label as all of its neighbors". Therefore, the problem can be modelled as follows:

$$\text{maximize}(\sum_{j \in L} \min(1, \sum_{i \in D} k_{i,j})), \tag{29}$$

$$s.t.$$

$$a_{i,j} = \begin{cases} 1 \text{ if } d(i,j) \leq \epsilon \\ 0 \text{ otherwise} \end{cases} \qquad \forall i,j \in D \tag{30}$$

$$k_{i,j} \in \{0,1\} \qquad \forall i \in D, \forall j \in L \tag{31}$$

$$r_i = \begin{cases} 1 \text{ if } \sum_{j \in D} a_{i,j} \geq minp \\ 0 \text{ otherwise} \end{cases} \qquad \forall i \in D \tag{32}$$

$$\sum_{j \in L} k_{i,j} = 1 \qquad \forall i \in D \tag{33}$$

$$r_h = 1 \wedge r_p = 1 \wedge a_{h,p} = 1 \wedge k_{h,j} = 1 \Rightarrow k_{p,j} = 1 \quad \forall h \in D, \forall r \in D, \forall j \in L \tag{34}$$

$$r_i = 0 \Rightarrow k_{i,j} = 1 \qquad\qquad\qquad \forall i \in D \tag{35}$$

where

$$j = \min(\{j \in L \setminus \{n+1\} \mid$$
$$\exists h : a_{h,i} = 1 \wedge r_h = 1 \wedge k_{h,j} = 1\} \cup \{n+1\})$$

In more detail this model can be described as follows. Boolean variables $k_{i,j}$ denote the color of a point (node i), by setting $k_{i,j} = 1$ if the point i has the color j. Variables $a_{i,j}$ indicate the presence or absence of an edge between two nodes. Variables r_i denote whether node i is a core point or not.

Assumed given are: (a) the set of points D, and (b) the ordered set of colors $L = \{1, \ldots, n\}$. Note that in the set of colors we have the color $n + 1$ that is an additional color used for coloring the noise points. The model imposes that a point has one and only one color, and that all the connected core points must have the same color (Eq. 34). Another requirement is that each point that is not a core point takes the same color of the core points that are connected to it. If it does not have any core point around it then this point takes the additional color $n + 1$ because it is a noise (Eq. 35). Such a constraint also captures the special case in which the point i can be connected to more than one core with different colors. In this case, the model assigns to i the color of the core that in the ordered set $C \setminus \{n + 1\}$ has a lower rank. Finally, the model is intended to maximize the number of different colors. Notice that a solution where all points have distinct colors does not satisfy Eq. 34 because connected points do not have the same color.

This constraint programming formulation makes it easy to extend the standard method with other constraints. In principle, any constraint that requires to merge clusters as identified by DBSCAN can be added to the model above. As an example, we could specify constraints on the minimum cluster size: in this case, clusters will need to be merged in order to obtain a required cluster size; by enforcing a diameter constraint on the resulting clusters, it can be ensured that the resulting clusters are not arbitrary combinations of clusters as identified by traditional DBSCAN.

Moreover, we can also extend the above problem by changing one of the constraints of the standard formulation. In the following, we will show that by changing a constraint of the DBSCAN formulation we obtain a problem that corresponds to one of the most famous algorithms for discovering communities in network data.

3.2   Label Propagation

When considering graph or network data, a task very similar to clustering is *community discovery*, which can be seen as a network variant of standard data clustering. The concept of a "community" in a (web, social, or informational) network is intuitively understood as a set of individuals that are very similar, or close, to each other, more than to anybody else outside the community [5]. This has often been translated in network terms into finding sets of nodes densely connected to each other and sparsely connected with the rest of the network. An interesting community discovery algorithm is the Label Propagation algorithm [17] that detects communities by spreading labels through the edges of the graph and then labeling nodes according to the majority of the labels attached to their neighbors, iterating until a general consensus is reached.

Before introducing a constraint programming model for this algorithm we recall the details of the iterative label propagation algorithm presented in [17].

Iterative Label Propagation (LP). Suppose that a node v has neighbors $v_1, v_2, \ldots, v_k$ and that each neighbor carries a label denoting the community that it belongs to. Then, v determines its community based on the labels of its neighbors. [17] assumes that each node in the network chooses to join the community to which the maximum number of its neighbors belong to. As the labels propagate, densely connected groups of nodes quickly reach a consensus on a unique label. At the end of the propagation process, nodes with the same labels are grouped together as one community. Clearly, a node with an equal maximum number of neighbors in two or more communities will take one of the two labels by a random choice. For clarity, we report here the procedure of the LP algorithm. Note that, in the following $C_v(t)$ denotes the label assigned to the node v at time (or iteration) t.

1. Initialize the labels at all nodes in the network. For any given node v, $C_v(0) = v$.
2. Set $t = 1$.
3. Arrange the nodes in the network in a random order and set it to V.
4. For each $v_i \in V$, in the specific order, let $C_{v_i}(t) = f(C_{v_{i1}}(t-1), \ldots, C_{v_{ik}}(t-1))$. Function f here returns the label occurring with the highest frequency among neighbors and ties are broken uniformly randomly.
5. If every node has a label that the maximum number of its neighbors has, or t hits a maximum number of iterations t_{max} then stop the algorithm. Else, set $t = t + 1$ and go to (3).

The drawback of this algorithm is the fact that *ties are broken uniformly randomly*. This random behavior can lead to different results for different executions and some of these results cannot be optimal. In Fig. 2 we show how given the same network as input of the LP algorithm we obtain four different results.

In the next section we propose a constraint programming model that solves this problem by providing the optimal solution.

3.2.1   Constraint Programming Model for Label Propagation

Let us now propose a constraint programming model for the community discovering problem based on label propagation. Our aim is to capture the basic idea that "each node takes the label of the majority of its neighborhood". Therefore, the model is the following:

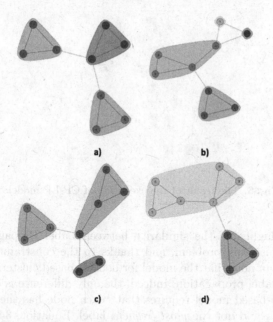

Fig. 2. The result of four executions of LP algorithm (Color figure online)

$$\text{maximize}(\sum_{j \in L} \min(1, \sum_{i \in N} k_{i,j})), \tag{36}$$

$$s.t.$$

$$a_{i,j} = \begin{cases} 1 \text{ if } (i,j) \in E \\ 0 \text{ otherwise} \end{cases} \tag{37}$$

$$k_{i,j} \in \{0,1\} \tag{38}$$

$$\sum_{j \in L} k_{i,j} = 1 \qquad\qquad \forall i \in N \tag{39}$$

$$n_{i,h} = \sum_{\forall j:a_{i,j}=1} k_{j,h} \qquad\qquad \forall i \in N, \forall h \in L \tag{40}$$

$$k_{i,l} = 1 \Rightarrow n_{i,l} = \max_{h \in L} n_{i,h} \qquad\qquad \forall i \in N, \forall l \in L \tag{41}$$

Here, variables $a_{i,j}$ indicate the presence or absence of an edge between two nodes. Variables $k_{i,j}$ denote the color (label) of a node in the network. Assumed given is (a) the set of nodes N, (b) the set of edges E, and (c) an ordered set of colors L. A node can be assigned one and only one color. Variables $n_{i,h}$ denote the number of neighbors of node i with assigned color h. The model assigns to the node i the color h if it is the most popular among its neighbors, as shown in Eq. 41. Such a constraint also captures the case of ties. In such a case, node i is assigned the color that has the lowest rank in the ordered set L. Finally, the model maximizes the number of different colors in the network, as shown in Eq. 36.

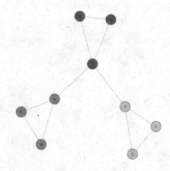

Fig. 3. The result of the execution of CP-LP model

This model highlights the similarity between Label Propagation and the Density-based clustering problem, and thanks to the constraint programming formulation we can note that the model for density-based clustering is a variant of the standard label propagation. Indeed, the only difference is due to the fact that the Density-based model requires that "each node has the *same* label of all its neighbors", and not the *most frequent* label. Equations 34 and 35 in the DBSCAN model and Eq. 41 in the LP model express this difference. By executing our model we obtain the optimal solution depicted in Fig. 3, where we consider as input the same network in Fig. 2.

4   Conclusions

In this chapter, we have presented how different well-established approaches to partition-based clustering can be modeled and optimized via constraints. In particular, we investigated two main families of partition-based methods, i.e. *direct* and *indirect*. In this perspective, the chapter has presented several examples where the clustering methods are explicitly modeled by constraints. In this way, it has parted from the more traditional *algorithmic* view on clustering. We discussed different optimization criteria and constraints, showed different modeling choices for *direct* methods and related the *indirect* methods of DBSCAN and label propagation through a constraint formulation.

References

1. Aggarwal, C.C., Reddy, C.K.: Data Clustering: Algorithms and Applications, 1st edn. Chapman & Hall/CRC, Boca Raton (2013)
2. Babaki, B., Guns, T., Nijssen, S.: Constrained clustering using column generation. In: Simonis, H. (ed.) CPAIOR 2014. LNCS, vol. 8451, pp. 438–454. Springer, Heidelberg (2014). doi:10.1007/978-3-319-07046-9_31
3. Basu, S., Davidson, I., Wagstaff, K., Clustering, C.: Advances in Algorithms, Theory, and Applications, 1st edn. Chapman & Hall/CRC, Boca Raton (2008)

4. Berthold, M.R., Borgelt, C., Hppner, F., Klawonn, F.: Guide to Intelligent Data Analysis: How to Intelligently Make Sense of Real Data, 1st edn. Springer, Heidelberg (2010)
5. Coscia, M., Giannotti, F., Pedreschi, D.: A classification for community discovery methods in complex networks. Stat. Anal. Data Min. 4(5), 512–546 (2011)
6. Dao, T., Duong, K., Vrain, C.: A filtering algorithm for constrained clustering with within-cluster sum of dissimilarities criterion. In: 2013 IEEE 25th International Conference on Tools with Artificial Intelligence, Herndon, VA, USA, 4–6 November 2013, pp. 1060–1067 (2013)
7. Dao, T.-B.-H., Duong, K.-C., Vrain, C.: A declarative framework for constrained clustering. In: Blockeel, H., Kersting, K., Nijssen, S., Železný, F. (eds.) ECML PKDD 2013. LNCS (LNAI), vol. 8190, pp. 419–434. Springer, Heidelberg (2013). doi:10.1007/978-3-642-40994-3_27
8. Dao, T.-B.-H., Duong, K.-C., Vrain, C.: Constrained clustering by constraint programming. Artif. Intell. (2015)
9. Davidson, I., Ravi, S.: The complexity of non-hierarchical clustering with instance and cluster level constraints. Data Min. Knowl. Disc. 14(1), 25–61 (2007)
10. Davidson, I., Ravi, S.S.: Clustering with constraints: feasibility issues and the k-means algorithm. In: Proceedings of the 2005 SIAM International Conference on Data Mining, SDM 2005, Newport Beach, CA, USA, 21–23 April 2005, pp. 138–149 (2005)
11. du Merle, O., Hansen, P., Jaumard, B., Mladenovic, N.: An interior point algorithm for minimum sum-of-squares clustering. SIAM J. Sci. Comput. 21(4), 1485–1505 (1999)
12. Ester, M., Kriegel, H.-P., Sander, J., Xu, X.: A density-based algorithm for discovering clusters in large spatial databases with noise. In: Simoudis, E., Han, J., Fayyad, U.M. (eds.) KDD, pp. 226–231. AAAI Press, Menlo Park (1996)
13. Gonzalez, T.F.: Clustering to minimize the maximum intercluster distance. Theor. Comput. Sci. 38, 293–306 (1985)
14. Hansen,P., Aloise, D.: A survey on exact methods for minimum sum-of-squares clustering, pp. 1–2, January 2009. http://www.math.iit.edu/Buck65files/msscStLouis.pdf
15. Jain, A.K., Murty, M.N., Flynn, P.J.: Data clustering: a review. ACM Comput. Surv. 31(3), 264–323 (1999)
16. Mueller, M., Kramer, S.: Integer linear programming models for constrained clustering. In: Pfahringer, B., Holmes, G., Hoffmann, A. (eds.) DS 2010. LNCS (LNAI), vol. 6332, pp. 159–173. Springer, Heidelberg (2010). doi:10.1007/978-3-642-16184-1_12
17. Raghavan, U.N., Albert, R., Kumara, S.: Near linear time algorithm to detect community structures in large-scale networks. Phys. Rev. E 76(2), 036106+ (2007)
18. Wagstaff, K., Cardie,C.: Clustering with instance-level constraints. In: Proceedings of the Seventeenth International Conference on Machine Learning (ICML 2000), Stanford University, Stanford, CA, USA, June 29–July 2 2000, pp. 1103–1110 (2000)

Showcases

The Inductive Constraint Programming Loop

Christian Bessiere[1]([✉]), Luc De Raedt[2], Tias Guns[2], Lars Kotthoff[3],
Mirco Nanni[4], Siegfried Nijssen[2,5], Barry O'Sullivan[3], Anastasia Paparrizou[1],
Dino Pedreschi[4], and Helmut Simonis[3]

[1] CNRS, University of Montpellier, Montpellier, France
bessiere@lirmm.fr
[2] DTAI, KU Leuven, Leuven, Belgium
[3] Insight, University College Cork, Cork, Ireland
[4] University of Pisa, Pisa, Italy
[5] LIACS, Universiteit Leiden, Leiden, The Netherlands

Abstract. Constraint programming is used for a variety of real-world
optimization problems, such as planning, scheduling and resource allocation problems. At the same time, one continuously gathers vast amounts
of data about these problems. Current constraint programming software
does not exploit such data to update schedules, resources and plans. We
propose a new framework, that we call the *Inductive Constraint Programming (ICON) loop*. In this approach data is gathered and analyzed
systematically in order to dynamically revise and adapt constraints and
optimization criteria. Inductive Constraint Programming aims at bridging the gap between the areas of data mining and machine learning on
the one hand, and constraint programming on the other hand.

This chapter is an extended abstract of

Christian Bessiere, Luc De Raedt, Tias Guns, Lars Kotthoff, Mirco Nanni,
Siegfried Nijssen, Barry O'Sullivan, Anastasia Paparrizou, Dino Pedreschi,
Helmut Simonis. *The Inductive Constraint Programming Loop.* CoRR,
abs/1510.03317, 2015. http://arxiv.org/abs/1510.03317

1 Introduction

Machine Learning/Data Mining (ML/DM) and Constraint Programming (CP)
are central to many application problems. ML is concerned with learning func-
tions/patterns characterizing some training data whereas CP is concerned with
finding solutions to problems subject to constraints and possibly an optimization
function.

Siegfried Nijssen can currently be reached at the Institute of Information and
Communication Technologies, Electronics and Applied Mathematics, UC Louvain,
Belgium.

Tias Guns can currently be reached at the Vrije Universiteit Brussel.

© Springer International Publishing AG 2016
C. Bessiere et al. (Eds.): Data Mining and Constraint Programming, LNAI 10101, pp. 303–309, 2016.
DOI: 10.1007/978-3-319-50137-6_12

The problem with current technology is that the problems of data analysis and constraint satisfaction/optimization have almost always been studied independently and in isolation. Indeed, there exist a wide variety of successful approaches to analysing data in the field of ML, DM and statistics, and at the same time, advanced techniques for addressing constraint satisfaction and optimization problems have been developed in the CP community. Over the past decade a limited number of isolated studies on specific cases has indicated that significant benefits can be obtained by connecting these two fields [EF01, XHHL08, DGN08, BHO09, KBC10, CJSS12], but so far a truly general, integrated and cross-disciplinary approach is missing.

CP technology is used to solve many types of constraint satisfaction and optimization problems, such as in power companies generating and distributing electricity, in hospitals planning their surgeries, and in public transportation companies scheduling buses. Despite the availability of effective and scalable solvers, current approaches are still unsatisfactory. The reason is that when using CP technology to solve these applications, the constraints and criteria, that is, the *model*, must be specified statically. However, in reality often this model needs to be revised over time. The revision can be needed to reflect changes in the environment due to external events that impact the problem. The revision can also be needed because the execution of the solution generated by the model has modified the characteristics of the problem. Finally the revision can be needed simply because the original model did not capture correctly the problem. Observing the impact of the solution allows us to correct or improve the model. Therefore, there is an urgent need for improving and revising a model over time *based on data* that is continuously gathered about the performance of the solutions and the environment they are used in. The CP community has extended the basic constraint satisfaction and optimization problems to better tackle changing environments. The *dynamic* constraint satisfaction approach [DD88] allows the addition/retraction of constraints from the initial model. This approach does not predict the changes from data, but rather the addition/retraction of constraints is performed by the user. The *online/stochastic* constraint programming approach [BH04, Wal02] offers a framework to deal with unknown future events, such as customer requests. It builds a finite set of *future scenarios*, e.g. using sampling from a known distribution, and the optimization problem is then defined over each of the scenarios. The framework does not capture ways of using data, other than for the prediction of possible scenarios of events.

In general, exploiting gathered data to modify and adjust any aspect of a model is difficult and labor intensive with state-of-the-art solvers. As a consequence, the data that is being gathered today in order to monitor the quality of the produced solutions and to help evaluating the effect of possible adjustments to the constraints or optimization criteria, is not fully exploited when changes in a schedule or plan are needed. Hence, schedules and plans that are produced are often suboptimal. This, in turn, leads to a waste of resources. Instead of using data passively, data should be actively analysed in order to discover and update the underlying regularities, constraints and criteria that govern the data.

In this chapter, we propose and formalize the new framework of inductive constraint programming. This framework is based on what we call the *Inductive Constraint Programming (ICON) loop*, which is an interaction between a machine learning component (ML) and a constraint programming component (CP). The ML component observes the world and extracts patterns. The CP component solves a constraint satisfaction or optimization problem using these patterns; its solution is applied to the world. We assume the world changes over time, possibly due to the impact of applying our solution. This process is repeated in a loop. Inductive constraint programming will serve the long-term vision of easier-to-use and more effective tools for resource optimization and task scheduling.

An introduction to Constraint Programming and Data Mining was already given earlier in this book; the focus of this chapter is on introducing the formalism behind the loop. Extensive examples of the loop can subsequently be found in the last chapters.

2   Inductive Constraint Programming Loop

The inductive constraint programming loop will cope with changes in the world by iteratively solving a learning problem and a constraint problem. The loop is composed of several components that interact with each other through writing and reading operations. A visualization of the loop is given in Fig. 1. We introduce each of the elements in the loop in turn.

CP Component. An important element of the CP component is the constraint network. A *constraint network* $N = (X, D, C, f)$ is composed of: a set X of variables taking values in domain D. These variables are subject to constraints in the set C. The optional evaluation function f takes as input an assignment on X and returns a cost for it. A solution (optionally *best* solution) of N is an assignment in D^X satisfying all the constraints in C (optionally minimizing f).

A *solver* Xsolve takes as input a constraint network and returns a solution/best solution or failure in case no solution satisfying all the constraints exists.

The CP component is composed of the constraint network $N = (X, D, C, f)$, the constraint solver Xsolve, and a Solutions repository. Xsolve generates solutions of N, or good/best solutions of N according to f, that it writes in the Solutions repository. In case Xsolve is not able to produce any solution to be applied to the world, the CP component notifies the ML component by sending information about the failure.

More details about CP can be found in the first chapter of this book.

ML Component. A *learning problem* $L = (E, H, t, loss)$ is composed of a set E of examples, a hypothesis space H, the target function t that one wants to learn, and a loss function $loss(E, h, t)$ that measures the quality of a hypothesis $h \in H$ w.r.t. dataset E and the target hypothesis t. The goal is to find a hypothesis that minimizes the loss.

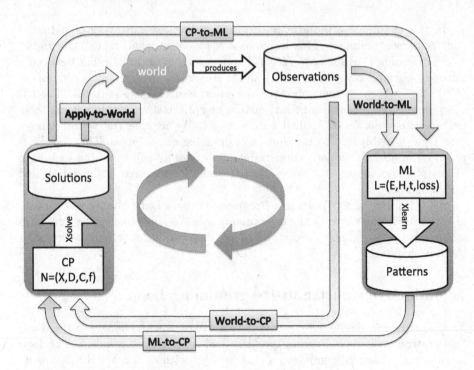

Fig. 1. The inductive constraint programming loop

For example, given real-valued data $E \subset \mathbb{R}^d$ and real-valued labels identified by target function t, where $\forall \mathbf{e} \in E : t(\mathbf{e}) \in \mathbb{R}$, the goal of linear regression is to learn a linear function $h_{\mathbf{c}} : E \to \mathbb{R}$ with coefficients $\mathbf{c}$ that minimizes the sum of squared errors between the predicted value and the observed value: $loss(E, h_{\mathbf{c}}, t) = \sum_{\mathbf{e} \in E} |h_{\mathbf{c}}(\mathbf{e}) - t(\mathbf{e})|^2 = \sum_{\mathbf{e} \in E} |\mathbf{e} \cdot \mathbf{c} - t(\mathbf{e})|^2$. Many other loss functions and hypothesis spaces have been defined in the literature.

The ML component is composed of the learning problem $L = (E, H, t, loss)$, the learner XLearn, and a Patterns repository. XLearn learns hypotheses t (typically one) and writes them in the Patterns repository.

More information about data mining and machine learning can be found in the second chapter of this book.

World. The World component is composed of a world W, an evaluation function $eval\_world$, and a Observations repository. The world W can have its own independent behavior, dynamically changing under the effect of time and the effect of applying solutions of the Solutions repository. The solutions are evaluated by the $eval\_world$ function and this feedback is stored in the Observations repository.

Now that we have defined the basis of the inductive constraint programming loop, we need to define the way the CP component, the ML component and the world interact with each other. They interact through a set of reading/writing functions.

An *inductive constraint programming loop* is composed of a world (W, *eval_world*), a CP component (N, Xsolve), and an ML component (L, XLearn). The loop uses the following channels of communication:

- function World-to-ML reads data and evaluations from the Observations repository and updates the learning problem L, that will be used by XLearn to learn a hypothesis h;
- function CP-to-ML is used to send feedback from the previous iteration of the CP component to the ML component, e.g. when Xsolve cannot find any satisfactory solution to be applied to the world;
- function World-to-CP reads data from the Observations repository that can be used to directly update the constraint network N used by Xsolve;
- function ML-to-CP reads patterns from the Patterns repository and updates the constraint network N used by Xsolve to produce solutions;
- function Apply-to-World takes solutions in the Solutions repository and applies them to the world, if possible.

The following pseudo code demonstrates how these communication channels are used in the inductive constraint programming loop:

Algorithm 1. Pseudo code of a loop cycle using the components.

```
function CYCLE(Observations, optional Solutions)
    repeat
        L_o ← World-to-ML(Observations)
        L_p ← CP-to-ML(Solutions)
        L ← constructL(L_o, L_p)
        Patterns ← applyXlearn(L)

        N_o ← World-to-CP(Observations)
        N_p ← ML-to-CP(Patterns)
        N ← constructN(N_o, N_p)
        Solutions ← applyXsolve(N)
    until Apply-to-World(Solutions)
end function
```

Initially, World-to-ML is used to gather training data to the ML component. These data can be feedback from previous executions of solutions of the CP component on the world. The solution of the previous cycle can also directly be used as well, through CP-to-ML. This is especially useful if the previous solution could not be applied to the world, for example because the learned patterns lead to an inconsistency. Using the output of World-to-ML and CP-to-ML, the learning problem L can then be constructed, specific to the learner at hand. Next, the learner is applied to L and patterns are obtained. These patterns can be weights of an objective function, constraints, or any other type of structural information that is part of the CP problem.

A similar process then happens for the CP component, the network is constructed using the output of World-to-CP and ML-to-CP, after which the solving method is used and solutions are obtained.

These solutions are then applied to the world using Apply-to-World. As mentioned before, it may be that the found solution (or non-solution) is not applicable to the world. In that case, a new iteration of the loop is started immediately which bypasses the world. Otherwise the solutions are applied to the world, after which a new cycle with new observations can be started.

We can observe that there is no direct link between the ML component and the world. Our framework is indeed devoted to solving combinatorial problems such as scheduling and routing, revising them based on feedback from the world; it does not aim to only classify or predict events in the world.

3   Conclusions

The key idea in the inductive constraint programming (ICON) loop is that the CP and ML components interact with each other and with the world in order to adapt the solutions to changes in the world. This is an essential need in problems that change under the effect of time, or problems that are influenced by the application of a previous solution. It is also very effective for problems that are only partially specified and where the ML component learns from observation of applying a partial solution, e.g. in the case of constraint acquisition.

The subsequent chapters will provide a number of examples of the use of the ICON loop.

References

[BH04]   Bent, R., Hentenryck, P.: Online stochastic and robust optimization. In: Maher, M.J. (ed.) ASIAN 2004. LNCS, vol. 3321, pp. 286–300. Springer, Heidelberg (2004). doi:10.1007/978-3-540-30502-6_21

[BHO09] Bessiere, C., Hebrard, E., O'Sullivan, B.: Minimising decision tree size as combinatorial optimisation. In: Gent, I.P. (ed.) CP 2009. LNCS, vol. 5732, pp. 173–187. Springer, Heidelberg (2009). doi:10.1007/978-3-642-04244-7_16

[CJSS12] Coquery, E., Jabbour, S., Saïs, L., Salhi, Y.: A SAT-based approach for discovering frequent, closed and maximal patterns in a sequence. In: Proceedings of the 20th European Conference on Artificial Intelligence (ECAI 2012), Montpellier, France, pp. 258–263. IOS Press (2012)

[DD88]   Dechter, R., Dechter, A.: Belief maintenance in dynamic constraint networks. In: Proceedings of the 7th National Conference on Artificial Intelligence (AAAI 1888), St. Paul, MN, pp. 37–42. AAAI Press/The MIT Press (1988)

[DGN08] De Raedt, L., Guns, T., Nijssen, S.: Constraint programming for itemset mining. In: Proceedings of the 14th ACM SIGKDD International Conference on Knowledge Discovery and Data Mining (KDD), Las Vegas, Nevada, pp. 204–212. ACM (2008)

[EF01]   Epstein, S.L., Freuder, E.C.: Collaborative learning for constraint solving. In: Walsh, T. (ed.) CP 2001. LNCS, vol. 2239, pp. 46–60. Springer, Heidelberg (2001). doi:10.1007/3-540-45578-7_4

[KBC10] Khiari, M., Boizumault, P., Crémilleux, B.: Constraint programming for mining n-ary patterns. In: Cohen, D. (ed.) CP 2010. LNCS, vol. 6308, pp. 552–567. Springer, Heidelberg (2010). doi:10.1007/978-3-642-15396-9_44

[Wal02] Walsh, T.: Stochastic constraint programming. In: Proceedings of the 15th Eureopean Conference on Artificial Intelligence (ECAI 2002), Lyon, France, pp. 111–115. IOS Press (2002)

[XHHL08] Xu, L., Hutter, F., Hoos, H.H., Leyton-Brown, K.: Satzilla: portfolio-based algorithm selection for SAT. J. Artif. Intell. Res. (JAIR) 32, 565–606 (2008)

ICON Loop Carpooling Show Case

Mirco Nanni[1](✉), Lars Kotthoff[3], Riccardo Guidotti[2], Barry O'Sullivan[4], and Dino Pedreschi[2]

[1] KDD Lab, ISTI-CNR, Pisa, Italy
mirco.nanni@isti.cnr.it
[2] KDD Lab, CS Department, University of Pisa, Pisa, Italy
{riccardo.guidotti,pedre}@di.unipi.it
[3] University of British Columbia, Vancouver, Canada
larsko@cs.ubc.ca
[4] University College Cork, Cork, Ireland
b.osullivan@cs.ucc.ie

Abstract. In this chapter we describe a proactive carpooling service that combines induction and optimization mechanisms to maximize the impact of carpooling within a community. The approach autonomously infers the mobility demand of the users through the analysis of their mobility traces (i.e. Data Mining of GPS trajectories) and builds the network of all possible ride sharing opportunities among the users. Then, the maximal set of carpooling matches that satisfy some standard requirements (maximal capacity of vehicles, etc.) is computed through Constraint Programming models, and the resulting matches are proactively proposed to the users. Finally, in order to maximize the expected impact of the service, the probability that each carpooling match is accepted by the users involved is inferred through Machine Learning mechanisms and put in the CP model. The whole process is reiterated at regular intervals, thus forming an instance of the general ICON loop.

1 Introduction

Carpooling, i.e., the act where two or more travellers share the same car for a common trip, is an old idea brought forward, among many others, to reduce traffic and its externalities. If a large proportion of travellers, especially daily commuters, would adopt carpooling, a substantial traffic reduction could indeed take place. However, experiences from many projects internationally have shown that it is extremely difficult to boost the adoption of carpooling to levels that significantly diminish traffic as a whole. There are many reasons why this happens: psychological, organizational, technological. As a matter of fact, we do not know much yet about the real carpooling potential that emerges from people's mobility—a very preliminary step towards designing the right mechanisms and incentives for a successful carpooling system. Nevertheless, we now have access to the data to observe individual mobility at microscopic level and for large populations of travellers, such as the digitised trajectories of vehicular travels recorded by GPS-enabled on-board devices. These forms of *big data* have been used in

© Springer International Publishing AG 2016
C. Bessiere et al. (Eds.): Data Mining and Constraint Programming, LNAI 10101, pp. 310–324, 2016.
DOI: 10.1007/978-3-319-50137-6_13

[12] to discover the mobility profiles of individual travellers, and to understand when two individuals have compatible matching needs, so that they can share part of their travels.

In the present work we pursue this approach further by exploiting the combination of mobility data mining, machine learning methods and constraint programming in an iterative process that follows the general "ICON loop" schema. Through the analysis of mobility data from a community of travellers in a given territory, we construct the *network of potential carpooling* for that community, where nodes correspond to the users and each link between user u and user v corresponds to the fact that u can take a lift from v, because there is a trip in v's profile that can serve u (u can be a passenger of driver v). Then, a globally optimal matching between potential drivers and potential passengers is performed, aimed to minimize the number of vehicles needed to perform all trips. Such assignments become potential suggestions for the users involved, which can agree or reject them. In order to better target suggestions with higher chances of being accepted, a user acceptance model is built and continuously updated through the analysis of previous iterations of the loop – i.e. previous assignments, labelled with the outcome of the suggestion (accept vs. reject) – and its predictions are used to evaluate the probability of success of each potential carpooling match. Such probabilities are then used in the optimization step, in order to achieve a maximum *expected* number of *accepted* suggestions.

The proposed approach has two main application levels. The first one consists in the actual implementation of a collective carpooling service, with real users involved in the whole process. The second one provides what-if analyses where the potential outcome of such a service – in terms of overall success of the service in itself, or its impact on the traffic – is measured through simulations. In this chapter, after describing the overall methodology, we will provide some examples of the latter kind of application, by simulating a carpooling service over a real dataset of mobility data and over various types of users – each representing a different attitude towards carpooling.

The following descriptions will start from a simplified view of the problem, where we basically assume that all users will accept any carpooling suggestion. Then, a refinement of the process will be illustrated, which takes into consideration the users' behaviours in terms of acceptance and rejection of carpooling suggestions.

2   Related Work

Carpooling is the second most popular way of commuting, and maybe one of the least understood. This probably explains such a large corpus of studies in the literature.

Many works have been devoted to study the carpooling phenomena. In [11], the authors describe the characteristics of carpoolers, distinguishing among different types of carpooler, and identifying the key differences between carpoolers. In [12], the methodology for extracting the mobility profiles used also in this work is introduced, and the criteria to match common routes. Something similar is illustrated in [5]. The authors extract home and work locations, and the social ties among the users for matching the users according to similar mobility patterns. [6] studies how to overtake the psycological barriers associated with riding with strangers and exploit it to find compatible matches for traditional groups of users and to find rides in alternative groups.

An approach widely followed for analyzing carpooling is the agent based model (ABM). In [2] an ABM is designed to optimize transports by the ride sharing of people who usually cover the same route. The information obtained from this simulator is used to study the functioning of the clearing services and business models. In [3] the authors use a multi-ABM to investigate opportunities among simulated commuters and by providing an online matching for those living and working in close areas. [4] present a conceptual design of an ABM for the carpooling application to simulate the interactions of autonomous agents and to analyze the effects of changes in factors related to the infrastructure, behavior and cost.

In other studies the authors try to find simulated or theoretical matches among users asking for a ride in a carpooling scenario and evaluate it in terms of simulated users' feedbacks. [9] develops the concept of real carpooling by allowing a large base of member passengers and drivers that declared their route to be matched against each other automatically using mobile phone calls. [8] considers simulated straight-line trajectories observing only origin and destination of trips and classifies users as eligible or ineligible for carpooling by minimizing the time of the trip. In [7] the authors build the users' network with edges labeled with the probability of negotiation success for carpooling to represent planned periodic trips. The probability values are calculated by a learning mechanism using the registered person features, the trip characteristics, and the negotiation feedback. The algorithm provides advice by maximizing the expected value for negotiation success.

3   Simple Carpooling Loop

A first formulation of a proactive carpooling service is summarized in Fig. 1. First, the system acquires the recent mobility history of the users that joined the service, in the form of trips (or trajectories) performed by car in that period. Then, such data is analyzed to extract, for each user, her mobility profile, described as the set of typical trips. At this point a network is built, which describes the possible pairs of trips (from different users) that could carpool, i.e. one of the users can become a passenger for the other. Allocating passengers to drivers is a non-trivial optimization problem, which is solved through a constraint model. The resulting allocation is then the input for the users of the service, who can choose to accept to adopt the suggested pairings or to reject them. The process is iterated from the beginning after a predefined observation period, at the end of which a new mobility history is collected for each user.

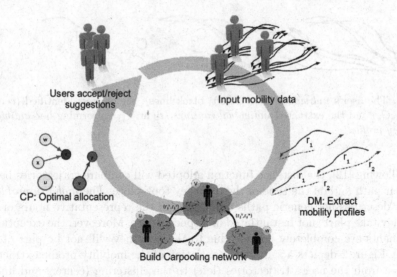

Fig. 1. Basic instantiation of the ICON loop for the carpooling scenario.

3.1   Inferring Mobility Profiles

Given a set of *users*, their mobility can be described by the set of trips performed in the period of analysis. Each trip, then, is defined by a trajectory, i.e. a sequence of spatio-temporal points:

Definition 1 (Trajectory). A *trajectory* T is a sequence of spatio-temporal points $T = \langle (x_1, y_1, t_1), \ldots, (x_n, y_n, t_n) \rangle$, where x_i and y_i $(1 \leq i \leq n)$ are the coordinates of the i-th point and t_i is its corresponding timestamp, with: $\forall 1 \leq i < n : t_i < t_{i+1}$.

The set of all the trajectories travelled by a user u makes her *individual history*.

Using the above definitions and following the profiling procedure proposed in [12], we can retrieve the systematic movements of a certain user u. The method consists in clustering the trajectories of the user by means of an ad hoc *distance function* that defines the concept of trajectory similarity to be adopted. In particular, a density-based clustering method is adopted, basically a variant of the generic OPTICS method [1] that breaks down large (i.e. too extensive) clusters into smaller compact ones. Two trajectories closer than a given threshold will be considered similar and contribute to the same mobility behaviour.

The result of the process is a partitioning of the original dataset of user's trajectories, from which we filter out the *clusters* with few trajectories (statistically non significant behaviors) and the trajectories that are noise (specifically detected by the clustering algorithm). Finally, for each valid cluster remaining, we extract a *representative trajectory*, which is called a *routine*. The set of all routines of a user is called her *mobility profile*.

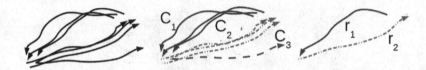

Fig. 2. The user's *individual history* (left: black lines), the clusters identified (center: C_1, C_2, C_3) and the extracted *individual routines* (right: r_1, r_2) forming her *individual mobility profile*.

Following [12], the distance function adopted will compare trajectories based on their path and on the time of the day they took place. The *mobility profile* of a user describes her generic paths followed, and the representative hours of the day they take place, not instantiated in a specific date. Moreover, the exceptional movements are completely ignored due to the fact they will not be part of the profile. Figure 2 depicts a sample instantiation of the mobility profile extraction process, from the user's trajectories (left) to the clustering (center) and finally to the resulting routines that form her mobility profile.

3.2 Building the Carpooling Network

Starting from the routines which constitutes the user mobility profiles, our first objective is to test whether a routine is *contained* in another. If a routine r_1 is contained in a routine r_2 then the user that systematically follows r_1 could leave her car at home and travel with the user that systematically follows r_2. The relation of routine containment is defined as follows:

Definition 2 (Routine Containment). Given two routines r_1 and r_2, two thresholds $spat_{tol}$ and $temp_{tol}$, we say that r_1 *is contained in* r_2, i.e. contained$(r_1, r_2, spat_{tol}, temp_{tol})$, if there exists a contiguous subsequence r_2' of r_2 such that:

$$dist(start(r_1), start(r_2')) + dist(end(r_1), end(r_2')) \leq spat_{tol} \wedge$$
$$|start\_time(r_1) - start\_time(r_2')| + |end\_time(r_1) - end\_time(r_2)'| \leq temp_{tol}$$

where *dist* represents the Euclidean distance, *start* and *end* extract resp. the first and last location of a routine, and *start_time* and *end_time* do the same for time.

Here, $spat_{tol}$ represents the maximum total distance that the user which *is served* could walk to reach the pick-up point, and to reach her final destination from the drop-off point; and $temp_{tol}$ is the maximum total amount of time that the user which *is served* is allowed to waste, as delay or anticipation w.r.t. her original trip, considering the departure and the arrival time. Figure 3 provides a visual depiction of the containment relation over a simple example. Clearly, routine containment is an asymmetric relation. For instance, when the routines compared have different lengths, the origin of the user which *serves* the other

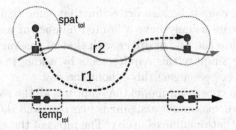

Fig. 3. Sample routines containment: r_1 *is contained* in r_2 because the starting and ending points of r_1 (circles) are spatially and temporally close enough to some points of r_2 (squares).

can be very far from the origin of the one who is *served*, and similarly for the destination point, in which case the second routine does not contain the first one.

A *carpooling network* represents all the links induced by the routine containment relation:

Definition 3 (Carpooling Network). A *carpooling network* $G = \langle N, E \rangle$ is a multigraph where N represents the set of all users, and E is the set of all labeled edges (u, v, r_i^u, r_j^v), where r_i^u is a routine of user $u \in N$, r_j^v is a routine of user $v \in N$, and r_i^u is *contained* in r_j^v.

Figure 4 shows a simple example of a carpooling network with three users. The network is a multigraph, since two nodes can be connected by several different edges – in our case, each edge is characterized by the pair of matching routines it represents.

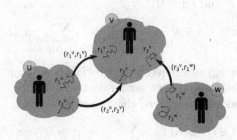

Fig. 4. Sample carpooling network involving three users and 7 routines.

3.3 Optimal Drivers-Passengers Matching

The carpooling network describes the set of all possible pairings of a passenger with a driver. Clearly, not all of them can take place in the real world, since that might violate some physical constraints. For instance, if a user is a passenger for another one, she cannot be also driver for that same trip, since she left the car at home. Also, cars usually have a limited number of seats. Finally, a user cannot freely alternate the role of driver with that of passenger, since she needs to pick her car – we simplify this requirements by asking that a user keeps the same status (driver or passenger) throughout the day.

Here we will see a constraint model that captures all the requirements needed to make a set of driver-passenger assignments correct, and therefore look for the admissible solution that maximizes utility. The input of the model are the edges in the carpooling network. Here we assume that each car can hold up to 5 people, including the driver. Everybody in the network must be either a driver or a passenger and assigned to a driver's car (maybe her own).

Let $G = (V, E)$ be the carpooling network. Let $trips(n)$ be the function that returns the trips t user n participates in and $lift(n, t)$ the function that returns the nodes n' that can give a lift to n on trip t and $lift'(n, t)$ its dual, i.e. the function that returns the nodes n' that can *receive* a lift from n on trip t.

Variables. We consider five groups of variables:

- A set of Boolean variables D: $\forall n \in V, t \in trips(n) : d_{nt} = 1$ iff node n is a driver on trip t.
- A set of Boolean variables DA: $\forall n \in V, t \in trips(n) : da_{nt} = 1$ iff node n drives alone on trip t.
- A set of Boolean variables P: $\forall n \in V, t \in trips(n) : p_{nt} = 1$ iff node n is a passenger in someone else's car on trip t.
- A set of Boolean variables U: $\forall n \in V, t \in trips(n), n' \in lift(n, t) : u_{ntn'} = 1$ iff n is a passenger with n' for trip t.
- An integer variable $sum = \sum D + \sum DA$.

Constraints. Any solution (i.e. variable assignment) should satisfy the following set of constraints. First, a set of *channelling constraints* define the relation between passengers P and trips U:

$$\forall n \in V, t \in trips(n) : p_{nt} = \sum_{n' \in lift(n,t)} u_{ntn'} \tag{1}$$

Notice that variables p_{nt} are requested to be Boolean, therefore each user can be passenger of at most one other user for each of her trips.

If somebody is taking passengers, they must be a driver:

$$\forall n \in V, t \in trips(n) : \sum_{n' \in lift'(n,t)} u_{n'tn} > 0 \iff d_{nt} = 1 \tag{2}$$

At most 4 passengers (5 people altogether) in a car:

$$\forall n \in V, t \in trips(n) : \sum_{n' \in lift'(n,t)} u_{n'tn} \leq 4 \tag{3}$$

Everybody has to be either driver or passenger:

$$\forall n \in V, t \in trips(n) : p_{nt} = 1 \oplus d_{nt} = 1 \tag{4}$$

If somebody is a driver and has no passengers, they drive alone:

$$\forall n \in V, t \in trips(n) : \left(d_{nt} = 1 \wedge \sum_{n' \in lift'(n,t)} u_{n'tn} = 0 \right) \iff da_{nt} = 1 \quad (5)$$

Everybody should be driver for all or none of their trips:

$$\forall n \in V : \sum_{t \in trips(n)} d_{nt} = 0 \vee \sum_{t \in trips(n)} d_{nt} = \|trips(n)\| \quad (6)$$

Objective. Finally, we look for the admissible solution that minimizes the variable *sum*, defined above, representing the number of drivers plus users driving alone.

Notice that the objective function *sum* not only counts the drivers to minimize the number of circulating vehicles. It also contains an additional contribution for those that travel alone (DA), meaning that solutions involving more users are preferred, thus introducing the diffusion of carpooling in the optimization criteria, which is one way to make the carpooling community larger and stronger, and the service more robust and successful in the long term.

4   Users' Preference Learning in Loop

The process described in the previous section aims to extract a set of carpooling assignments that turns as many users as possible into passengers, since each one represents a car saved from circulating. However, in a real world not all such assignments would be accepted and implemented by the users involved. Indeed, several factors might induce a user to reject a suggested match, such as incompatibilities with the assigned mate, specific need of using her own vehicle (therefore preventing her from being a passenger), or excessive delays/effort to reach and get on board of the driver's vehicle. As a result, a tentative assignment that simply maximizes the number of suggested pairings (as in the model shown in Sect. 3.3) might actually suggest the wrong ones, resulting in massive rejection on the side of the users.

As solution to this problem, here we propose to learn from previous iterations of the loop, i.e. to extract from the feedback provided by the users (basically, whether they accepted the suggestions they received, or not) a model to estimate the success probability of future matches. Such probabilities are then used in the matching phase.

This leads to a modification of the loop, as shown in Fig. 5, which is enriched with some components (highlighted in yellow in the figure). During each iteration, the answers of the users are stored to form a training dataset (top-left of the figure), from which to learn a success model for matches (lower-right) that is deployed to add weights to the carpooling network (bottom). Then, the constraint matching model will be modified accordingly (left).

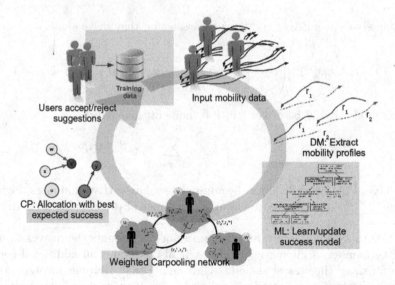

Fig. 5. Instantiation of the ICON loop for the carpooling scenario, including users preference learning. The changes over the basic instance of Fig. 1 are highlighted.

4.1   Preference Learning

In this work we aim to associate to each candidate match – i.e. pair of routines, each associated to its respective user – a probability of success. To this purpose, we employ a very simple model, called Probability Estimation Tree, that basically extends traditional decision trees by returning on each leaf a class distribution instead of a crisp class selection. In our case, each leaf will return the probability of a success and its complementary probability of failure. While several improvements exist to build more reliable estimates – e.g. by adapting the tree construction procedure or by smoothing the probabilities on the leaves – we will adopt the simple solution of extracting a decision tree through standard C4.5 and then compute the associated probabilities on the leaves, without any post-processing.

Predictive Features. The features available to predict the success of a match, include three categories:

Quality of the match. In this group we have measures that describe the ease -of sharing the vehicle for a trip. In particular, we include: the distance to walk for the pick-up and drop-off, the delay caused to the passenger, time spent traveling together, whether the two routines start or end together. In addition, also the distance between the residence locations and the working places of the users are included.

Personal features. For both the users involved, we consider age, gender, marital status, occupation, whether they smoke, have children or animals.

Past usage of carpooling. Here we collect the statistics about the feedbacks of the single user (how often he accepted to carpool as a passenger or as a driver) and the pair of users involved in the match (how frequently they accepted to travel together, if they already happened to be matched).

Training Set. At each iteration the system suggests a set of matches to (part of) the users. For each of them, we store an input instance of the training set, containing all the features associated to the match (predictive features) and the feedback received from the users involved (target variable, having values "success" or "failure"). This way, the training set grows at each iteration. An important fact to observe is that only the matches selected by the drivers-passengers matching phase will generate an instance for the training set, since they are the only ones which were submitted to the users. Therefore, only the routines and their associated users that have been involved in such matches contribute to the preference learning task.

Learning and Deploying the Model. At each iteration the preference model is learned using the most recent training set available. Then, as soon as the carpooling network has been built, the model is used to compute a weight for each edge in the network. In particular, in order to apply the preference model we need to compute all the predictive features associated to each edge on the network, based on the information we have about the users involved and the routines that are matched. The result is an estimate for the success probability of each match in the network, to be used in the drivers-passengers assignment phase, which is presented (in revised form) in the next section.

4.2   Preference-Aware Matching Model

The model presented in Sect. 3.3 is revised in order to change the optimization criterion. In particular, instead of maximizing the number of matches (which correspond to the number of passengers, as well as to the cars saved from circulating) now we aim to maximize the *expected* number of *successful* matches. That can be obtained by simply maximizing the value $\sum_{e \in S} p(e)$, where S is the set of matches returned as solution, and $p(e)$ is the success probability of match e (recall that in the previous step of the loop we obtained an estimate of such value through machine learning).

Then, the new model is basically the same we had in Sect. 3.3, except that (i) now we do not need the variables in DA (tracking users that drive alone, whose count was part of the objective function) as the corresponding channeling constraints; (ii) we assume to have a function $prob_{est}(t)$ for each trip/match t, that provides its success probability; and (iii) the objective function is modified into the following:

$$\max \left(\sum_{n \in V, t \in trips(n)} prob_{est}(t) \cdot p_{nt} \right) \tag{7}$$

5   Simulation of a Carpooling Service

The carpooling loop illustrated so far is designed having in mind two kinds of interactions with users: (i) all users provide data about their mobility at regular intervals; (ii) the users involved in a carpooling match answer by accepting or rejecting the suggestion. As we mentioned at the beginning of this chapter, beside running a real carpooling service, the system developed so far can be used to perform simulations aimed to study various facts about carpooling. For instance, that can be used to simply evaluate the potential impact of carpooling on areas with different features (large urban areas vs. rural ones, areas with one single attractor vs. areas with several ones, etc.), or to understand what would be the impact of specific attitudes of users, such as discrimination based on gender and age or preference towards local travel mates. In the following we describe a possible way of implementing such kind of simulations, also providing a concrete example based on real mobility data and simulated information about the users–including their attitude towards carpooling.

5.1   Simulating Mobility

As a proxy of the mobility for potential users of the carpooling service, we use real GPS traces collected for insurance purposes by *Octo Telematics S.p.A* [10]. The complete dataset contains 9.8 million car travels performed by about 159,000 vehicles active in a geographical area focused on Tuscany, Italy in a period of one month in 2011.

Fig. 6. (Left) The trajectories used for the simulation and (Right) corresponding mobility profiles.

For the purposes of this chapter, we selected a subset of 100 users that move around the city of Pisa. Figure 6 depicts the trajectories of such users, together with the mobility profiles extracted from them (See Sect. 3.1).

In the simulation, the carpooling loop is re-iterated every week, each time recomputing the mobility profiles based on the trips performed in the last 14 days. That basically creates a sliding window on the mobility history of the users, having width of 14 days, which moves one week forward at each step. In these experiments we keep also the trips performed by users during the time they acted as passengers. More sophisticated simulations might omit them, possibly keeping as profile the trips that made her carpool.

The mobility data is used to build the carpooling network, and also to extract all the mobility-based predictive features needed for the preference learning task. That also includes the inference of home and work locations (defined here simply as the two most frequent start locations), used to compute some of the features.

5.2   Simulating Users' Preferences

In order to coherently simulate the behaviour of our users, we need (i) to associate them with a set of individual data (gender, age, etc.), and then (ii) provide a model that evaluates any carpooling proposal based on the three categories of variables described in Sect. 4.1.

The first task was achieved by randomly generating a personal profile for each user, based on the features distribution provided by the national bureau of statistics and other external sources. For the second task, instead, a set of rules have been defined, which compute a score that combines:

- the similarity of the personal profiles of the users (same age, gender, etc.);
- the ease of carpooling together, computed as linear combination of spatial and temporal distance between the matching routines, plus a bonus if they start or end close to each other (basically, the two users can start or end the trip together);
- whether the two users shared a trip in the past, and therefore already know each other.

The weights of the different components can be easily modified, allowing to simulate a rather wide range of behaviours, such as those interested in social compatibility (high weight to the first group of features) or those only focused on efficient and comfortable transportation (second group of features). Also, such rules can be replaced with alternative ones, in order to evaluate the effect of more complex attitudes of the users.

5.3   Results

Here we summarize some sample results that show the typical behaviour of the system. In particular, we adopt a schema of users' preferences that emphasizes the social compatibility. The simulation was run for 5 iterations. Figure 7(left) shows the impact of carpooling at each step, comparing the number of matches suggested to the users and those that were actually accepted.

The results show that, after an initial phase of instability, the system improved the number of successful matches at each iteration. Also the number of suggestions to the users remains smaller than what we had in the first iteration – where the model was equivalent to the simpler version described in Sect. 3.3 – basically showing that the success probabilities are better estimated in later iterations, due to the larger size of the training set. These results also provide a

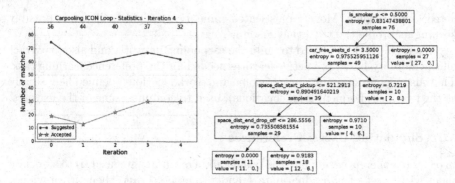

Fig. 7. Results of a sample simulation: number of suggested vs. accepted matches (left); sample users' preference model learned at the second iteration of the loop (right).

comparison between the two models, simple vs. success probability-aware, showing that the latter improves performances significantly. Figure 7(right) shows a sample preference model learned at the second iteration – thus computed over the training data collected at the end of the first iteration. The model is characterized by a mixture of personal features (is_smoker and number of car seats) and trip features (distance to walk for pick-up and for drop-off), as well as by a relatively low purity of the leaves. The models learned during the successive iterations (not shown here for space reasons, since they are significantly larger than the one just described) get more and more sophisticated and accurate. Also, the setup of the learning task ensured that the model never results in extreme overfitting. Below we can have a view of the evolution of such models looking at a ranking of most relevant features, ranked by the cumulative entropy gain yielded by each attribute:

Iteration 0
is_smoker_p : 0.51763342041
car_free_seats_d : 0.196822768067
space_dist_end_drop_off : 0.161445930025
space_dist_start_pickup : 0.124097881498
time_dist_start_pickup : 0.0
last_accepted_pair : 0.0
l1_l1_dist : 0.0
age_d : 0.0
gender_p : 0.0
has_children_p : 0.0

. . .

Iteration 4
last_accepted_pair : 0.300609683595
%_accepted_pair : 0.18422352604
gender_d : 0.121782490916
is_smoker_d : 0.096830535215
l1_l1_dist : 0.0947711528021
is_smoker_p : 0.0921934235296
age_p : 0.0549409842076
gender_p : 0.0396236591312
time_dist_start_pickup : 0.00874162379163
car_free_seats_d : 0.00628292077177

We see, in particular, that later models (i) make large use of the right features, i.e. those that determine social compatibility of individuals, which most influence the acceptance probability; and (ii) they also start exploiting the outcomes of previous iterations (e.g. the two top ones, describing when the pair successfully carpooled and their percentage of success), basically recognizing the successful pairs of the past, which constitute very good candidate for a successful pair in the present iteration.

6   Conclusions

In this work we developed a tool to implement or simulate a proactive carpooling service that combines induction and optimization mechanisms to maximize the impact of carpooling within a community. The system opens a variety of ways to explore, both at the social and technical level. Among them, we mention the following questions that could be studied: what is the impact, in terms of improved traffic, of specific attitudes towards carpooling, especially focusing on those that can be emphasized through appropriate incentives? Does the geographical location and scale of the service influence the performances – in other terms: should carpooling be organized locally, or at a regional/national level? What are the best models and algorithms for preference learning? Also, should the system select only users most likely to succeed (as it happens in the present version) or should it also include other ones, with the purpose of getting better training sets and therefore improve the preference learning component (that looks to have relations with active learning schemata)?

While the experiments described here were conducted over a specific set of users' preferences, the approach can be easily applied over several different scenarios, in order to evaluate which kind of behaviors can affect the success of carpooling the most. Also, some parameters of the model, such as the number of available seats in a car or the distance that users are willing to walk for carpooling, can be explored in order to understand which are critical (and therefore actions might be taken at the public level to influence them in the real world), and which are not.

Finally, while our simulations were based on vehicle data, other data sources, such as smartphone GPS traces, might well replace them, and possibly overcome some of their limitations – e.g. the lack of data when a user is not using the car, for instance because she is carpooling.

References

1. Ankerst, M., Breunig, M.M., Kriegel, H.P., Sander, J.: OPTICS: ordering points to identify the clustering structure. In: ACM Press, pp. 49–60. ACM Press (1999)
2. Armendáriz, M., Burguillo, J., Peleteiro, A., Arnould, G., Khadraoui, D.: Carpooling: a multi-agent simulation in Netlogo. In: Proceedings of ECMS (2010)
3. Bellemans, T., Bothe, S., Cho, S., Giannotti, F., Janssens, D., Knapen, L., Körner, C., May, M., Nanni, M., Pedreschi, D., et al.: An agent-based model to evaluate carpooling at large manufacturing plants. Procedia Comput. Sci. 10, 1221–1227 (2012)
4. Cho, S., Yasar, A.-U.-H., Knapen, L., Bellemans, T., Janssens, D., Wets, G.: A conceptual design of an agent-based interaction model for the carpooling application. Procedia Comput. Sci. 10, 801–807 (2012)
5. Cici, B., Markopoulou, A., Frias-Martinez, E., Laoutaris, N.: Assessing the potential of ride-sharing using mobile and social data: a tale of four cities. In: Proceedings of the 2014 ACM International Joint Conference on Pervasive and Ubiquitous Computing, pp. 201–211. ACM (2014)

6. Correia, G., Viegas, J.M.: Carpooling and carpool clubs: clarifying concepts and assessing value enhancement possibilities through a stated preference web survey in Lisbon, Portugal. Transp. Res. Part A: Policy Pract. **45**(2), 81–90 (2011)
7. Knapen, L., Keren, D., Yasar, A.-U.-H., Cho, S., Bellemans, T., Janssens, D., Wets, G.: Estimating scalability issues while finding an optimal assignment for carpooling. Procedia Comput. Sci. **19**, 372–379 (2013)
8. Lerenc, V.: Increasing throughput for carpool assignment matching. US Patent App. 13/329,899, 19 December 2011
9. Massaro, D.W., Chaney, B., Bigler, S., Lancaster, J., Iyer, S., Gawade, M., Eccleston, M., Gurrola, E., Lopez, A.: Carpoolnow-just-in-time carpooling without elaborate preplanning. In: WEBIST, pp. 219–224 (2009)
10. 2014. http://www.octotelematics.com/it
11. Teal, R.F.: Carpooling: who, how and why. Transp. Res. Part A: Gen. **21**(3), 203–214 (1987)
12. Trasarti, R., Pinelli, F., Nanni, M., Giannotti, F.: Mining mobility user profiles for car pooling. In: Proceedings of the 17th ACM SIGKDD International Conference on Knowledge Discovery and Data Mining, pp. 1190–1198. ACM (2011)

ICON Loop Health Show Case

Barry Hurley[1], Lars Kotthoff[2], Barry O'Sullivan[1], and Helmut Simonis[1(✉)]

[1] Insight Centre for Data Analytics, University College Cork,
Cork, Ireland
h.simonis@4c.ucc.ie
[2] University of British Columbia, Vancouver, Canada

Abstract. In this document we describe the health show case for the ICON project. This corresponds to Task 6.2 in WP 6 of the Description of Work for the project. The description provides a high-level abstraction, detailed description of the interfaces between modules, and a description of sample data.

1 Overview

A hospital collects data on its patients in a patient management system (PMS). Based on this data, ideally an operational schedule for the hospital would be created; for instance, in an ideal hospital no beds would be unoccupied and patients would not have to be on a waiting list for a long period. Due to the dynamic nature of the hospital, planning is however difficult. By analysing the historical data of the hospital, we can discover weaknesses in its planning policies; this in turn can lead to recommendations for improved planning, which can be implemented in improved planning systems.

This show case considers a scenario where the machine learning component learns to predict the duration of various tasks in a some standard workflows. Predictions are made based on the types of tasks and workflow involved, and various patient attributes such as mobility, age, diagnosis, etc. Subsequently, these duration predictions are passed to a constraint programming module which schedules the execution of the tasks while obeying workflow precedences and resource consumption constraints such as a nurse, consultants, x-ray, dialysis machines and so on. The schedule is subsequently simulated with actual task durations. Inaccuracies in the duration prediction may lead to knock-on effects where tasks are delayed awaiting a resource to be freed or a preceding task to finish. Such delays can have a detrimental effect on the ability to release a patient during the scheduling window. Figure 1 presents a high level overview of this flow.

Data are exchanged by files, describing the data valid at a given time in the world. The world maintains its own data for historical information, and for the evaluation of the results produced.

© Springer International Publishing AG 2016

C. Bessiere et al. (Eds.): Data Mining and Constraint Programming, LNAI 10101, pp. 325–333, 2016.
DOI: 10.1007/978-3-319-50137-6_14

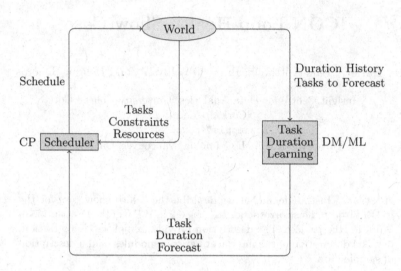

Fig. 1. High level overview of health show case

2 World

Figure 2 shows a more detailed structure of the assumed world.

2.1 Workflow

The workflow description provides a job schema for the different types of patient in each of the wards. The same workflow in different wards may use different resources, or have other task durations. Workflows are defined by events (start, end, milestone), and by task prototypes (activities), linked by precedence constraints. We implement a subset of XPDL to capture and-split and and-join operations, where multiple tasks can be executed in parallel and the following activity can start when all tasks are finished.

Our resource model assumes constant resource availability and resource consumption. A task may require multiple resources, each over its complete duration. Resources can be disjunctive (resource availability one, each task uses one unit of resource), or cumulative (resource availability is a positive integer, resource use of tasks are integer values as well.

Start and end event mark the beginning and end of a workflow. We can also define milestones with an attached due date. Not meeting the due date of the milestone contributes to the objective cost value.

3 CP Component

3.1 Simple Model

A first model as a finite domain constraint program can be formulated with the following notation:

World

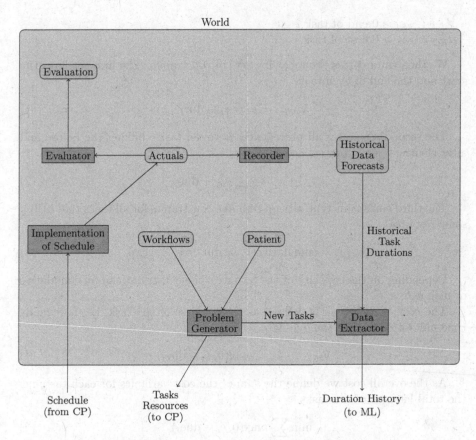

Fig. 2. Detailed view of world for health show case

- T, the set of all tasks, indexed by i
- P, the set of all precedences, indexed by p
- Q, the set of all patients, indexed by k
- R, the set of all resources, indexed by r

We also use the following constants:

- dur_i forecasted duration of task i
- rel_i release date of task i
- due_i due date of task i
- cap_r capacity of resource r
- res_{ir} resource use of task i on resource r
- pat_i patient to whose workflow the task belongs

We introduce finite domain variables for the start, the end and the cost of each task:

- $\forall_{i \in T}: \quad s_i \in [rel_i, \infty]$ start of task i

- $\forall_{i \in T}:$ $e_i \geq 0$ end of task i
- $\forall_{i \in T}:$ $c_i \geq 0$ cost of task i

We then can express the following sets of constraints. The first set links the start and the end of each task.

$$\forall_{i \in T}:   e_i = s_i + \mathrm{dur}_i \tag{1}$$

The second set states all precedences between tasks, linking the before and after elements of the precedence data.

$$\forall_{p \in P}:   s_{a_p} \geq s_{b_p} + \mathrm{dur}_{b_p} \tag{2}$$

The third constraint type sets up resource constraints for all tasks that utilise some resource.

$$\forall_{r \in R}:   \mathrm{cumulative}(< s_i, \mathrm{dur}_i, \mathrm{res}_{ir} >, \mathrm{cap}_r) \tag{3}$$

Depending on the capacity of the resource, the constraint can be cumulative or disjunctive.

The cost of each task is defined as the lateness of the task, i.e. how many time units it ends after the due-date.

$$\forall_{i \in T}:   c_i = \max(0, e_i - \mathrm{due}_i) \tag{4}$$

As the overall cost we define the sum of the cost variables for each task, i.e. the total lateness of the tasks.

$$\min \sum_{i \in T} \max(0, e_i - \mathrm{due}_i) \tag{5}$$

If we are able to schedule all tasks before their due-date, then all patients can be discharged on the day, if the schedule can be implemented without delay. As the durations are forecasted, changes in the duration may mean that in the implemented schedule some tasks will extend beyond the cutoff time.

3.2 Model 2 - Accepting Patients

In our previous model, the tasks for a patient may stretch beyond the cutoff time, this means that the patient can not be discharged in time, and therefore it is pointless to perform the earlier tasks of the workflow, as they will have to be repeated on the next day. These tasks use up resources, which may stop other jobs from finishing in time. A better model introduces a decision variable for each job (Patient), which states if the patient will be discharged on the current day. We can then state that all tasks must finish before their due-date, and use an optional resource use of 0 to exclude rejected jobs from the resource constraints.

The model then takes the following form:

We introduce finite domain variables for the state, the end and the cost of each task:

- $s_i \in [\mathrm{rel}_i, \infty]$ start of task i
- $e_i \in [0, \mathrm{due}_i]$ end of task i
- $x_k \in \{0, 1\}$ patient k is accepted for discharge on the current day

The objective is changed to maximise the number of patients that will be discharged

$$\max \sum_{k \in Q} x_k \tag{6}$$

The resource constraints (Eq. 3) are modified to deal with a variable resource height for each task, depending on the acceptance of the patient.

$$\forall_{r \in R} : \quad \mathrm{cumulative}(< s_i, \mathrm{dur}_i, x_{\mathrm{pat}_i} * \mathrm{res}_{ir} >, \mathrm{cap}_r) \tag{7}$$

The precedence (Eq. (2)) and linkage (Eq. 1) constraints are not changed.

There are two potential issues with the objective function. The first is that it is not clear how good the propagation of the 0/1 variable in the cumulative can work, and how symmetries are affecting the cost function.

The second concern is that we are no longer interested in the robustness of the schedule. If a duration forecast is wrong, this has a different effect of the task is early during the day, or it is the last task in the schedule. In the first case, a delay may be easily recovered, or a knock-on effect may delay not just one, but multiple patients. If the task ends just before the cutoff time, a delay will push the task beyond the deadline. We should therefore consider how we can improve the robustness of the schedule.

A final comment affects the search routine. By introducing both 0/1 and finite domain variables, we can no longer rely on an automated strategy based on domain size, we have to interleave the assignment of the acceptance variables for a patient with the scheduling of all tasks belonging to the patient.

4   ML Component

The input of the ML component consists of two inputs, one containing historical data, the other the new data for which the task duration is to be predicted.

The format of the inputs is nearly identical, except that the regression problem data does not contain the duration field. Each record is one line.

```
task_id (integer)            patient_age (integer)
date (Integer)               patient_sex (0/1)
weekday(integer 0-6)         patient_mobility (0-5)
ward (integer)               patient_attribute_1 (number)
workflow (integer)           ...
workflow_task (integer)      patient_attribute_n (number)
patient (integer)            duration (integer)
```

The attributes may consist of numerical values (blood pressure, heart rate, temperature), 0/1 categories (use of specific medications like blood thinner, which may require extra care), or medical conditions besides the one treated for (diabetes, high blood pressure, pacemaker), which affect the duration of certain tasks.

The output of the ML component is a file containing the duration forecasts for all tasks in the regression problem data.

5 Evaluation

We compare the effectiveness of workflow schedules produced based on two ML components, one which is trained once initially versus a looping model which is retrained as each day passes, thus having access to the more recent, representative data which should, in theory, be able to adapt to recent trends. This hypothesis is tested when new procedures come into affect, effectively changing the distribution of task durations, with some now taking longer to complete and others shorter. Both ML models will start with the same 60 days of historical training data, and the looping model maintains a sliding-window of the same quantity of training data.

The predicted task durations from the two ML components are used by the CP component to schedule task start times while obeying resource capacity, and precedence constraints. The two schedules are then passed back to the world and simulated based on the actual task durations, mismatches in the duration prediction may lead to tasks being delayed in practice, with a combinatorial knock-on effect. We contrast the two schedules to one based on actual task durations provided by an oracle.

Figure 3 shows the cumulative makespan penalty, that is the difference in makespan between the simulated schedules based on the two ML components

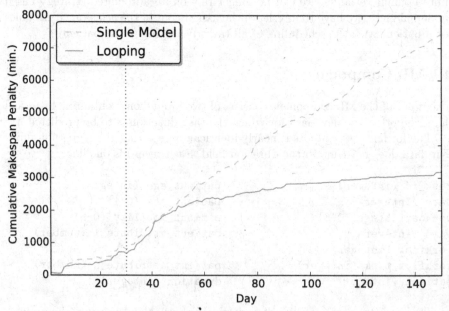

Fig. 3. Scheduling makespan comparison between looping model and a trained once model.

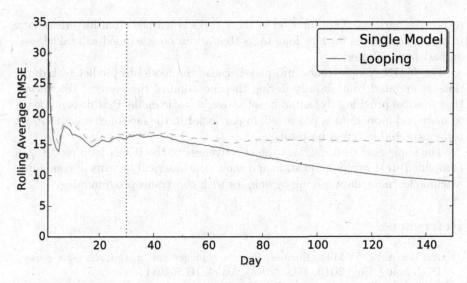

Fig. 4. Prediction accuracy comparison between looping model and a trained once model.

versus scheduling based on the actual task durations provided by an oracle. The vertical line marks the day where new procedures come into affect, changing the distribution of task durations.

In the initial 30 days, before the new procedures, both ML components produce very similar predictions resulting in schedules with comparable makespans. Once the new procedures come into effect, the prediction accuracy of both models deteriorate and the makespan penalty increases more rapidly. However, the looping model slowly adapts as a sufficient quantity of the new data becomes available. With around half of its training data coming from the new distribution, the performance of looping model diverges from the trained-once model, consistently producing more accurate predictions and subsequently, better schedules.

Further evidence for this divergence is provided when we compare the rolling average root mean squared error (RMSE) for the two ML models, Fig. 4. The relative accuracy of the looping model steadily improves as a larger portion of its training data is drawn from more recent, representative days. The model which has been trained once, does not adapt well to the new distributions.

6   Conclusions

We have presented the health show case for the ICON project. It represents an instantiation of the ICON loop, where a machine learning/data mining component and a constraint programming component interact with the outside world and each other.

The show case simulates the workings of part of a simplified hospital; in particular it is concerned with the scheduling of tasks that need to be performed to process patients. The duration of these tasks is learned from historical data to provide estimates for how long tasks that are to be scheduled will take based on patient attributes.

The ICON "loop" comes into play because the model to predict task durations is updated continuously during the operation of the system. We showed that this can provide substantial benefits over a static model that does not learn as more and more data is processed, in particular in the face of changes that are to be expected in a real hospital.

The show case demonstrates the effectiveness of the ICON loop model in a (simplified) real-world application and shows the practical benefits of combining machine learning/data mining techniques with constraint programming.

References

1. van der Aalst, W.M.P.: Business process management: a comprehensive survey. ISRN Softw. Eng. **2013**, 37 p. (2013). Article ID 507984
2. Beliën, J., Demeulemeester, E.: A branch-and-price approach for integrating nurse and surgery scheduling. Eur. J. Oper. Res. **189**(3), 652–668 (2008). http://dx.doi.org/10.1016/j.ejor.2006.10.060
3. Cardoen, B., Demeulemeester, E., Beliën, J.: Operating room planning and scheduling: a literature review. Eur. J. Oper. Res. **201**(3), 921–932 (2010). http://dx.doi.org/10.1016/j.ejor.2009.04.011
4. Decker, K., Li, J.: Coordinated hospital patient scheduling. In: Demazeau, Y. (ed.) Proceedings of the Third International Conference on Multiagent Systems, ICMAS 1998, 3-7 July 1998, Paris, France, pp. 104–111. IEEE Computer Society (1998)
5. Hannebauer, M., Müller, S.: Distributed constraint optimization for medical appointment scheduling. In: Proceedings of the Fifth International Conference on Autonomous Agents, AGENTS 2001, pp. 139–140. ACM, New York (2001). http://doi.acm.org/10.1145/375735.376026
6. Hansson, J., Tolf, S., Øvretveit, J., Carlsson, J., Brommels, M.: What happened to the no-wait hospital? A case study of implementation of operational plans for reduced waits. Qual. Manag. Health Care **1**(21), 34–43 (2012)
7. Lenz, R., Elstner, T., Siegele, H., Kuhn, K.A.: Application of information technology: a practical approach to process support in health information systems. JAMIA **9**(6), 571–585 (2002). http://dx.doi.org/10.1197/jamia.M1016
8. Mans, R.S., Schonenberg, M.H., Song, M., Aalst, W.M.P., Bakker, P.J.M.: Application of process mining in healthcare – a case study in a Dutch hospital. In: Fred, A., Filipe, J., Gamboa, H. (eds.) BIOSTEC 2008. CCIS, vol. 25, pp. 425–438. Springer, Heidelberg (2008). doi:10.1007/978-3-540-92219-3_32
9. Müller, R., Rogge-Solti, A.: BPMN for healthcare processes. In: Eichhorn, D., Koschmider, A., Zhang, H. (eds.) Proceedings of the 3rd Central-European Workshop on Services and their Composition, Services und ihre Komposition, ZEUS 2011, Karlsruhe, Germany, 21–22 February 2011, CEUR Workshop Proceedings, vol. 705, pp. 65–72. CEUR-WS.org (2011). http://ceur-ws.org/Vol-705/paper9.pdf
10. Russell, N., Aalst, W.M.P., Hofstede, A.H.M., Edmond, D.: Workflow resource patterns: identification, representation and tool support. In: Pastor, O., Falcão e Cunha, J. (eds.) CAiSE 2005. LNCS, vol. 3520, pp. 216–232. Springer, Heidelberg (2005). doi:10.1007/11431855_16

11. Schaus, P., Hentenryck, P., Régin, J.-C.: Scalable load balancing in nurse to patient assignment problems. In: Hoeve, W.-J., Hooker, J.N. (eds.) CPAIOR 2009. LNCS, vol. 5547, pp. 248–262. Springer, Heidelberg (2009). doi:10.1007/978-3-642-01929-6_19

12. Svagård, I., Farshchian, B.A.: Using business process modelling to model integrated care processes: experiences from a European project. In: Omatu, S., Rocha, M.P., Bravo, J., Fernández, F., Corchado, E., Bustillo, A., Corchado, J.M. (eds.) IWANN 2009. LNCS, vol. 5518, pp. 922–925. Springer, Heidelberg (2009). doi:10.1007/978-3-642-02481-8_140

13. Wolf, A.: Constraint-based modeling and scheduling of clinical pathways. In: Larrosa, J., O'Sullivan, B. (eds.) CSCLP 2009. LNCS (LNAI), vol. 6384, pp. 122–138. Springer, Heidelberg (2011). doi:10.1007/978-3-642-19486-3_8

14. Wolf, A., Geske, U., Finsterbusch, A., Rothe, M.: Constraintbasierte behandlungsplanung in der dialyse. In: Fähnrich, K., Franczyk, B. (eds.) Informatik 2010: Service Science - Neue Perspektiven für die Informatik, Beiträge der 40. Jahrestagung der Gesellschaft für Informatik e.V. (GI), Band 2, 27.09. - 1.10.2010, Leipzig. LNI, vol. 176, pp. 711–716. GI (2010). http://subs.emis.de/LNI/Proceedings/Proceedings176/article6167.html

ICON Loop Energy Show Case

Barry Hurley, Barry O'Sullivan, and Helmut Simonis[✉]

Insight Centre for Data Analytics, University College Cork, Cork, Ireland
`helmut.simonis@insight-centre.org`

Abstract. This chapter demonstrates the effectiveness of the ICON loop when applied to energy cost optimization in a data centre. The objective is to schedule the execution of customer tasks such that the overall energy cost is minimised. This is complicated by the fact that the real-time energy price is not known a-priori, therefore machine learning techniques are employed to produce a forecast price vector ahead of time. In practice such a forecast needs to adapt to changes in the world affecting the pricing model over time. Therefore, the model needs to adapt in an iterative process, realised by employing the ICON loop approach.

1 Overview

The aim of this showcase is to improve the energy efficiency of data centres through the integration of machine learning (ML) and constraint programming (CP). Consider a cloud grid computing service, where customers contract to run computing services (tasks) throughout the day. Tasks are scheduled within their execution time window and assigned to machines within the data centre. Each task requires certain amounts of resources of different types (e.g. CPU, memory, IO) during their execution. The cumulative resource limit for each resource type of all tasks scheduled on a machine can not exceed the resource capacity of that machine. The objective is to schedule the tasks in such a way that the overall cost of energy used is minimised.

However this is complicated by the fact that large electricity consumers, like a data centre, may use a time-variable electricity tariff, which follows the wholesale market price. This price is not known in advance; in Ireland for example, the price is not known until four days after the event. The price is influenced by variable consumer demand over time, wind energy production, which varies with weather conditions, and the availability of generating plants. The price therefore fluctuates significantly throughout the day (on some days by more than a factor of ten), which provides an opportunity to reduce the energy use during peak price periods and instead perform the work during cheaper periods, as long as the time windows for each task are respected. This requires a forecast of the price ahead of time, so that the tasks can be scheduled accordingly. As the forecast will not be 100% accurate, it might lead to a schedule were tasks are run at times when the forecast price is low, but the actual price is much higher. This increases the eventual cost of the schedule, which is always computed with the actual price.

© Springer International Publishing AG 2016
C. Bessiere et al. (Eds.): Data Mining and Constraint Programming, LNAI 10101, pp. 334–347, 2016.
DOI: 10.1007/978-3-319-50137-6_15

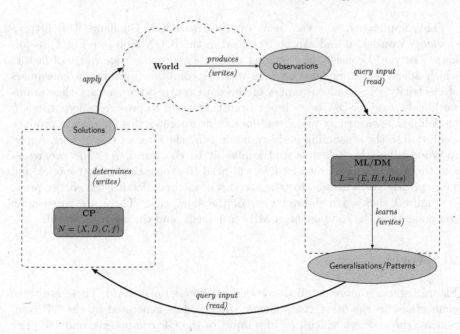

Fig. 1. The inductive constraint programming loop

The model assumes that the duration and resource requirements of all tasks are given accurately by the customers. In order to avoid run-away processes using up all available resources, we assume that tasks that exceed resource limits or their duration significantly are removed by the supervisor process.

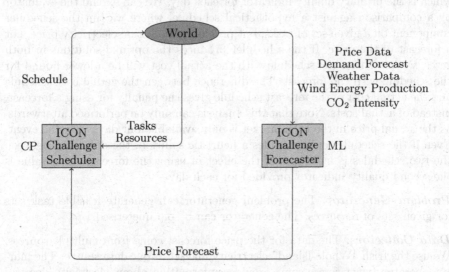

Fig. 2. High level overview of energy show case

This application was the focus of the first ICON Challenge [19] http://challenge.icon-fet.eu and can be modelled in the ICON loop (see Fig. 1) as follows. The world consists of a number of components: the wide range of factors which affect the energy market, like weather conditions, producers/consumers of electricity, etc.; and customers of the data centre who contract the various workloads. read-for-DM takes input from this world to produce a hypothesis h modelling the electricity price. read-for-CP incorporates this forecast to produce a solution to the scheduling problem minimising the forecast energy cost. Apply-to-World takes this schedule and applies it to the world. As time progresses and the world changes, read-for-DM will need to evolve the forecast model, and subsequently the schedule, to take account of factors affecting the energy price.

Figure 2 shows a high-level view of the show case. There are three main components, a CP component, a ML component, and the assumed world.

2   World

Figure 3 shows a more detailed overview of the assumed world. There are three connections to the other components; the schedule generated by the CP component, the task set which provides input of the CP component, and the price data, which are input to the ML component.

2.1   Evaluator

The evaluator takes a generated schedule, and evaluates its cost against the actual price of electricity. This produces the cost of the implemented schedule, which is the primary quality indicator for each day. We can extend the evaluator by a comparison against a hypothetical schedule, where we run the scheduler component on a given set of tasks, but providing the actual electricity price, not a forecast of the price. If the scheduler produces the optimal solutions in both cases, we know that the schedule with the actual cost will be a lower bound for the schedule with the forecast. The difference between the actual cost schedule and the actual cost of the forecast schedule gives the penalty for using a forecast, instead of actual costs. Note that this analysis can only be performed afterwards, as the actual price in the Irish market is only available four days after the event. Even if the scheduler only provides a heuristic solution, the difference between the two schedules is indicative of the effect of using the forecast. This value is the second quality indicator provided for each day.

Problem Generator. The problem generator can generate feasible task sets for given sets of resources. The generator can be parameterised.

Data Collectors. The data for the price forecast come from multiple sources. We use the Irish "Whole-Island" electricity market as our data source. The market operator (www.sem-o.com) provides information about electricity demand and prices for Ireland in half-hour resolution. The national grid operator Eirgrid http://www.eirgrid.com/operations/systemperformancedata/ provides

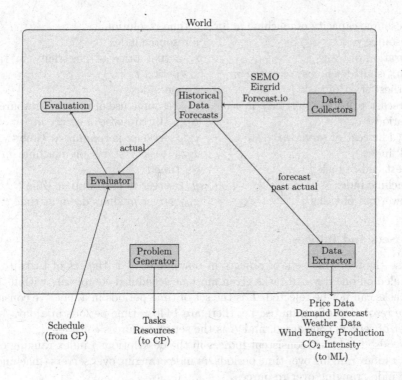

Fig. 3. Detailed view of the energy show case world.

data about wind energy production and CO_2 intensity of the generation. Different services provide weather forecasts and historical weather data, including Weather Underground and forecast.io.

Data Extractor. The data extractor parses the data store for the different collected data, cleans the data, rejecting infeasible data elements, and produces the required price data.

3 Optimisation Component

In this section we introduce a formal description of the CP component of the show case. The component takes as input a set of tasks to schedule, and a price forecast for electricity, and produces a schedule for all tasks. We give a description of the scheduling problem being solved, and present the format of the input and output files.

3.1 Notation

The following entities, sorted alphabetically, are defined in the model.

c_{mr} resource capacity of machine m for resource r

d_j duration of tasks j

down$_m$ shut-down cost of server m

e_j earliest start of task j

f_t forecast price of electricity in time period t

idle$_m$ idle cost of server m

j task index

l_j latest end of task j

m machine index

p_j power use of tasks j

q time resolution

r resource index

r_t actual price of electricity in time period t

t time index

u_{jr} resource use of task j of resource r

up$_m$ startup cost of server m

v_{mt} server m is running at time t

x_{jmt} task j starts on machine m at time t

y_{mt} server m starts up at time t

z_{mt} server m shuts down at time t

3.2 Sets and Indices

We use the following sets of objects in our model: J is the set of tasks to be scheduled on one day. All tasks given must be scheduled completely within that day, tasks can not be rejected. T is the set of time-periods in a day. We consider a time resolution of q minutes, i.e. there are $1440/q$ time periods in a day. M is the set of servers considered, and R is the set of resources considered.

We also try to use consistent indices in the description: j index ranging over tasks, t index ranging over time periods, m index ranging over servers (machines), and r index ranging over resources.

3.3 Constants

q The time resolution in the scheduling model. The value is an integer expressed in minutes, i.e. there are $|T| = 24 * 60/q + 1$ time periods in a day.

d_j The duration of task j, the values are positive integers less than or equal to $|T|$. The duration gives the length of the task such that the end is the start plus the duration. The task is not considered active at its end-time.

e_j The earliest start of task j, the values are integers between 0 and $|T|$.

l_j The latest end of task j, the values are integers between 0 and $|T|$. This latest end corresponds to a latest start of $l_j - d_j$.

u_{jr} The resource use for task j for resource r, this is a positive integer value.

p_j The power use of task j, which is constant during execution of the task. This is a non-negative integer value.

c_{mr} The capacity of server m for resource r, this is a positive integer.

idle$_m$ The idle power of running server m for one time period, this is a non-negative integer number.

up$_m$ The start-up cost of server m. The cost of starting up machine m once. This cost is not dependent on the energy cost at this time. The value is a non-negative floating point number.

down$_m$ The shut-down cost of server m. The cost of shutting down a server once. This cost is not dependent on the energy cost at the shutdown time. The values is a non-negative floating point number.

f_t This is the forecast price of one unit of energy in time period t. This is a floating point number, and can be zero or even negative.

r_t The actual price of one unit of energy in time period t. This is a floating point number, and can be zero or even negative. This information cannot be used to make scheduling decisions, and is only used to evaluate a resulting schedule.

3.4   Model

We now describe the variables, constraints and objective of the model.

Variables.

x_{jmt} Task j starts at time t on machine m. A 0/1 integer variable indicating when and where a task is run. As each task must be scheduled, exactly one of the variables linked to a task must be set to one.

v_{mt} Server m is running at time t. If any task is running on a machine at some time t, the machine must be active.

y_{mt} Machine m starts up at time t. Initially, all machines are off.

z_{mt} Machine m shuts down after time period t. The machine is active at time t, but not at time $t+1$. All machines must be switched off at the end of the scheduling horizon.

Constraints.
We first need to enforce that each task is scheduled on one machine, exactly once:

$$\forall_{j \in J} : \quad \sum_{m \in M} \sum_{t \in t} x_{jmt} = 1 \tag{1}$$

No task can be scheduled before its earliest start:

$$\forall_{j \in J} \forall_{m \in M} \forall_{t < e_{ij}} : \quad x_{jmt} = 0 \tag{2}$$

No task can be scheduled to end after its latest end:

$$\forall_{j \in J} \forall_{m \in M} \forall_{t + d_{ij} > l_{ij}} : \quad x_{jmt} = 0 \tag{3}$$

The resource requirements of all tasks scheduled on the same machine at the same time must fit within the capacity of the machine:

$$\forall_{m \in M} \forall_{r \in R} \forall_{t \in T} : \quad \sum_{j \in J} \sum_{t - d_j < t' \leq t} x_{jmt'} u_{jr} \leq c_{mr} \tag{4}$$

The following constraints all link the different types of variables related to machines.

If a machine is starting at time t, then we consider it running at this time.

$$\forall_{m \in M} \forall_{t \in T} : \quad y_{mt} \Rightarrow v_{mt} \tag{5}$$

If a machine is starting up at time t, then it was not running at time $t - 1$.

$$\forall_{m \in M} \forall_{t \in T} : \quad y_{mt} \Rightarrow v_{mt-1} = 0 \qquad (6)$$

If a machine is shutting down at time t, then it is still running at this time.

$$\forall_{m \in M} \forall_{t \in T} : \quad z_{mt} \Rightarrow v_{mt} \qquad (7)$$

If a machine is shutting down at time t, then it is not running at time $t + 1$.

$$\forall_{m \in M} \forall_{t \in T} : \quad z_{mt} \Rightarrow v_{mt+1} = 0 \qquad (8)$$

If a task is starting at time t on some machine m, then that machine must be active at least while the task is running, i.e. for all time points from t to $t + d_j - 1$.

$$\forall_{j \in J} \forall_{m \in M} \forall_{t \in T} \forall_{t \leq t' < t+d_j} : \quad x_{jmt} \Rightarrow v_{mt'} \qquad (9)$$

If a machine is running at time t, then it was either already running at time $t - 1$ or it starts up at time t.

$$\forall_{m \in M} \forall_{t \in T} : \quad v_{mt} \Rightarrow v_{mt-1} \vee y_{mt} \qquad (10)$$

If a machine is running at time t, then it is either also running at time $t + 1$, or it shuts down at time t.

$$\forall_{m \in M} \forall_{t \in T} : \quad v_{mt} \Rightarrow v_{mt+1} \vee z_{mt} \qquad (11)$$

Objective Function. The objective is to minimise the total cost of operation, consisting of the energy cost running all tasks c_J, the cost of running the servers when they are active c_M, the startup cost of the servers c_{up} and the shutdown cost of the servers c_{down}. This means we minimise the following function

$$\text{cost} := \min c_J + c_M + c_{up} + c_{down} \qquad (12)$$

The energy cost of running all tasks is given by the sum

$$c_J := \sum_{j \in J} \sum_{m \in M} \sum_{t \in T} x_{jmt} \left(\sum_{t \leq t' < t+d_j} p_j r_{t'} q / 60 \right) \qquad (13)$$

Note that we have to convert the power use for the task into an energy value by multiplying with the duration of the time period (in hours). As we don't know the actual cost of electricity r_t when creating the schedule, we may decide to use the forecast price f_t in the optimisation instead. This may mean that an optimal solution for the forecast is not optimal for the actual price. But the final evaluation of the solution quality will be based on the actual price, which is only known after the fact.

The energy cost of running the servers (ignoring the cost of the tasks) is given by

$$c_M := \sum_{m \in M} \sum_{t \in T} v_{mt} \text{idle}_m r_t q / 60 \qquad (14)$$

The start-up cost is given by the sum

$$c_{\text{up}} := \sum_{m \in M} \sum_{t \in T} y_{mt} \text{up}_m \qquad (15)$$

The shut-down cost is given by the sum

$$c_{\text{down}} := \sum_{m \in M} \sum_{t \in T} z_{mt} \text{down}_m \qquad (16)$$

4   Machine Learning Component

In the forecast problem, we have to predict the actual electricity price for one day into the future based on historical and forecast data. The historical data is available from September 2011 onwards. The following fields are defined:

DateTime String, defines date and time of sample
Holiday String, gives name of holiday if day is a bank holiday
HolidayFlag integer, 1 if day is a bank holiday, zero otherwise
DayOfWeek integer (0–6), 0 Monday, day of week
WeekOfYear integer, running week within year of this date
Day integer, day of the date
Month integer, month of the date
Year integer, year of the date
PeriodOfDay integer, denotes half hour period of day (0–47)
WindForecast the forecast wind production for this period
LoadForecast the national load forecast for this period
PriceForecast the price forecast for this period
Temperature the actual temperature measured at Cork airport
Windspeed the actual wind speed measured at Cork airport
CO2Intensity the actual CO2 intensity in (g/kWh) for the electricity produced
ActualWind the actual wind energy production for this period
ActualLoad the actual national system load for this period
ActualPrice the actual price of this time period, the value to be forecast

The last four fields are only available for historical data, i.e. they can not be used to make the forecast. Also note that a model for price prediction is described in further detail in [9,12].

5   Evaluation

This section evaluates the core-benefit and reliance of the ICON loop model in the context of the energy showcase. Specifically, its ability to adapt to a market price increase of 10%. Such a change is not unrealistic as can be seen by the increasingly high volatility of the electricity market [12]. We consider real-world data from the Irish electricity market consisting of system demand forecasts, wind-generation forecasts, market operator price forecasts, along with weather forecasts. The prediction feature set is the same as that used in the ICON Energy Challenge which are a subset of those from [12].

To evaluate the effectiveness of the approach, instances of the scheduling problem described in the previous sections where generated at two day intervals over a period of a year, with 50 tasks, 1 machine, and 2 resources. Subsequently, these were each solved to optimality using the mixed integer programming model, described in Sect. 3.4, using one of three energy price profiles. First, the baseline optimal solution is to schedule the energy usage based on the actual price figures. In reality, this is impractical since it is not known at the time, in the Irish electricity market it is not known until four days after [12], however it will act as the ultimate baseline.

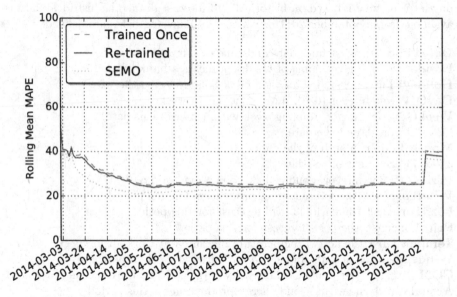

Fig. 4. Rolling mean absolute percentage error of two forecasting models, alongside the market operator's forecast (SEMO).

Second, a linear regression model which is trained on an initial set of historical data. Finally, a model which is retrained each day after new price data becomes available. The latter model, following the ICON loop should have the ability to adapt to changes in the market. Note that both models are trained on the same quantity of historical data (2 years), the difference being the looping model integrates the more recent, representative data. The cost for both forecast model is then reported based on the actual energy cost of implementing such a schedule.

The two regression models produce statistically significantly different price predictions (>95% confidence) over the course of the one year period considered. Figure 4 shows a rolling average of the mean absolute percentage error (MAPE) of the two regression models and the market operator's (SEMO) own forecast. The spike towards the end is due to a very low actual price (<0.3 cent/kWh) affecting the calculation. The difference appears small but the looping model produces more accurate predictions on average, over the course of the year.

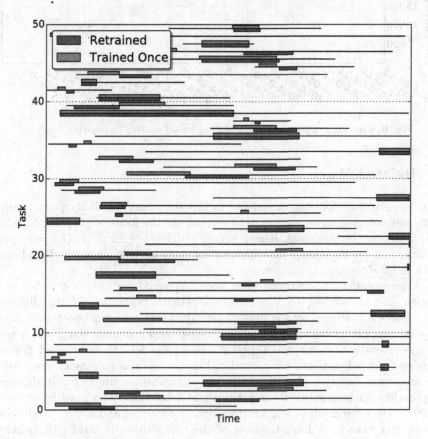

Fig. 5. An example gantt chart of two schedules based on two price forecast profiles.

Differences between the two price prediction models can result in significantly different schedules from the CP model. Figure 5 shows an example schedule produced by the two different price forecasts. The black lines plot the release window of the tasks and the solid boxes show when the task was scheduled to start in the optimal solution based on each forecast, the length of the solid box represents the duration. Understandably, the different price profiles lead to divergent schedules, often with tasks being scheduled at opposite ends of their release windows.

Figure 6 plots the cumulative penalty of scheduling based on the prediction models versus the optimal actual-price schedule, over the period of one year. Scheduling using the looping model produces statistically significantly lower penalty over the single trained-once model (>98% confidence).

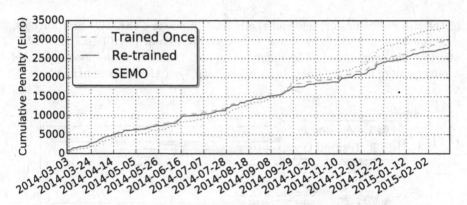

Fig. 6. Cumulative penalty of scheduling based on two forecasting models.

6   Related Work

The two problem domains considered in this showcase, electricity price prediction, and energy management in data centres, are the focus of intensive research activities over the last years. Recent survey articles, [20] for electricity price prediction, and [4] for energy management for data centres, show the breadth of work in these areas.

A large variety of methods have been suggested for electricity price prediction [20]. According to Weron, a competitive evaluation of the different approaches is hindered by the limited availability of common benchmark data, missing supporting data like weather forecasts and economic data, and a lack of agreed evaluation criteria. Different time-spans for the forecasting period, ranging from a few hours for operational use, to several months or even years for long-term capacity planning problems, may also require radically different approaches. An international competition on Global Energy Forecasting was held in 2012 [11], comprising of a hierarchical load forecasting and a wind production forecasting branch. A later instance of the competition in 2014 [10] included a probabilistic hourly price forecast branch [2], as well as a solar production forecasting branch. As discussed in [9], the absolute error in the forecast may not be he most important criterion to minimize when using forecasted prices for making scheduling decisions. Identifying price spikes accurately [8] may be more important, as their impact on the total energy cost is very high.

Energy cost management for datacentres [4] can focus on two objectives, reducing energy use and reducing energy cost. We can try to reduce the total energy use by improving the efficiency of servers and cooling systems, but also by work load consolidation inside and between datacentres [18]. This can take the form of moving virtual machines between servers to use a minimal number of machines at full power, while powering down unused resources [15], or, conversely, by distributing tasks between servers so that hot spots are avoided, and the overall cooling effort can be reduced [7]. Exploiting temperature and humidity differences between datacentre locations allows us to maximize the effect of free cooling, avoiding the expensive use of air handling units.

Wholesale electricity prices vary significantly over time and location, often by more than a factor of ten. We can therefore also try to reduce energy cost by reducing use in locations and times where prices are high, and increase use where and when prices are low [6]. Some computational tasks, especially in high performance computing [1], can be moved in time, while other services, like web-stores, are linked to fixed, time variable demand curves. Moving demand geographically has an impact on latency, where services like gaming and video, need relatively low latency connections, and can therefore only be moved in a limited area [17].

Scheduling of tasks in a computation-heavy data centre is discussed in [3,5], using Constraint Programming techniques for the EURORA supercomputer. In this work, tasks can not only be scheduled in time, but their duration can be changed as well by allocating different numbers of CPU cores and graphics cards to their execution.

The paper most closely related to the use case discussed here, modifying the computational load based on time-variable electricity prices, is [14]. It considers pre-empting certain tasks when the price is high, and resuming them when the price decreases again.

7   Conclusions

We have presented the ICON Energy Show Case, tightly integrating machine learning and constraint programming in a scenario for an energy-efficient data-centre. The objective is to minimize the data-centre's energy cost by scheduling tasks based on the time-variable electricity tariff. Machine learning plays a crucial role in trying to produce a good forecast of the electricity price, however the accuracy of the price prediction does not necessarily correlate with a good overall energy cost. We compare an ICON Loop model where the machine learning component is retrained each day and its ability to adapt to factors affecting the electricity price, versus a static model which is not able to adapt to changes. The looping model demonstrates significant benefits and is able to adapt to market changes.

An additional application of the work presented in this chapter has been applied to the optimisation of energy costs in a large mining company. Specifically, Boliden Tara Mines Ltd. consumed 184.7 GWh of electricity in 2014, equating to over 1% of the national demand of Ireland. Two prediction tasks are undertaken, both employing machine learning techniques. Firstly, a forecast of the real-time energy price is produced, and secondly a prediction of the highly-variable pumping demand is made. Based on these forecasts, an optimisation model produces an operational schedule of the pumps to minimise energy costs. An evaluation using real-world electricity prices and detailed sensor data demonstrates significant savings of up to 10.72% over the year compared to the existing control systems [13].

References

1. Aikema, D., Kiddle, C., Simmonds, R.: Energy-cost-aware scheduling of HPC workloads. In: 2011 IEEE International Symposium on a World of Wireless, Mobile and Multimedia Networks (WoWMoM), pp. 1–7, June 2011
2. Barta, G., Nagy, G.B.G., Kazi, S., Henk, T.: GEFCOM 2014—probabilistic electricity price forecasting. In: Neves-Silva, R., Jain, L.C., Howlett, R.J. (eds.) KES-IDT 2015. SIST, vol. 39, pp. 67–76. Springer, Heidelberg (2015). doi:10.1007/978-3-319-19857-6_7
3. Bartolini, A., Borghesi, A., Bridi, T., Lombardi, M., Milano, M.: Proactive workload dispatching on the EURORA supercomputer. In: O'Sullivan, B. (ed.) CP 2014. LNCS, vol. 8656, pp. 765–780. Springer, Heidelberg (2014). doi:10.1007/978-3-319-10428-7_55
4. Beloglazov, A., Buyya, R., Lee, Y.C., Zomaya, A.Y.: A taxonomy and survey of energy-efficient data centers and cloud computing systems. Adv. Comput. **82**, 47–111 (2011). http://dx.doi.org/10.1016/B978-0-12-385512-1.00003-7
5. Borghesi, A., Collina, F., Lombardi, M., Milano, M., Benini, L.: Power capping in high performance computing systems. In: Pesant, G. (ed.) CP 2015. LNCS, vol. 9255, pp. 524–540. Springer, Heidelberg (2015). doi:10.1007/978-3-319-23219-5_37
6. Castiñeiras, I., Mehta, D., O'Sullivan, B.: Energy cost minimisation of geographically distributed data centres. In: 4th IEEE International Conference on Cloud Networking, CloudNet 2015, Niagara Falls, ON, Canada, 5–7 October 2015, pp. 279–284. IEEE (2015). http://dx.doi.org/10.1109/CloudNet.2015.7335322
7. Chisca, D.S., Castiñeiras, I., Mehta, D., O'Sullivan, B.: On energy- and cooling-aware data centre workload management. In: 15th IEEE/ACM International Symposium on Cluster, Cloud and Grid Computing, CCGrid 2015, Shenzhen, China, 4–7 May 2015, pp. 1111–1114. IEEE (2015). http://dx.doi.org/10.1109/CCGrid.2015.141
8. Christensen, T., Hurn, A., Lindsay, K.: Forecasting spikes in electricity prices. Int. J. Forecast. **28**(2), 400–411 (2012). http://www.sciencedirect.com/science/article/pii/S0169207011000550
9. Grimes, D., Ifrim, G., O'Sullivan, B., Simonis, H.: Analyzing the impact of electricity price forecasting on energy cost-aware scheduling. Sustaina. Comput.: Inform. Syst. **4**(4), 276–291 (2014). http://www.sciencedirect.com/science/article/pii/S221053791400050X
10. Hong, T.: Energy forecasting: past, present, and future. Foresight Int. J. Appl. Forecast. **32**, 43–48 (2014). https://ideas.repec.org/a/for/ijafaa/y2014i32p43-48.html
11. Hong, T., Pinson, P., Fan, S.: Global energy forecasting competition 2012. Int. J. Forecast. **30**, 357–363 (2014)
12. Ifrim, G., O'Sullivan, B., Simonis, H.: Properties of energy-price forecasts for scheduling. In: Milano, M. (ed.) [16], pp. 957–972. http://dx.doi.org/10.1007/978-3-642-33558-7_68
13. Kinsella, A., Smeaton, A.F., Hurley, B., O'Sullivan, B., Simonis, H.: Optimizing energy costs in a zinc and lead mine. In: Proceedings of the Thirtieth AAAI Conference on Artificial Intelligence 2016 (2016)
14. Lucanin, D., Brandic, I.: Take a break: cloud scheduling optimized for real-time electricity pricing. In: 2013 International Conference on Cloud and Green Computing, Karlsruhe, Germany, September 30 - October 2, 2013, pp. 113–120. IEEE Computer Society (2013). http://dx.doi.org/10.1109/CGC.2013.25

15. Mehta, D., O'Sullivan, B., Simonis, H.: Comparing solution methods for the machine reassignment problem. In: Milano, M. (ed.) [16], pp. 782–797. http://dx.doi.org/10.1007/978-3-642-33558-7_56
16. Milano, M. (ed.): CP 2012. LNCS, vol. 7514. Springer, Heidelberg (2012). doi:10.1007/978-3-642-33558-7
17. Qureshi, A., Weber, R., Balakrishnan, H., Guttag, J., Maggs, B.: Cutting the electric bill for internet-scale systems. SIGCOMM Comput. Commun. Rev. 39(4), 123–134 (2009). http://doi.acm.org/10.1145/1594977.1592584
18. Rahman, A., Liu, X., Kong, F.: A survey on geographic load balancing based data center power management in the smart grid environment. IEEE Commun. Surv. Tutor. 16(1), 214–233 (2014)
19. Simonis, H., O'Sullivan, B., Mehta, D., Hurley, B., Cauwer, M.D.: Energy-cost aware scheduling/forecasting competition. Technical report, Insight Centre for Data Analytics, University College Cork, May 2014
20. Weron, R.: Electricity price forecasting: a review of the state-of-the-art with a look into the future. Int. J. Forecast. 30(4), 1030–1081 (2014). http://www.sciencedirect.com/science/article/pii/S0169207014001083

Author Index

Printed in the United States
By Bookmasters